Surveying Southern British Columbia

Surveying Southern British Columbia

A photojournal of Frank Swannell, 1901–1907

Jay Sherwood

Caitlin Press

01 02 03 04 05 18 17 16 15 14

Caitlin Press Inc.
8100 Alderwood Road,
Halfmoon Bay, BC V0N 1Y1
www.caitlin-press.com

Text and cover design by Vici Johnstone.
Printed in Canada.

Front cover image: "Hyiu Mowitch" (plenty of venison). Oil prospector John Watt is standing in the back. Swannnell surveyed leases in the Flathead Valley for him. Image G-03251 courtesy of Royal BC Museum, BC Archives.

Back cover image: Working in mosquito season. Swannell has a head net and thick gloves to protect him from biting insects. Image H-04700 courtesy of Royal BC Museum, BC Archives.

Dedication page: Surveyors en route to Fort Steele. Swannell took this photograph of the survey crew in 1902. Image I-33669 courtesy of Royal BC Museum, BC Archives.

Caitlin Press Inc. acknowledges financial support from the Government of Canada through the Canada Book Fund and the Canada Council for the Arts, and from the Province of British Columbia through the British Columbia Arts Council and the Book Publisher's Tax Credit.

Canada Council for the Arts | Conseil des Arts du Canada

BRITISH COLUMBIA ARTS COUNCIL
We acknowledge the support of the Province of British Columbia through the British Columbia Arts Council

Library and Archives Canada Cataloguing in Publication

Sherwood, Jay, 1947-, author
Surveying southern BC : a photojournal of Frank Swannell, 1901–7 / Jay Sherwood.

Includes 150 photographs and Frank Swannell's courtship letters to Ada Driver.
Includes bibliographical references and index.
ISBN 978-1-927575-51-2 (pbk.)

1. Surveying—British Columbia—History—20th century.
2. Frontier and pioneer life—British Columbia—Pictorial works.
3. British Columbia—History—20th century—Pictorial works.
4. Swannell, F. C. (Frank Cyril). 5. Surveyors—British Columbia—Biography. I. Title.

FC3845.S679S54 2014 971.1'03 C2014-904062-8

Contents

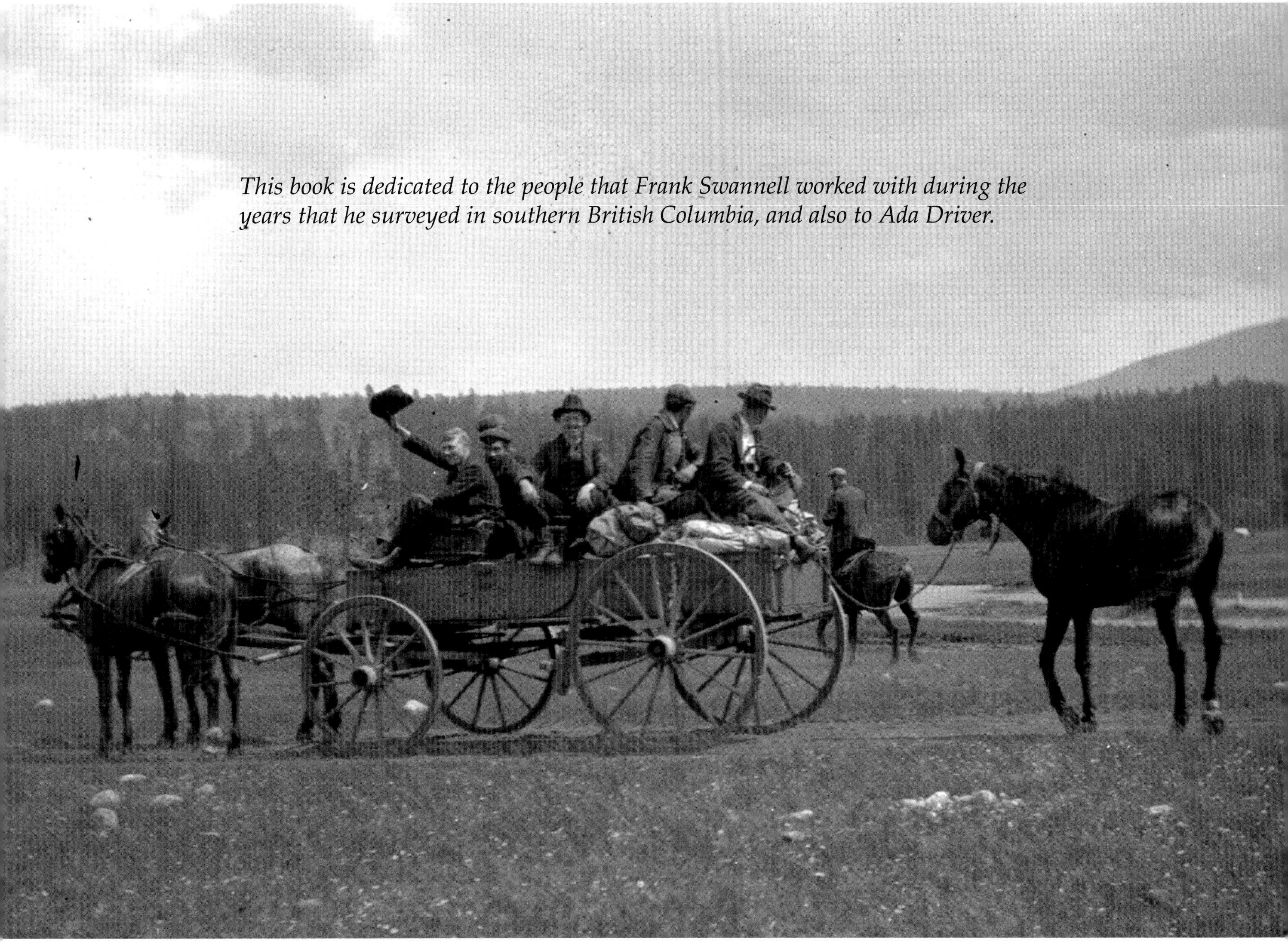

This book is dedicated to the people that Frank Swannell worked with during the years that he surveyed in southern British Columbia, and also to Ada Driver.

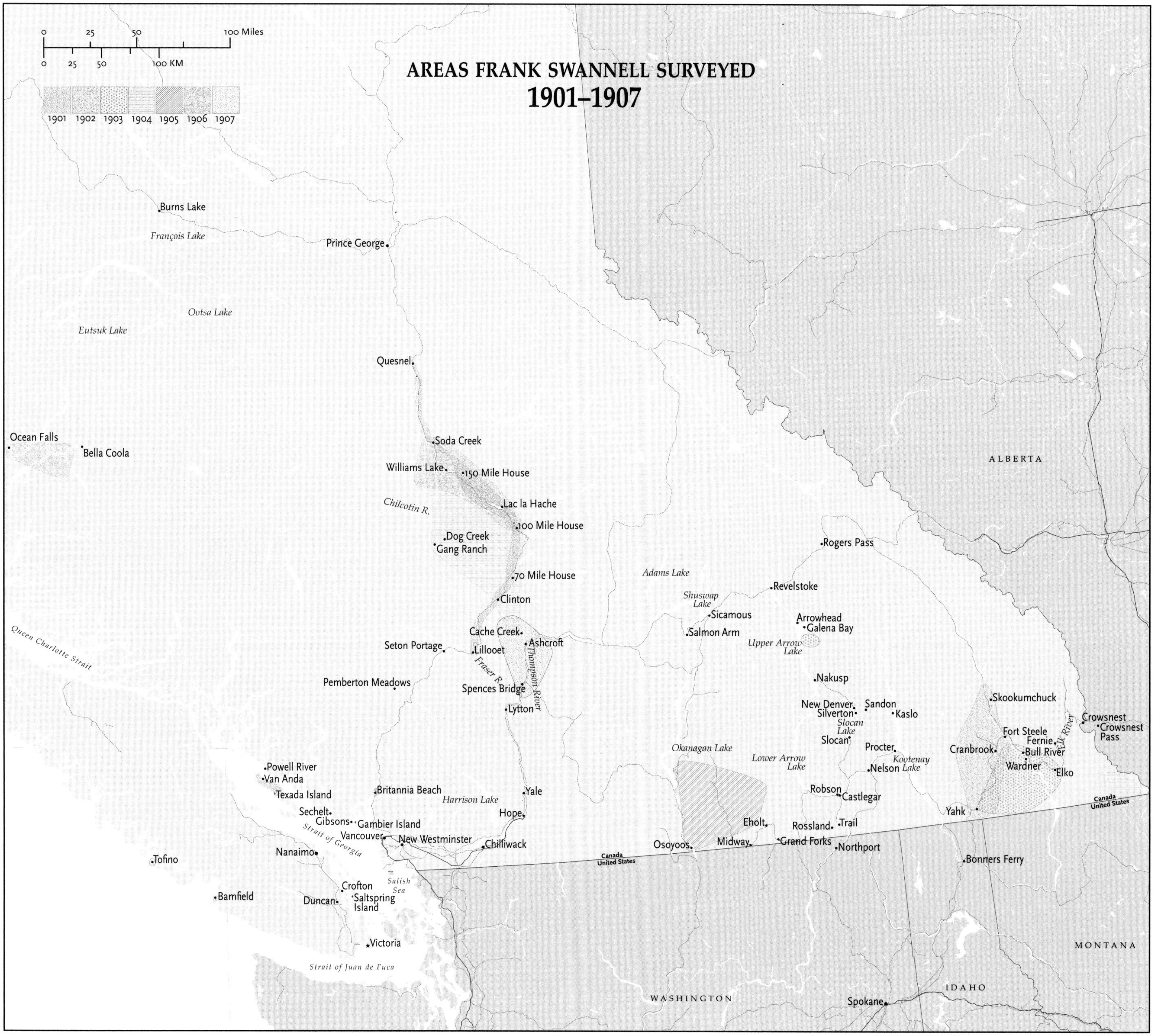
AREAS FRANK SWANNELL SURVEYED
1901–1907
0 25 50 100 Miles
0 25 50 100 KM
1901 1902 1903 1904 1905 1906 1907
Burns Lake
François Lake
Prince George
Ootsa Lake
Eutsuk Lake
Quesnel
Ocean Falls
Bella Coola
Soda Creek
Williams Lake
150 Mile House
Chilcotin R.
Lac la Hache
100 Mile House
Dog Creek
Gang Ranch
70 Mile House
Clinton
Cache Creek
Ashcroft
Seton Portage
Lillooet
Fraser R.
Thompson River
Spences Bridge
Lytton
Pemberton Meadows
Queen Charlotte Strait
Powell River
Van Anda
Texada Island
Britannia Beach
Harrison Lake
Yale
Sechelt
Gibsons
Gambier Island
Hope
Vancouver
New Westminster
Chilliwack
Strait of Georgia
Nanaimo
Tofino
Bamfield
Crofton
Salish Sea
Saltspring Island
Duncan
Victoria
Strait of Juan de Fuca
Adams Lake
Shuswap Lake
Sicamous
Salmon Arm
Revelstoke
Rogers Pass
Arrowhead
Galena Bay
Upper Arrow Lake
Nakusp
New Denver
Silverton
Sandon
Kaslo
Slocan Lake
Slocan
Okanagan Lake
Lower Arrow Lake
Procter
Kootenay Lake
Nelson
Robson
Castlegar
Eholt
Rossland
Trail
Osoyoos
Midway
Grand Forks
Northport
Canada
United States
Skookumchuck
Fort Steele
Fernie
Bull River
Cranbrook
Wardner
Elko
Elk River
Crowsnest
Crowsnest Pass
Yahk
Bonners Ferry
ALBERTA
MONTANA
IDAHO
WASHINGTON
Spokane

Introduction

Surveying Southern British Columbia, the fourth and final book of the Frank Swannell series, is a prequel to the three previous books that covered his career during the years that he worked on contract for the BC government. From 1901 to 1907 Swannell surveyed for Gore & McGregor, a Victoria-based surveying company that was one of the most prominent in the province at that time. The firm received contracts throughout the southern part of BC, so during his years with Gore & McGregor, Swannell surveyed from the west coast of Vancouver Island to the Rockies and from the southern Canada–US boundary north as far as Quesnel in the Cariboo and the Bella Coola area on the coast. In the process he worked on many projects that were important in the history of British Columbia at the beginning of the twentieth century.

A note in Swannell's diary for April 4, 1901, stated: "Bo't camera." This simple statement marked the beginning of an important aspect of his work. Swannell, who had started his surveying career in 1899, realized that he was seeing areas of British Columbia that few people visited, and he wanted to have a record of his work, the places where he surveyed and the people he met. From the beginning Swannell showed that he had a photographer's eye for composition and detail. Although he did not take as many pictures as in later years, his photographs from 1901 to 1907 provide a fine record of many remote areas of the southern part of the province. Several of Swannell's early photographs have been used in BC history books.

What is missing from Swannell's early work is the written record. There are diaries for some of the time from 1901 to 1907, but these are primarily one-line entries. Swannell's lack of a detailed written record during

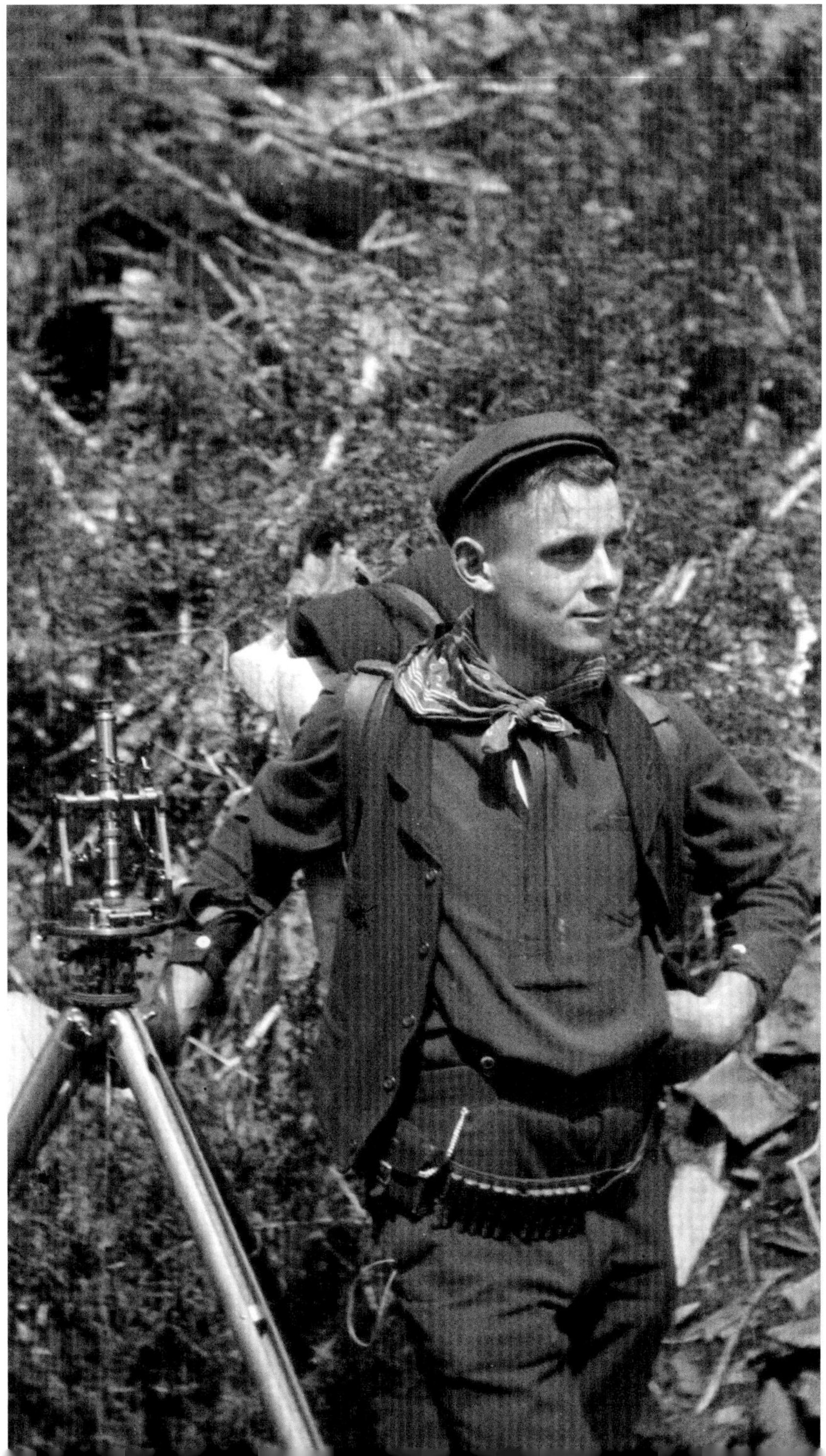

Frank Swannell at the beginning of the 1902 field season when he was running the transit for T.S. Gore. Gore & McGregor were beginning the survey of the land grants for the BC Southern Railway. In his right hand Swannell is holding his field book for recording the information. The pistol and ammunition are for shooting small game, particularly grouse. Image I-58381 courtesy of Royal BC Museum, BC Archives.

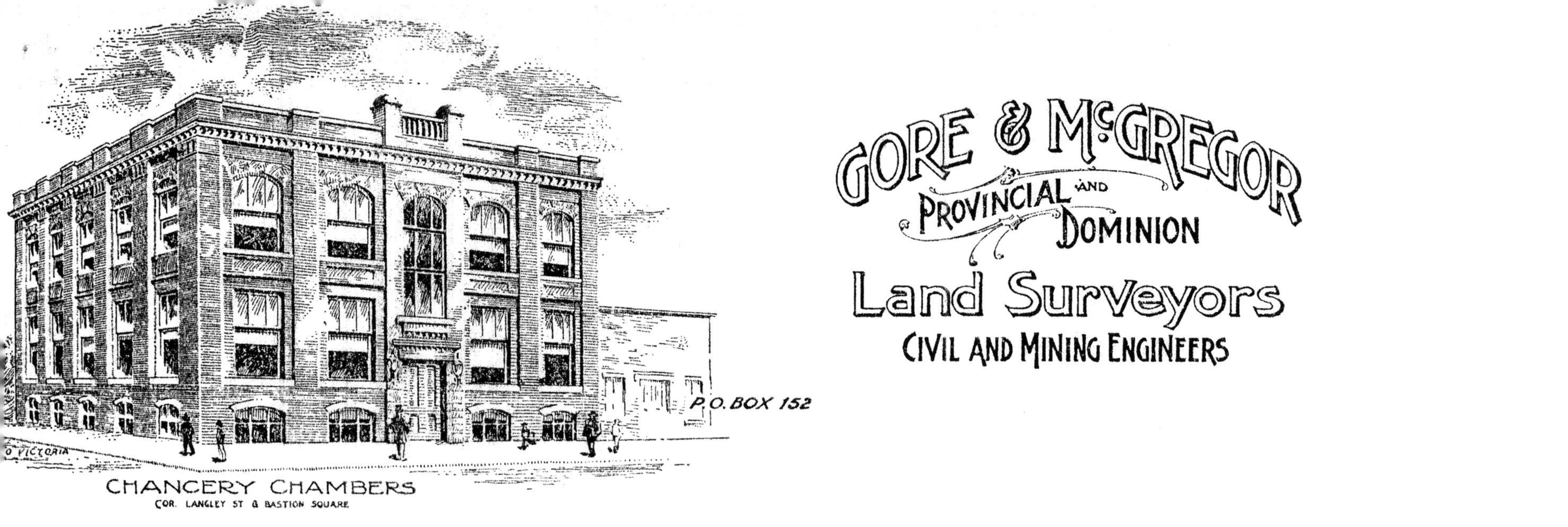

❖ Gore & McGregor letterhead courtesy Glenbow Archives.

these years is rather surprising considering that he was interested in journalism and was in contact with magazine editors regarding writing articles.

However, another important event occurred for Swannell in 1901. One evening in the spring he and his friend George Copley went to a dance where they saw two young ladies without partners. It took the two men a while to summon the courage to ask the ladies for a dance. Then they had to decide which one they were going to ask. Their choice proved to be the right one, for each married the young lady he first danced with that evening. Swannell's partner was Ada Driver, while Copley's was Nora Mabel Liddell. From 1901 until his marriage in 1904, Swannell wrote letters to Ada while he was out surveying. Many of these letters, or fragments of these letters, have survived. They are much different in style from his later writing, and in them a clearly romantic Frank Swannell provides detailed descriptions of his surveying work and adventures.

Frank Swannell was born in Hamilton, Ontario, in 1880, the eldest child (and the only one to live past infancy) of Frederick Swannell and his second wife, Mary Forster. When he was young the family moved to Toronto. In 1897, during the summer after high school graduation, Swannell came to western Canada for the first time to work in the wheat fields of Manitoba. After the summer Swannell enrolled in a two year mining engineering program at the University of Toronto. He came to British Columbia in the summer of 1898, working for the surveying firm of Drewry & Twigg, who were based in New Denver, a growing mining community in the West Kootenay. Most of their work involved mineral claims, but they also did some railway surveys.

After receiving his mining engineering diploma in 1899, Swannell intended to go to the Yukon. When he reached Victoria he realized that he did not have sufficient finances for the venture, and besides, the main gold rush was already over. Swannell obtained a surveying position with Gore & McGregor working on a right-of-way survey for the Columbia and Western Railway. He must have liked the work and way of life and impressed surveyor Thomas Gore with his ability, since Gore made special arrangements for Swannell and two other men to take their preliminary surveying exam in Greenwood, the area where they were working, instead of Victoria. Swannell passed this exam during the fall of 1899 and Gore accepted him as an articled student for his three-year apprenticeship.

Under Gore's tutelage, Swannell began learning the skills he needed to become a surveyor. In 1900 he worked on mining claims at Britannia, where the large copper mine on Howe Sound would soon develop. In October of that year Swannell's half-brother, Frederick, who lived in Nelson and was about fifteen years older, wrote a letter to him. He said, "I hope you are still satisfied with your choice of a profession and with the men under whom you are studying." He observed that "most surveyors and geologists are men

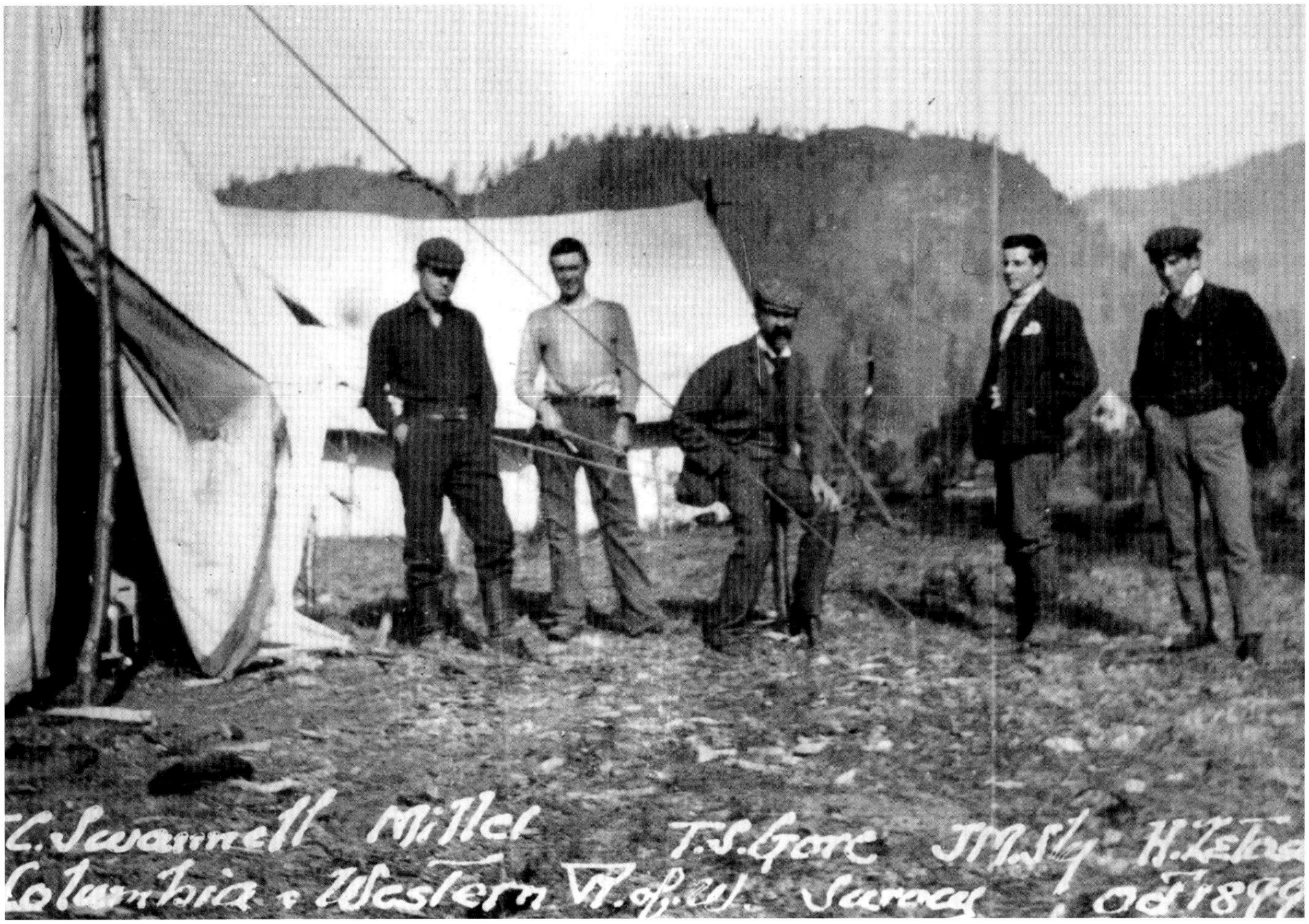

❖ Frank Swannell (left) and T.S. Gore (centre) on the Columbia and Western Railway right-of-way survey in 1899. Swannell did not have a camera at that time, so one of the other members of the crew probably gave him the pictures that he had for this survey. Image I-33523 courtesy of Royal BC Museum, BC Archives.

of good intellect and manners, and I think we can continually rub a little knowledge and polish from oft the people we associate with." Frederick also cautioned Frank: "I myself would not advise you to adopt literature as a profession, but if you are able to write sketches and articles, occasionally, as side issues, it would be very helpful to you both from a financial standpoint, and from the habit of observing men and affairs which it would naturally engender."

In January 1901 Swannell commenced keeping a diary with brief entries. With this and his new camera, Swannell was ready to record his work on important surveying projects throughout southern BC for the next seven years.

Note: In the early twentieth century Canadians used the imperial system of measurement. A foot is 30 centimetres and a mile is 1.6 kilometres. In direct quotations in this book the imperial measurement is used with a metric equivalent provided in brackets. The terms Indian and Siwash were commonly used to describe First Nations people at that time. By all accounts Swannell had a good relationship with the First Nations people he met, so the use of these words in his writing is not derogatory.

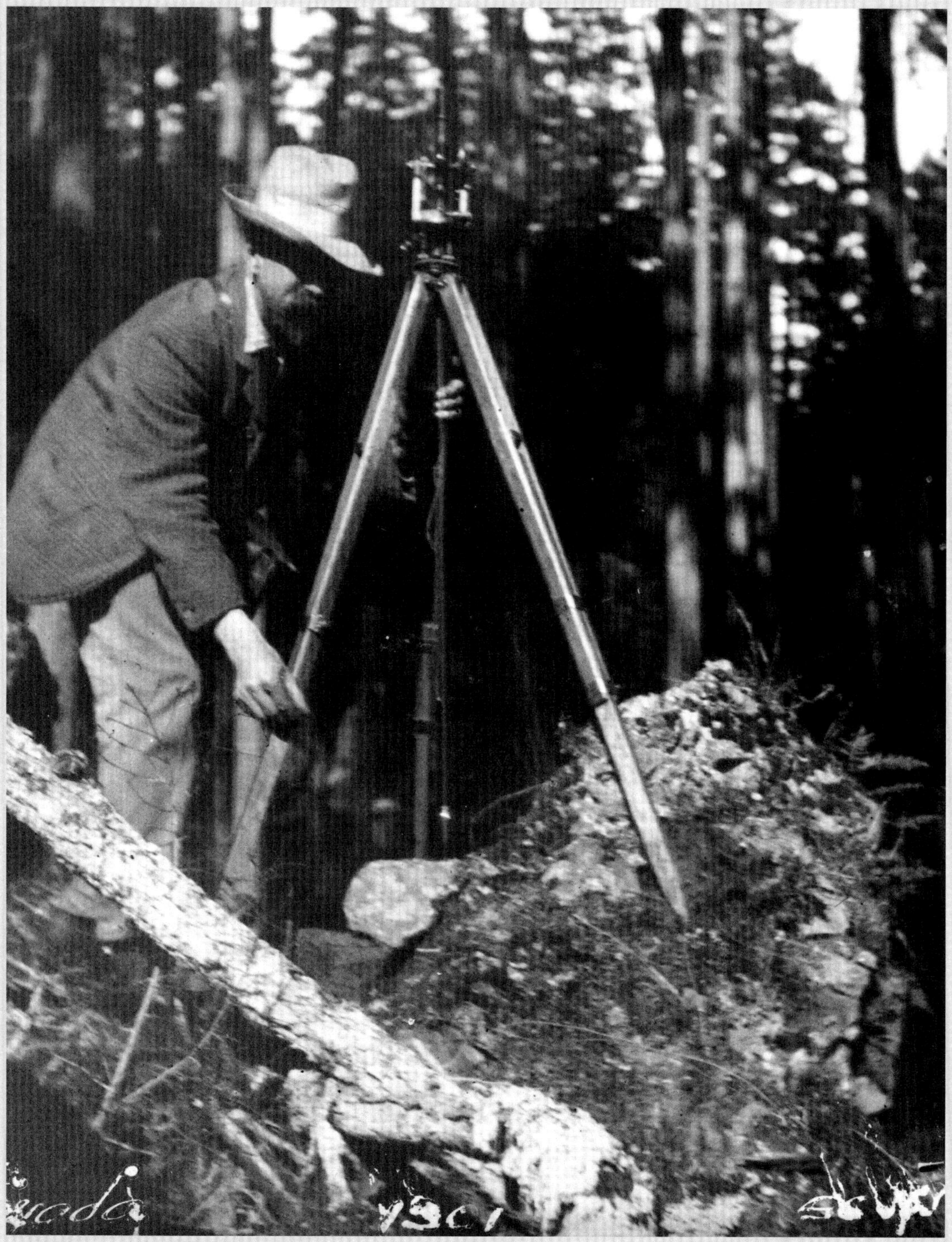

❖ Swannell took this photograph of T.S. Gore setting up the transit at a survey station on Texada Island in May 1901. This was the first trip on which Swannell took his camera. Image I-58060 courtesy of Royal BC Museum, BC Archives.

Thomas Gore

Thomas Sinclair Gore, LS,* was a prominent British Columbia land surveyor. He followed in the footsteps of his older brother, William Sinclair Gore, who was also a surveyor. William surveyed in the United States for the Chicago, Burlington and Quincy Railway for several years in the 1860s. When he returned to Canada he became the country's first certified Dominion land surveyor in 1873. After surveying federal land in the Prairie provinces for a few years, William came to British Columbia in 1875. From 1878 to 1891 he was the Surveyor General of British Columbia. Then William was promoted to Deputy Commissioner of Lands, a position he held until 1905.

Thomas surveyed for the same railway as William in the United States during the 1870s. He also became a Dominion land surveyor and surveyed in the Prairie provinces along with serving in the Dominion Land Surveyors Intelligence Corps during the 1885 Riel Rebellion. Thomas came to BC in 1888. In 1894 he formed a surveying company with Hugh Burnet, LS, and James McGregor, PLS #1. Burnet left around 1900. Gore & McGregor continued their partnership until the former retired from surveying in 1910. Both Gore and McGregor were considered excellent, experienced surveyors, but undoubtedly William's leading role in surveying in British Columbia was an important factor in Thomas Gore receiving many major contracts during the years that Swannell worked for the firm.

*From 1871 – 1891, registered BC land surveyors received the designation LS. From 1891-1905 the surveyors were called PLS (Provincial Land Surveyor) and received a number based on chronological order. James McGregor was the first PLS. Since 1905, surveyors in the province have been commissioned BCLS (British Columbia Land Surveyor) also in chronological order.

Columbia and Western Railway

During the 1890s and early twentieth century, the American Great Northern Railway and the Canadian Pacific Railway (CPR) engaged in a fierce, protracted struggle for control of the economic wealth being produced in the mines of the West Kootenay and Boundary regions of the province, along with the coal fields of Crowsnest Pass. Augustus Heinze, an American mining engineer and one of the famous "copper kings" of Butte, Montana, established a smelter at Trail in the West Kootenay and in 1896 received a charter from the British Columbia government for the Columbia and Western Railway. Initially he built a narrow-gauge railway from his Trail smelter to the Rossland mines and a standard railway north to Robson on the Columbia River, where it could connect with the sternwheelers travelling on the Arrow Lakes. The Columbia and Western Railway charter also included a provision to construct a 160-kilometre line to the Boundary mining district to the west. The provincial government was to provide a land grant of 20,000 acres per mile (5,030 ha per km) and the federal government would pay a cash subsidy of $3,200 per mile ($2,000 per km). Heinze did not have sufficient capital to build the railway into the Boundary district, so he sold both the smelter and the railway charter to the CPR in 1898. The 1898 CPR Annual Report stated that construction of the Columbia and Western line was necessary to protect the interests of the company in the region.

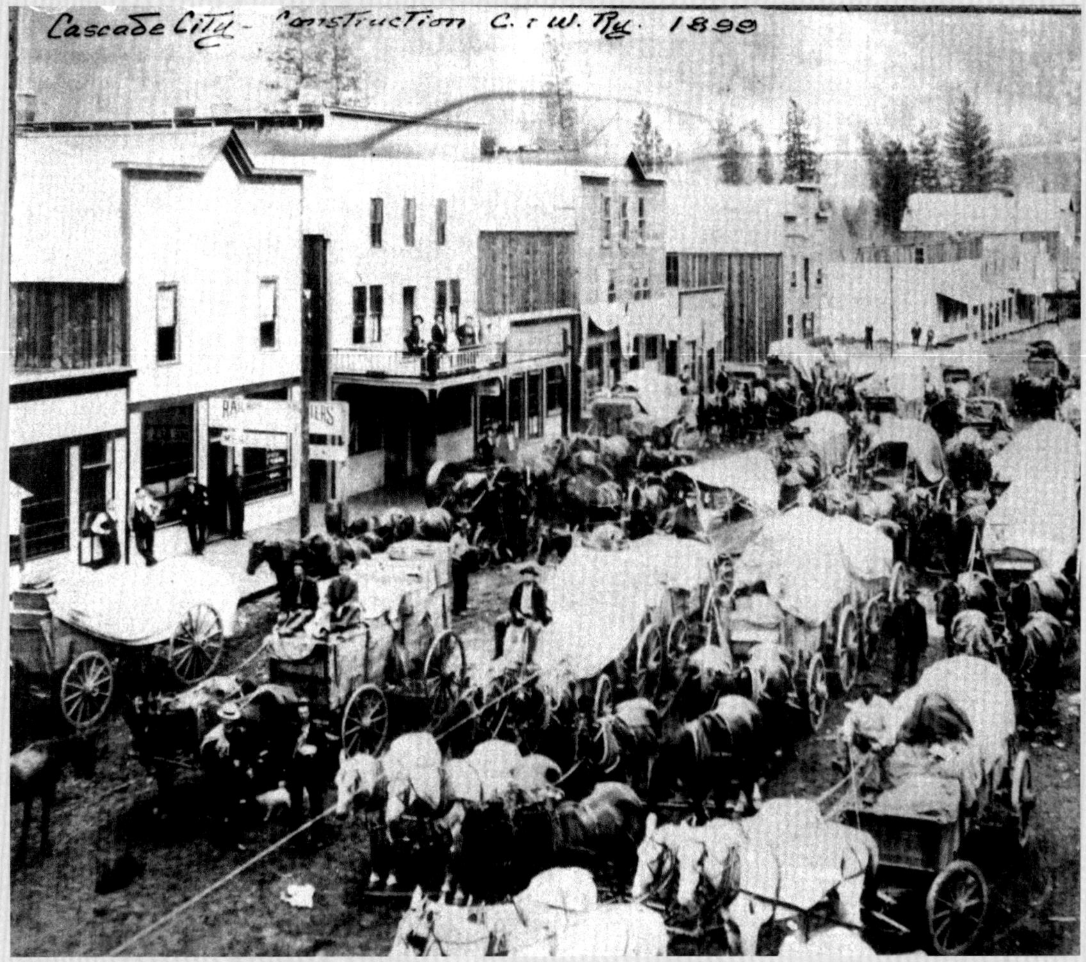

❖ Cascade City: construction of the Columbia and Western Railway. The community was located at the railway's crossing of the Kettle River. Cascade City was a centre for local mining activities and there was a hydroelectric dam and powerhouse on the river. As the photograph shows, it was a busy place during the construction of the railway. At the end of September 1899, not long after this picture was taken, there was a fire that destroyed a large part of Cascade City. Image A-04093 courtesy of Royal BC Museum, BC Archives.

The CPR completed the railway from Robson to Midway by early 1900. There were two mountain ranges to cross, steep grades, five tunnels, a large amount of blasting and several trestles to construct. At the height of the project there were over five thousand people working on the railway line. Thomas Shaughnessy, the president of the CPR, stated, "The road is by all means the most expensive the CPR has ever constructed. The country throughout was exceedingly difficult and the engineering difficulties very great." The railway right-of-way survey that Gore did in 1899 marked the limits of the actual width allotted for the railway (about 15 metres on either side of the centre of the track) and was connected to existing surveys wherever possible.

The rail line provided access to the rich Boundary mining district for the smelter at Trail and it proved profitable. Gore & McGregor later surveyed some of the land grants that the CPR received for construction of the Columbia and Western Railway.

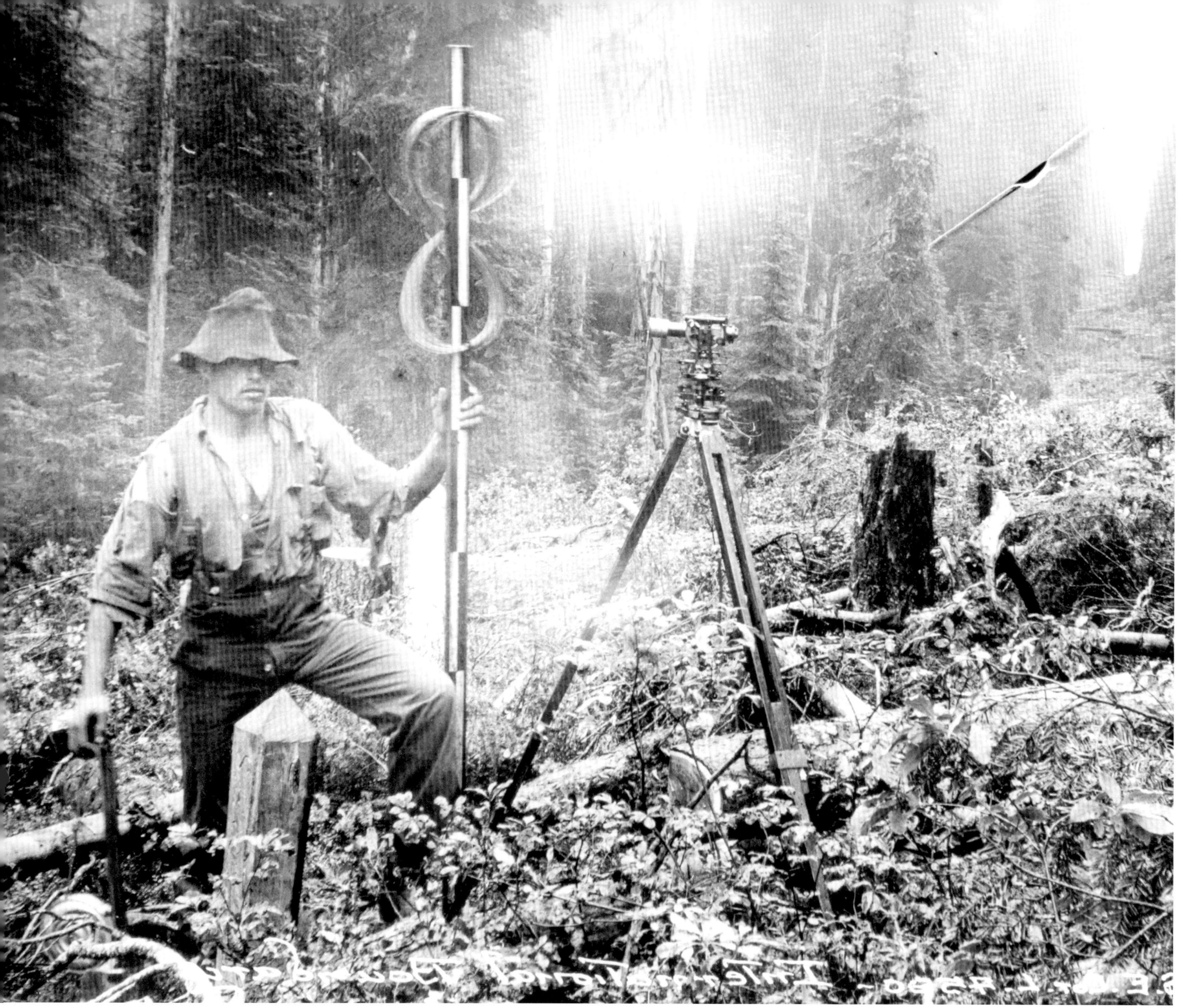

❖ R.M. Benson surveying at the Canada–United States boundary. The cutline for the boundary is visible in the background. Benson was a member of Swannell's crew in 1902, 1903 and 1908. In this posed photograph he is holding in his left hand a survey rod, which was used to help locate position on the survey line, along with a steel chain (in a figure-eight formation) on top. Beside him is a transit, while a post at a surveying station along the boundary is in front. G-09198 courtesy of Royal BC Museum, BC Archives.

Surveying in British Columbia in the Early Twentieth Century

Compared to his later work for the BC government, which involved more complex surveying techniques, most of Swannell's surveying for Gore & McGregor was basic lot work, much of it for resource development projects. The lots surveyed were usually rectangular or square, with some extending for several kilometres. The basic surveying equipment was a steel chain, used to measure distance, and a transit, which was used to keep the survey line straight and to measure the angle when the line changed direction. The surveyor or one of his assistants ran the transit.

On the lot surveys the line followed a set direction, so any trees that were in the line of sight had to be cut down. The transitman, who operated the transit, frequently had to remain at a surveying station for a long period of time while the crew chopped out the trees that obstructed his view. In the summer, the transitman often had to contend with mosquitoes and blackflies. All of the clearing was done by axe, so most of the crews that Swannell worked on had several axemen. When a large tree was on the line it took several men to chop it down. If only the branches of a tree were on the line, the crew would try to remove them by climbing up the tree or standing on the shoulders of one of the men on the crew. The axemen would chop out the trees that were on the line ahead and clear the survey line until they reached the location where the next station would be set.

The large lots for the BC Southern Railway land grants extended to over fifty kilometres in a continuous direction, and the survey line had to be cleared to a width of 1.8 metres. Sighting through the transit, the transitman would signal when the survey rod

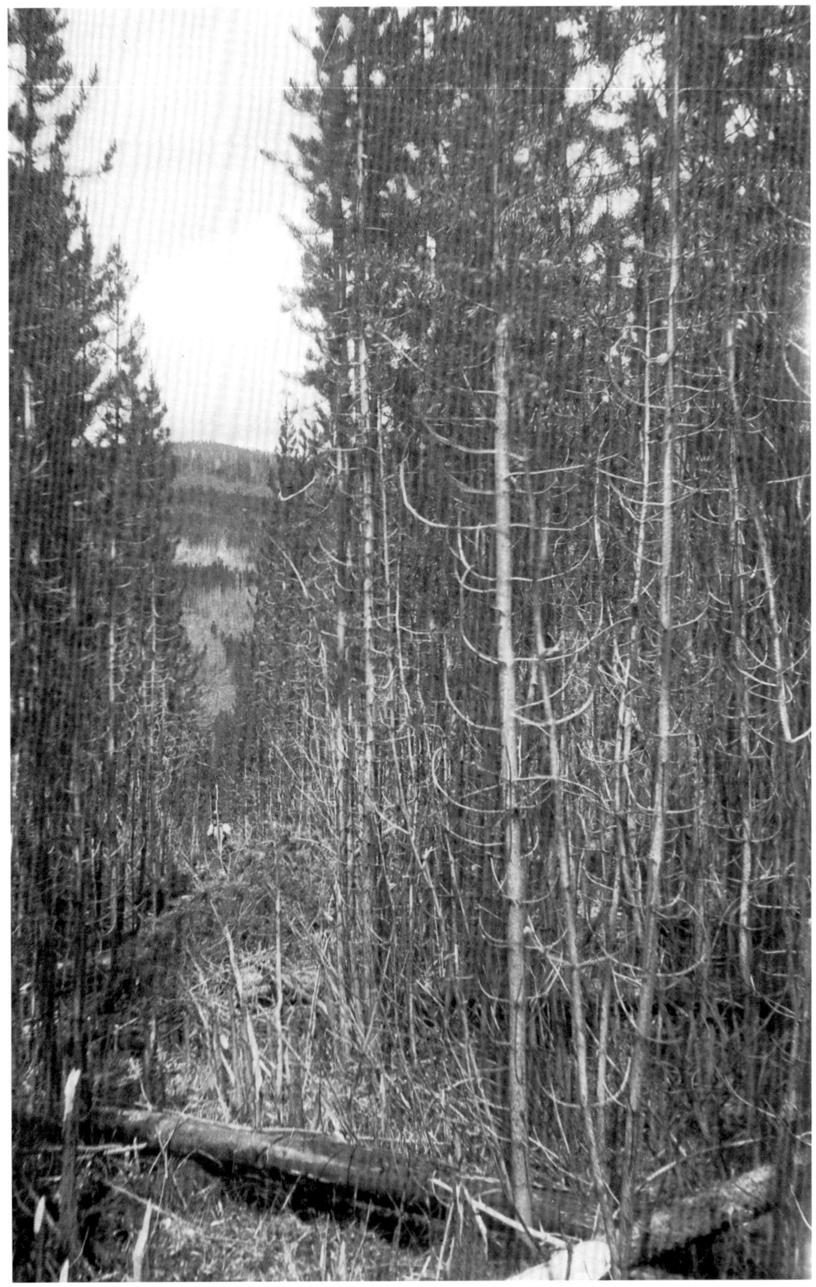

This survey line is going through a dense stand of young jack pines. At the top of a small hill about a third of the way up the picture, one of the men on the crew is standing on the line with a survey rod. Image I-57998 courtesy of Royal BC Museum, BC Archives.

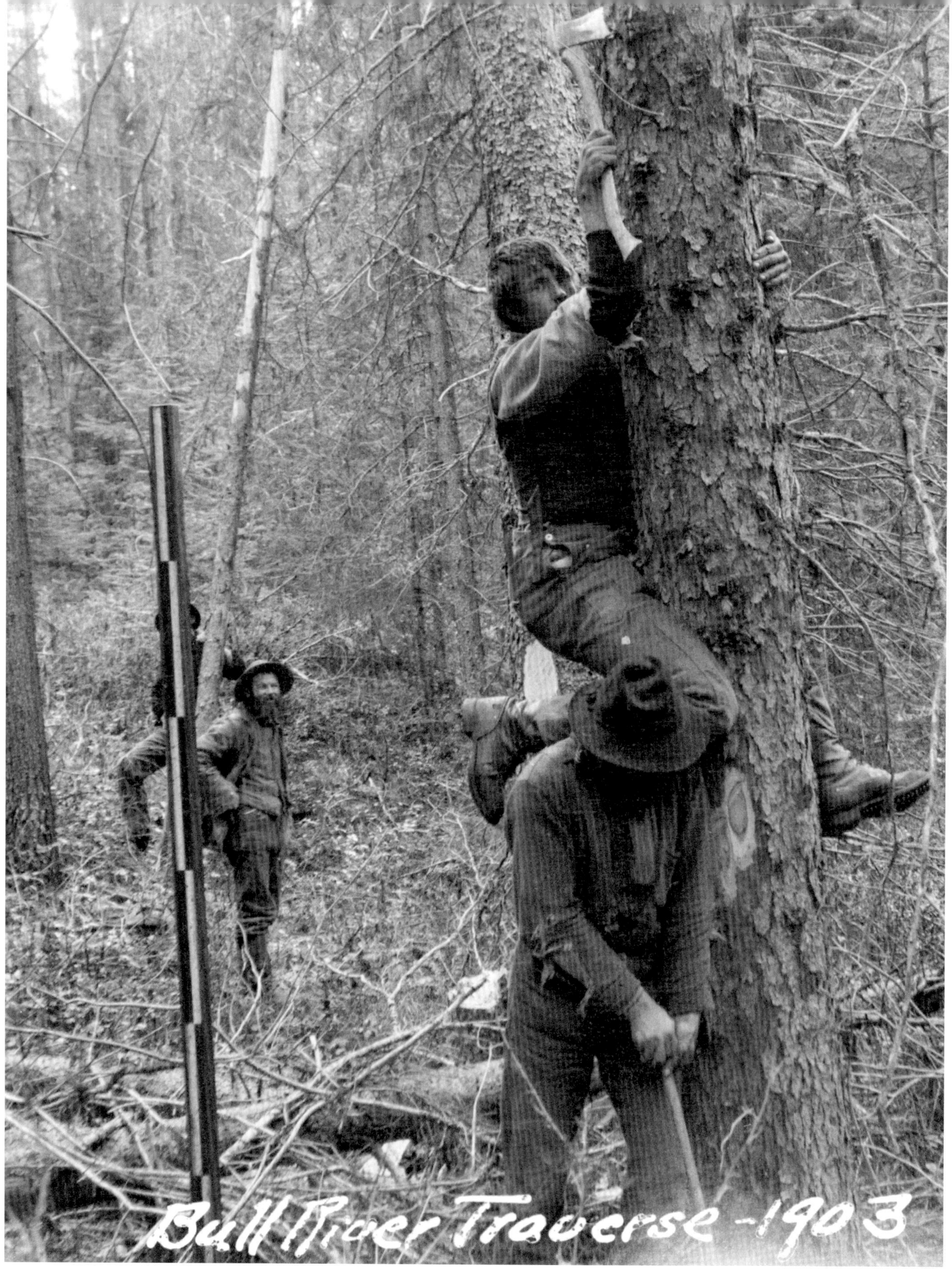

❖ Limbing a tree. Image I-58099 courtesy of Royal BC Museum, BC Archives.

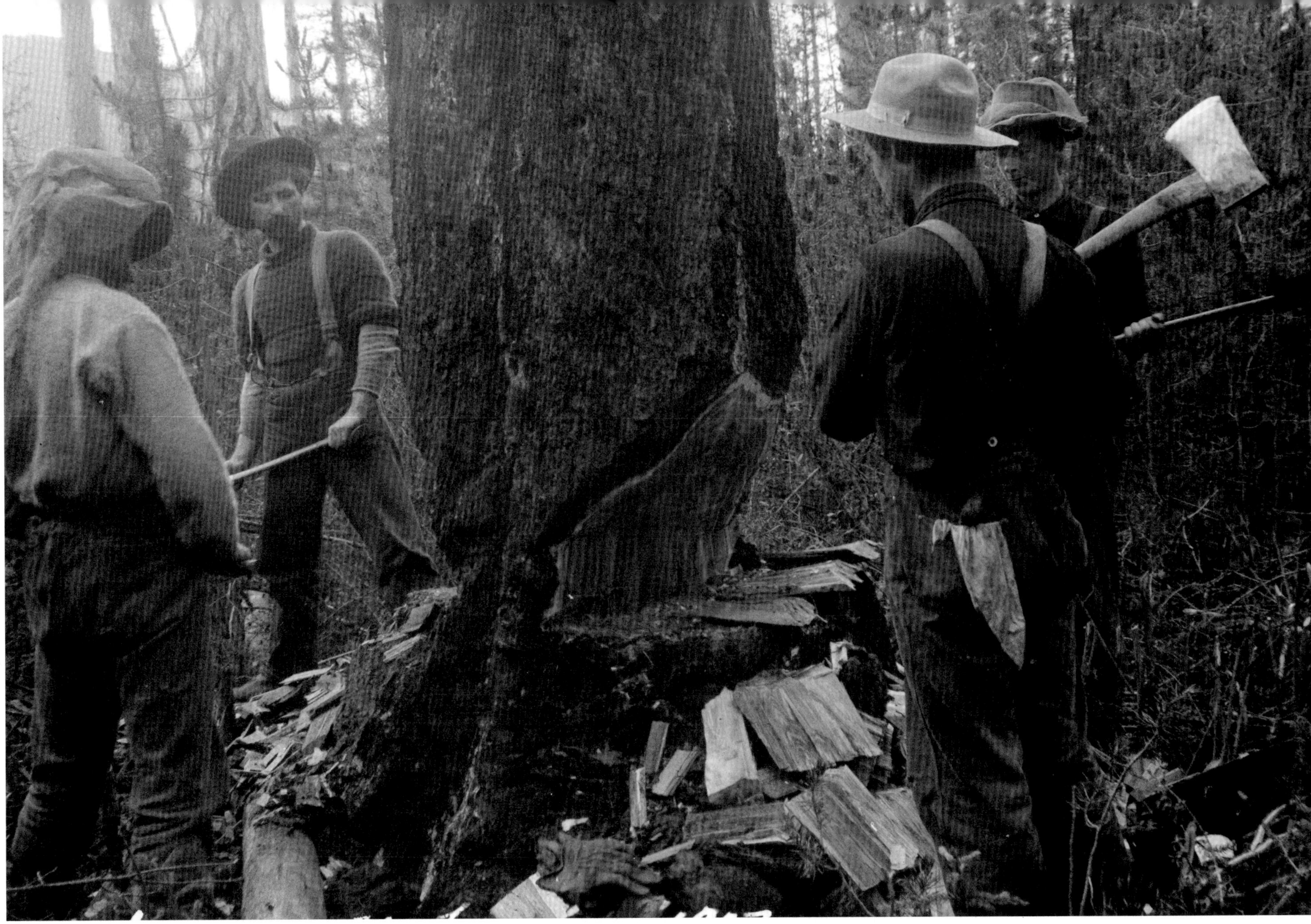

was on the exact location of the line, and a survey station would be set at that point. Usually the station was placed on a high point with a clear view ahead, or at an important geographical feature like a large creek or river. Then the distance between the stations would be measured. At that time surveying distances were measured in chains and links. There are eighty chains in a mile (5,280 feet or 1,600 metres), so each chain is 66 feet (20 metres). Each chain is divided into one hundred links of 8 inches (20 centimetres). The transitman would then move his instrument up to the next station, sight back to the previous station and continue the line forward in the same direction.

At every mile (1.6 kilometres), a mile post was set if possible. When stones were available, a mound was built to protect the station and make it easier to locate. Bearing trees were used to reference the station, and they would be blazed, facing toward the station, 4 feet (1.2 metres) above the ground. The slope distance from the blaze to a point that was also 1.2 metres above the station would be measured. The blaze on the bearing tree would be carved with the letters

Cutting down a large tree. Image I-58095 courtesy of Royal BC Museum, BC Archives.

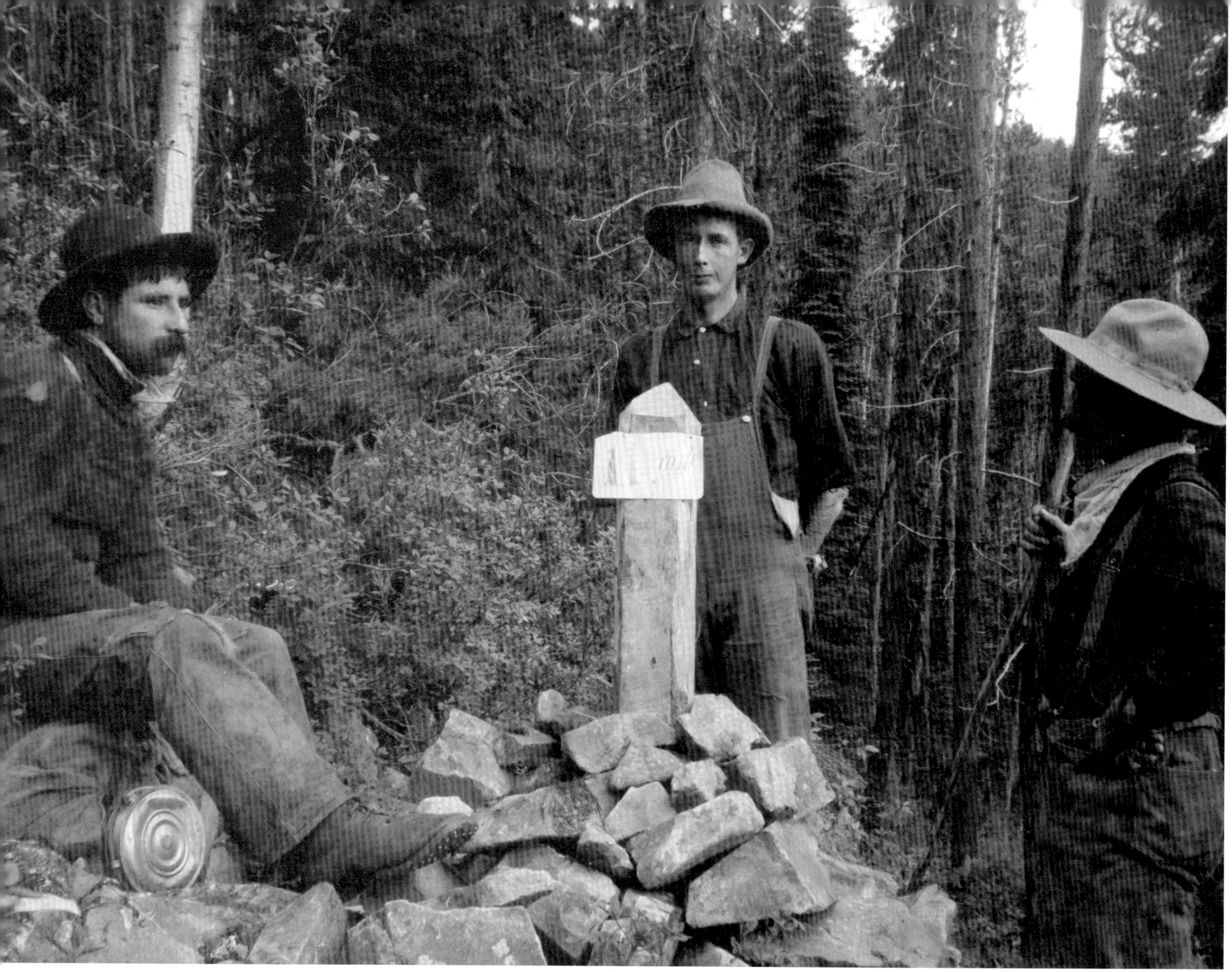

❖ At the 11-mile (17.6-kilometre) post on the east boundary of Lot 4591. The post, the stone mound around it and one bearing tree (behind the man on the left) are visible. Image I-58090 courtesy of Royal BC Museum, BC Archives.

BT (bearing tree) and the distance to the station. In his field notes the surveyor would record the species and diameter of the bearing tree, along with the distance and direction from the station to the bearing tree. A minimum of two, and preferably three, bearing trees would be established at a surveying station, and if possible would subtend an angle at the station that was close to 90 degrees. The bearing trees could be used to locate the station or even re-establish it if the post no longer existed.

Swannell took many photographs of the survey work, life in camp, and the geography of the areas where he worked from 1901 to 1907.

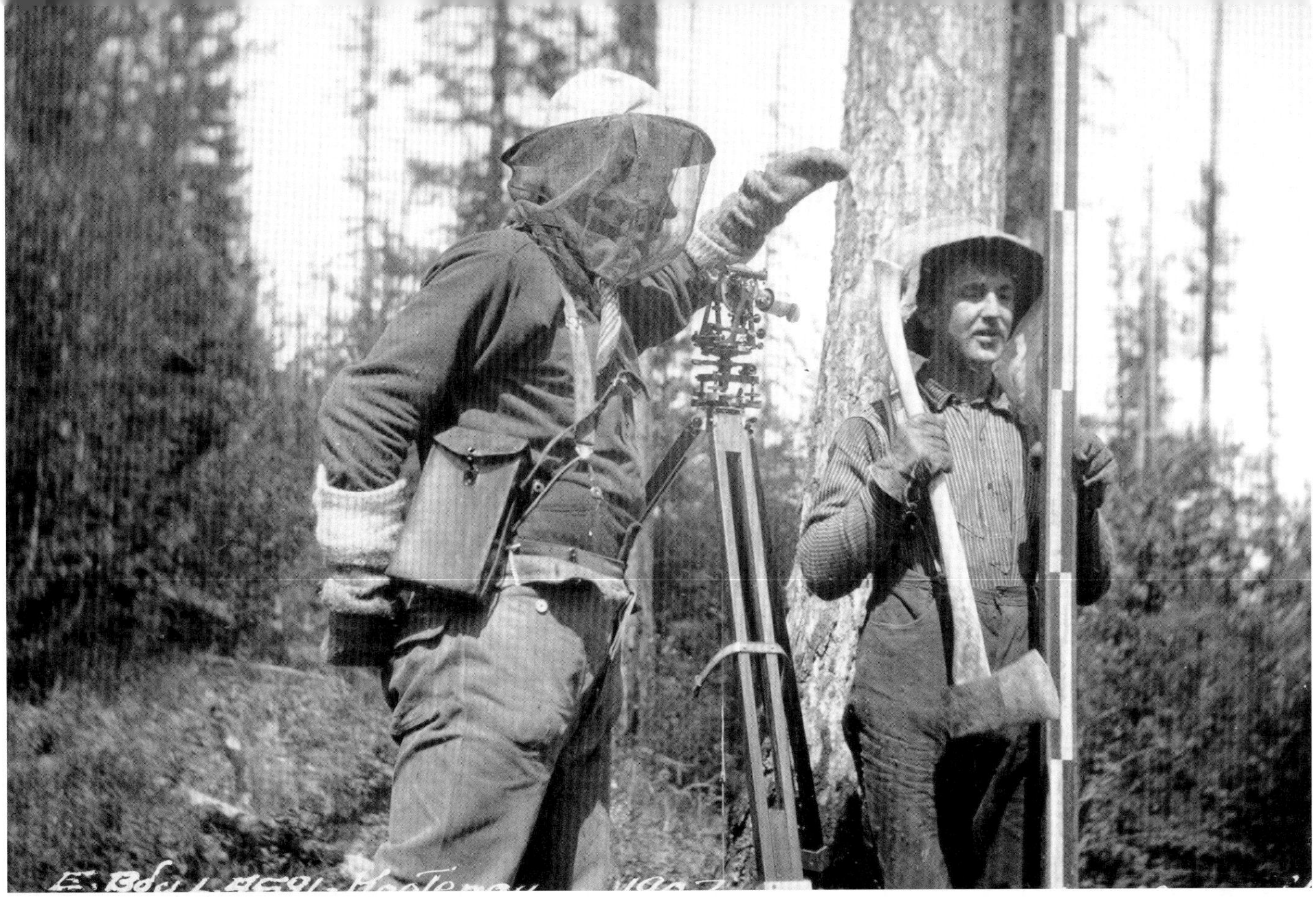

❖ Working in mosquito season. Swannell has a mosquito head net and thick gloves to protect him from biting insects. One of the crew members has a survey rod to help mark the location of the line and an axe to clear any brush or trees that obscured Swannell's view. Image H-04700 courtesy of Royal BC Museum, BC Archives.

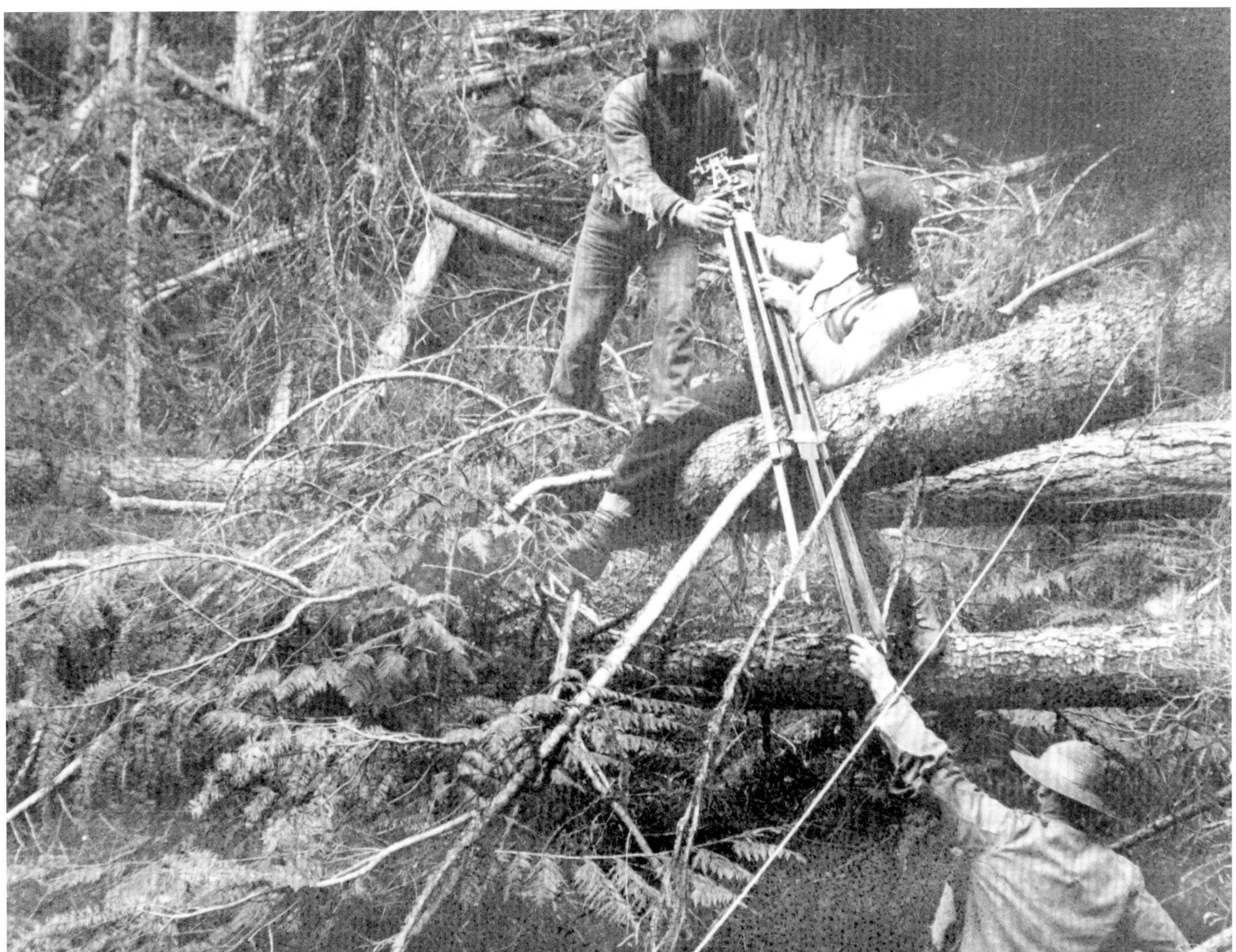

❖ Although the survey line was cleared, occasionally there were windfalls that had to be negotiated when moving from station to station, and the transitman had to be assisted. Particular care was needed to prevent damage to the transit. Image G-06435 courtesy of Royal BC Museum, BC Archives.

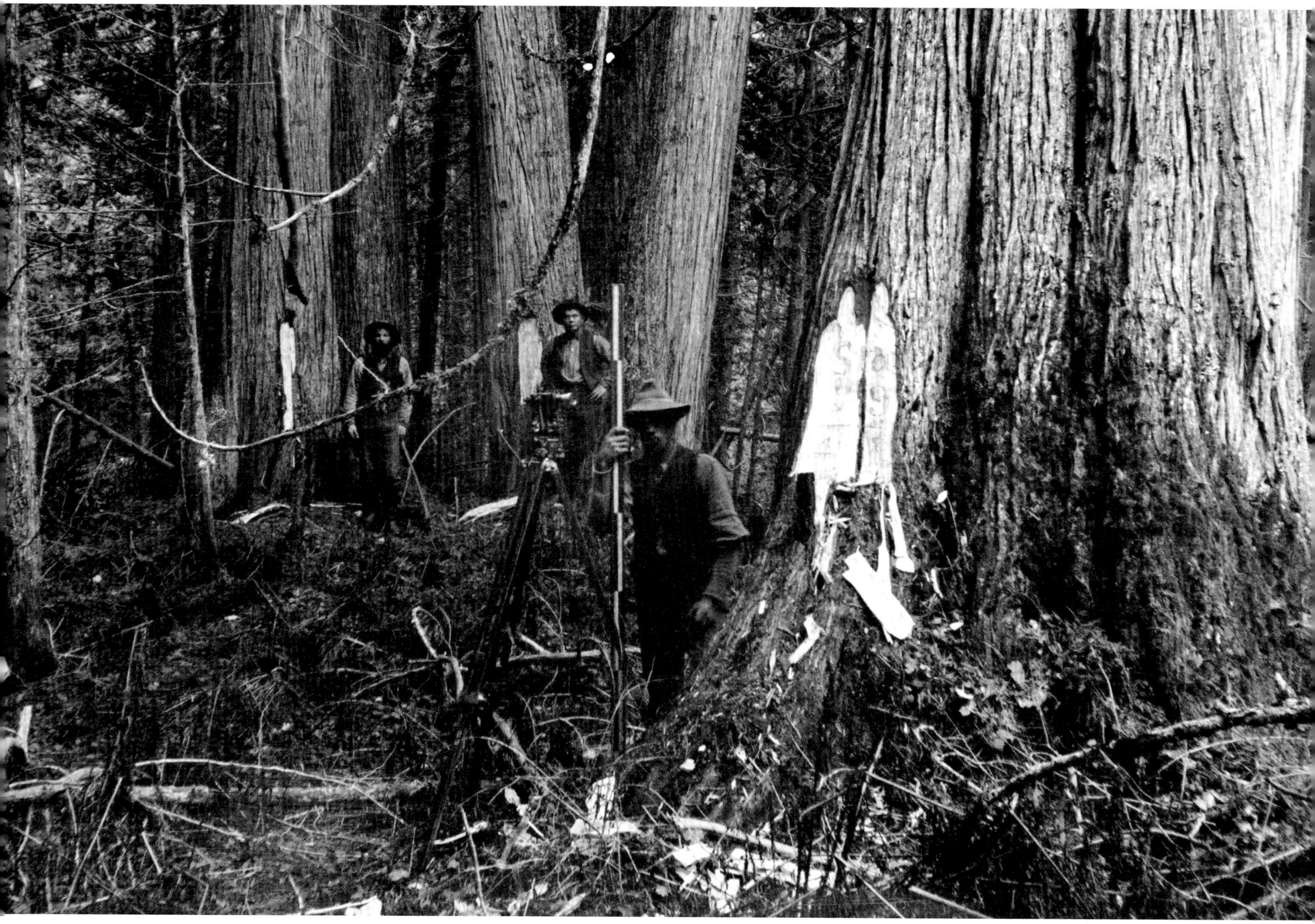

❖ Sand Creek traverse 1903: heavy cedar. Swannell has three members of his crew each standing beside one of the bearing trees at this station. The survey station is probably close to the place from where this photograph was taken. Although the three trees are not close to a 90-degree angle, they were probably the only suitable bearing trees nearby. Image I-58233 courtesy of Royal BC Museum, BC Archives.

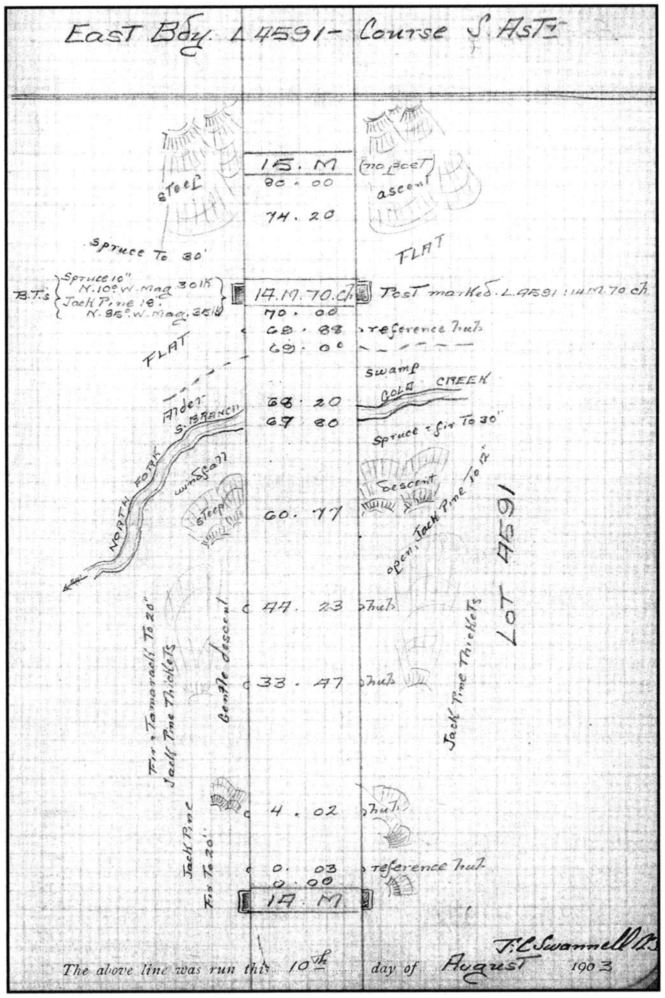

❖ This is a page from Swannell's 1903 field book for Lot 4591, heading south from Mile 14 to Mile 15. The page reads from the bottom up. The distances in the middle are marked in chains and links and denote items like the crossing of Gold Creek, from 67.80 to 68.20 chains. On the side Swannell noted the different types of terrain and trees that he found during this mile. Because of the steep terrain Swannell was not able to set a post exactly at 15 miles. Instead it was set at 14 miles 70 chains. At the left of the station Swannell noted the two bearing trees: a 10" (25 cm) spruce that was N10°W mag (magnetic direction) and 30 links from the station, and an 18" (45 cm) jack pine that was N85°W and 35 links away. The two stations subtend an angle of 75 degrees. Map courtesy Land Title Survey Authority.

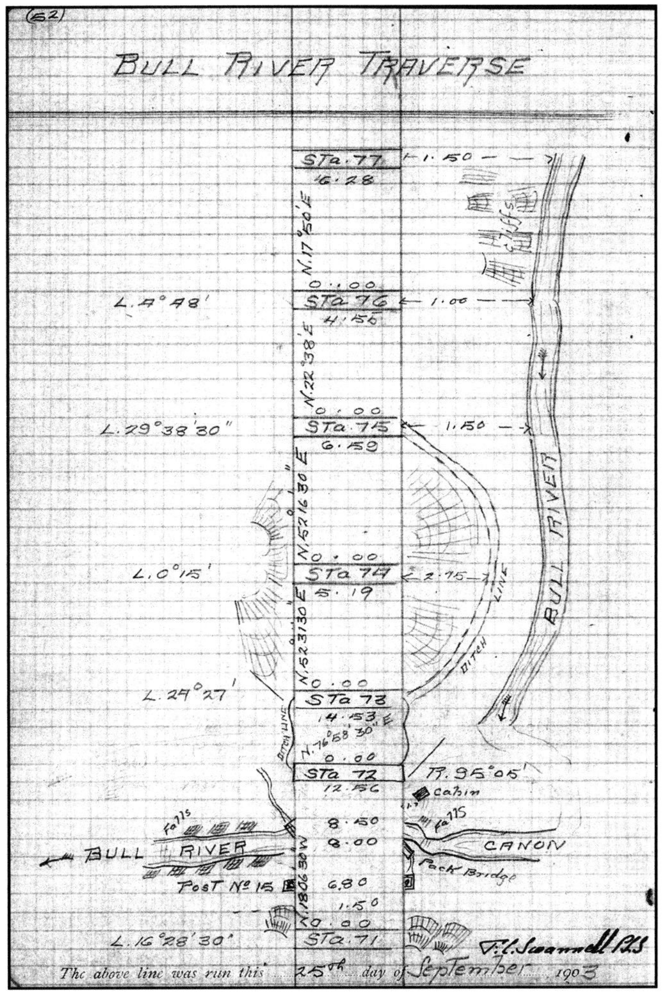

❖ Occasionally Swannell ran a traverse survey. This type of survey follows a general direction instead of a fixed line. In 1903 Swannell's crew ran a traverse survey up the Bull River valley that was over 40 kilometres long, making a connection between a station on the railway line and a point on the line of the lot that they were surveying. This page from the traverse survey is again read from the bottom up. The stations are numbered, and the distance between them is directly below the station. The actual astronomic direction computed by Swannell is written below the distance on the left side of the middle section. The deflection left or right from a 180 degree line, using the previous station as a back sight, is recorded beside the station number. For example, Swannell recorded that the distance from Station 76 to 77 was 6.28 chains and that the direction was N17°50'E. This was a deflection to the left of 4°48' (marked on the left of Station 76) from the direction of N22°38'E at the previous station. The only station on this page that has a deflection to the right is Station 72, where the direction changes from N18°06'30" W to N76°58'30"E. This change of 95°05' is marked to the right of Station 72. Swannell noted where the traverse crossed the Bull River and intersected the ditch line. He sketched the location of a nearby cabin, the falls on the river and the approximate distance (in chains) from the traverse line to the Bull River as they proceeded upstream. Map courtesy Land Title Survey Authority.

❖ A Tsilhquot'in First Nations man and his son. Image I-33188 courtesy of Royal BC Museum, BC Archives.

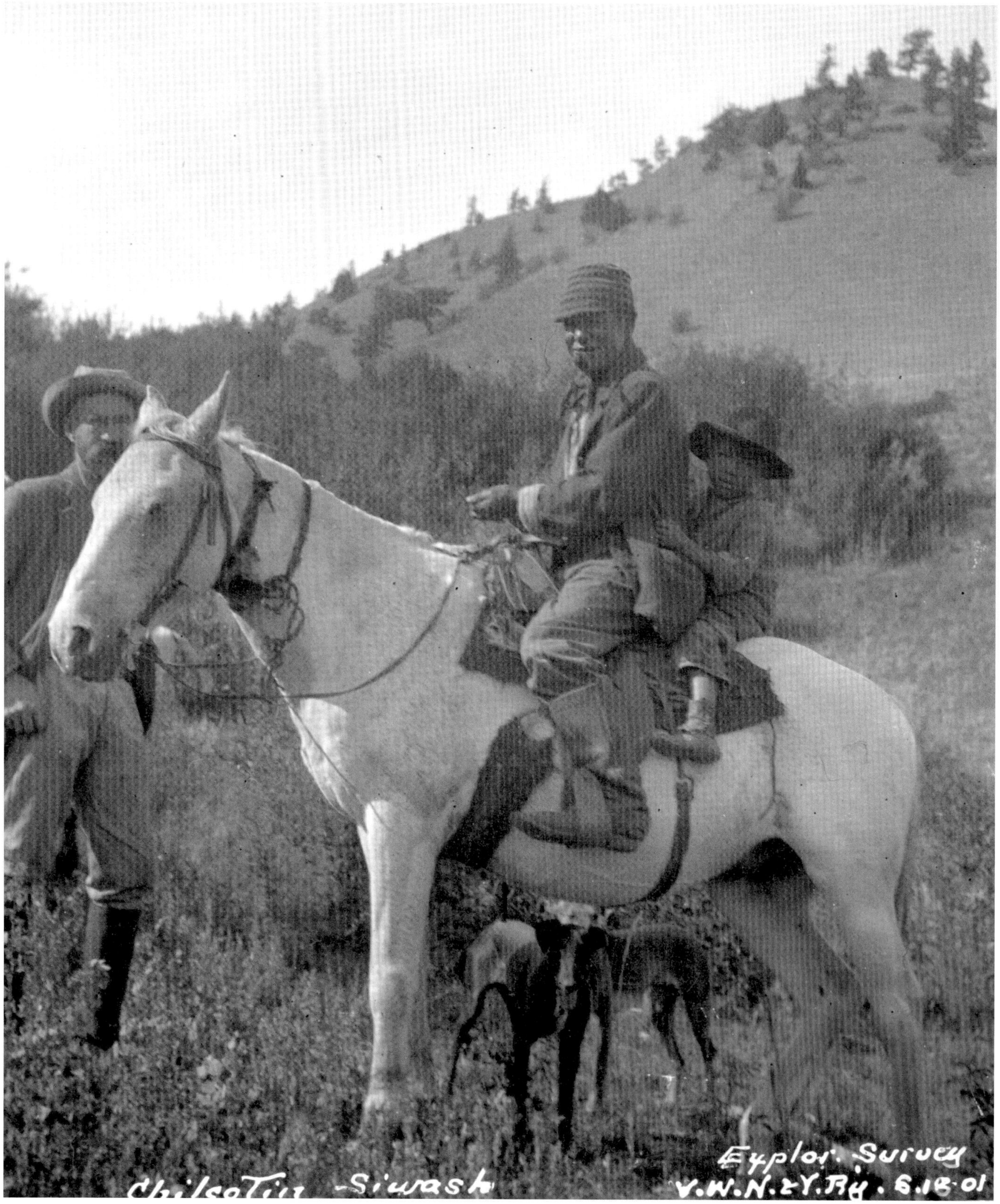

1901

During the winter of 1900–1901, Swannell completed the office work for the surveys from the previous summer, copying field notes, making calculations and learning the skills that would enable him to become one of BC's premier cartographers.

In his diary Swannell recorded that he celebrated the new year in Seattle before returning to Victoria on the steamship *Rosalie* the next day. A few weeks later he enlisted with the 5th (BC) Regiment of the Royal Canadian Artillery, a reserve unit, and he served for one year. Canada had soldiers in the South African War at the time. Swannell noted the death of Queen Victoria on January 22.

Swannell had an interest in and talent for languages. When he came to Victoria he already knew German. During 1901 he started taking French lessons and began using French phrases in his diary. Swannell tried to find someone to correspond with who knew Danish. He also purchased a Chinook dictionary and started learning that language so he could converse with the First Nations people he met on his surveys. Phrases of Chinook appeared in some of his diary entries and letters.

Swannell's first surveying in the spring of 1901 occurred from May 6 to 11 when he went with Gore and his crew to Saltspring Island. On the first day, one of the axemen sliced three of his fingers. The next day, the crew "retraced old line run in 1875. Found broken glass at corners." The early surveys used wooden posts to mark the corners of lots. Since these posts quickly deteriorated in the damp weather of the Pacific coast, the surveyors often left more permanent items like glass or metal near the survey post. When they ran another line the following day they found glass again near the post. May 9 was a special day for Swannell because he

Mining on Texada Island

Texada Island, the largest in the Strait of Georgia, was the scene of mining activity beginning in the early 1870s. Initially the main ore was iron, followed by a small gold rush in the 1880s. In the late 1890s a boom occurred, with several mines containing gold and copper ore being developed. In 1899 a smelter was constructed near Van Anda, the largest community on the island. In 1901, when Swannell surveyed Texada Island, there were 124 mining claims staked on the island, over half of the claims registered in the Nanaimo district that year. Several of the mining claims that Swannell surveyed on Texada eventually developed into mines.

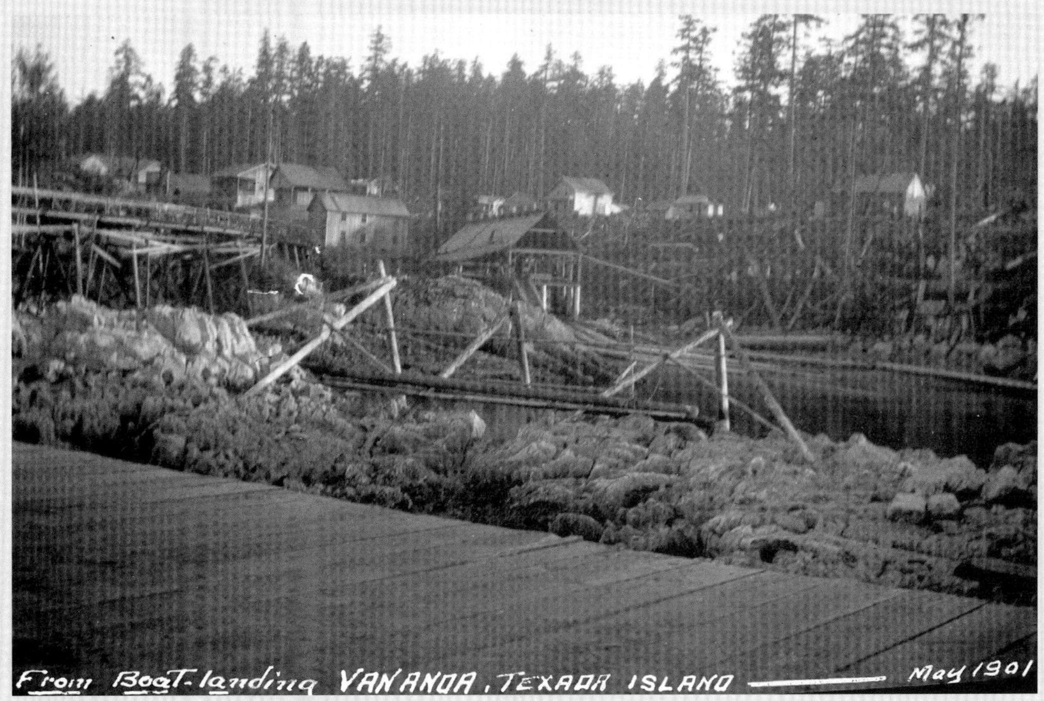

❖ Van Anda was the largest mining community on Texada Island in 1901. Image I-33546 courtesy of Royal BC Museum, BC Archives.

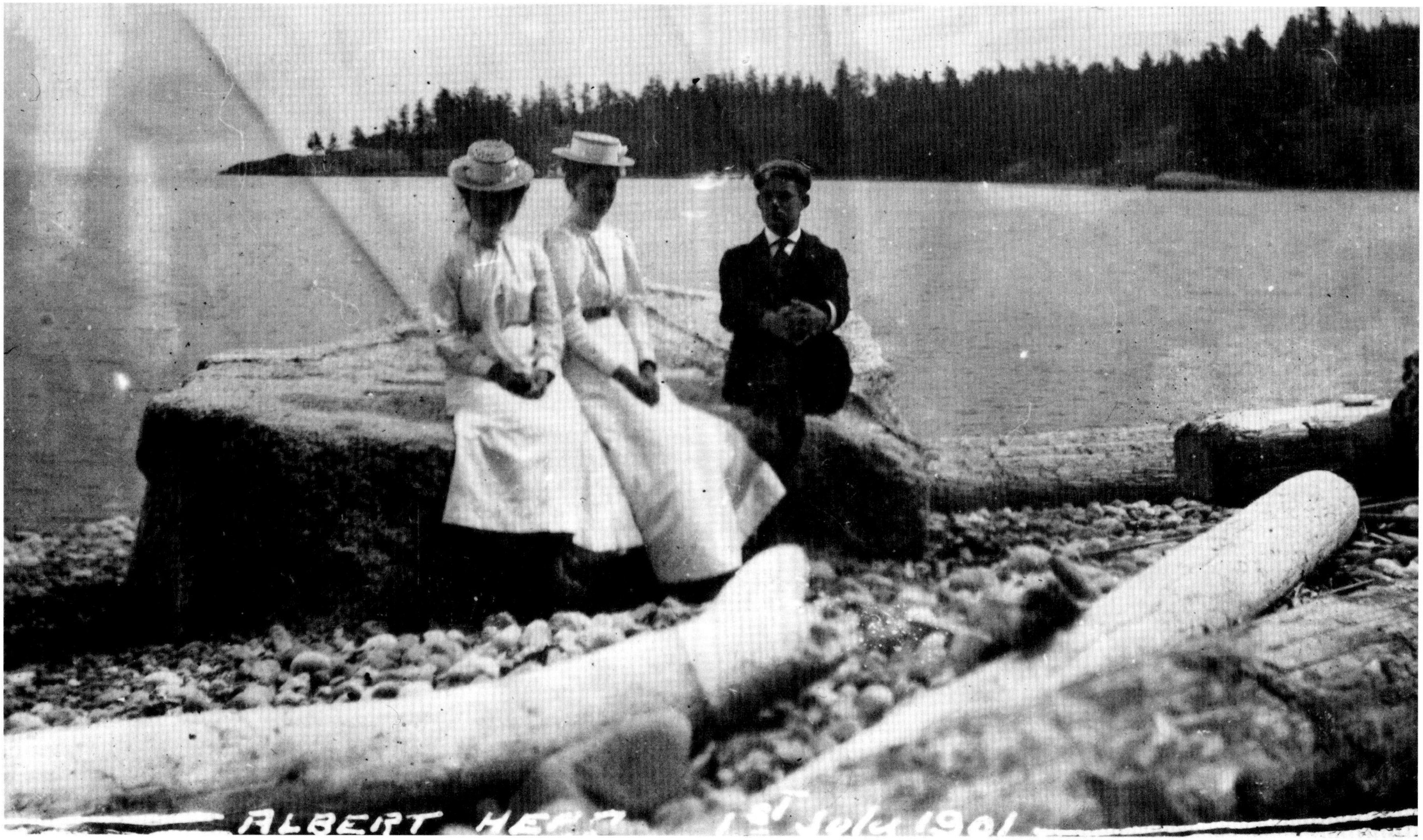

❖ Above: George Copley used Swannell's camera to take this photograph of a picnic near Victoria on Dominion Day (Canada Day today). Nora Mabel Liddell is on the left, Ada is in the middle and Frank is on the right. Image J-00314 courtesy of Royal BC Museum, BC Archives.

❖ Opposite page: Canoe used for the Pacific cable station survey. Image I-33640 courtesy of Royal BC Museum, BC Archives.

operated the transit by himself for the first time. He continued running the instrument the next day until the survey was completed. On May 11 Swannell and the crew returned to Victoria.

Two days later the men left for Vancouver on the SS *Joan* where they boarded the SS *Saga* on May 14, bound for Texada Island. The following day they began surveying the LeRoi mining claim. Once again Swannell operated the transit. But his new position also brought the possibility and responsibility for mistakes in his readings or calculations. When Gore and his crew next surveyed the Boulder Nest mining claim, Swannell noted one day that he had made a mistake in his calculation and, on another day, that he "skipped level measurement on end line which threw corner 41′ [12.5 m] out." Gore was patient with his young, articling pupil. Despite Swannell's errors he continued to let him operate the transit, and the survey proceeded without any further mistakes. In his diary Swannell noted that Gore gave him the opportunity to survey one mining claim himself and that it checked to .8 feet accuracy [24 cm]. One evening Swannell "went trolling, caught 10 lb [4.5 kg] dogfish." By May 25 Gore's crew had finished surveying five mining claims on Texada, and the following day they returned to Vancouver on the SS *Saga*. Swannell noted: "Siwashes with

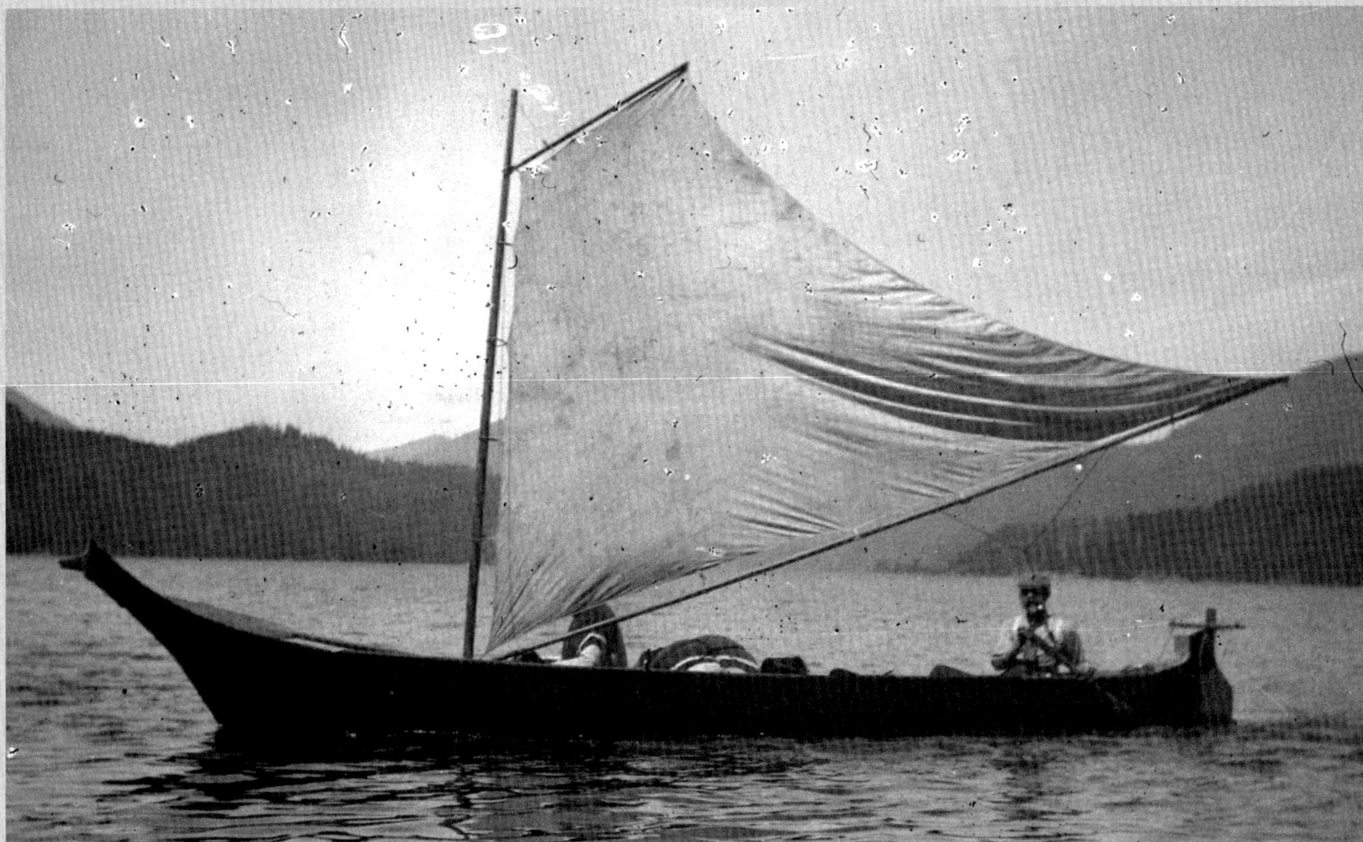

Pacific Cable Survey

At the beginning of the twentieth century, the telegraph was still the main form of long-distance communication. In the 1860s regular telegraph service between North America and Europe had been established through a transatlantic cable line. During the latter part of the nineteenth century Britain wanted to establish a telegraph system that would span the globe and link its empire with rapid communication. By 1900 there was one major section left to complete—from Canada across the Pacific Ocean to Australia and New Zealand. Great Britain, Canada, Australia and New Zealand formed the Pacific Cable Board to plan the logistics and finances of establishing this cable line that would be over 11,000 kilometres long. Sir Sandford Fleming of Canada was one of the chief architects of this project. The location selected for the eastern terminus of the cable was at Bamfield on the west coast of Vancouver Island. Although remote, it was the most suitable place, meeting the conditions necessary to successfully lay this long, 8,000-ton telegraph cable. In the summer of 1901 Gore & McGregor received the contract to survey the site for the cable station.

The cable was laid during the late summer and fall of 1902. From Bamfield the cable went to Fanning Island (Tabuaeran) south of Hawaii, then on to Norfolk Island between Australia and New Zealand. Here two cables were laid: one to New Zealand and the other to Australia. Australia already had a connection across the Indian Ocean to India and Europe. The first worldwide telegraph message was sent on November 1 by Sir Sandford Fleming from Ottawa to the Canadian Governor General in the same city. This British telegraph system that linked the globe was called the All Red Line, and it was very successful, both in establishing rapid communication throughout the British Empire and for political purposes. It also became a financial success.

The cable station at Bamfield operated for over fifty years. Today there is a national historic marker at Bamfield to recognize the significance of the Pacific Cable Survey. It states: "The movement for this British state-owned cable was initiated by Sir Sandford Fleming in 1879 and carried to a successful conclusion, nearly a quarter of a century later, mainly through his persistent efforts. The cable, laid at the joint expense of Great Britain, Canada, Australia, and New Zealand, was completed and officially opened on 31st October, 1902."

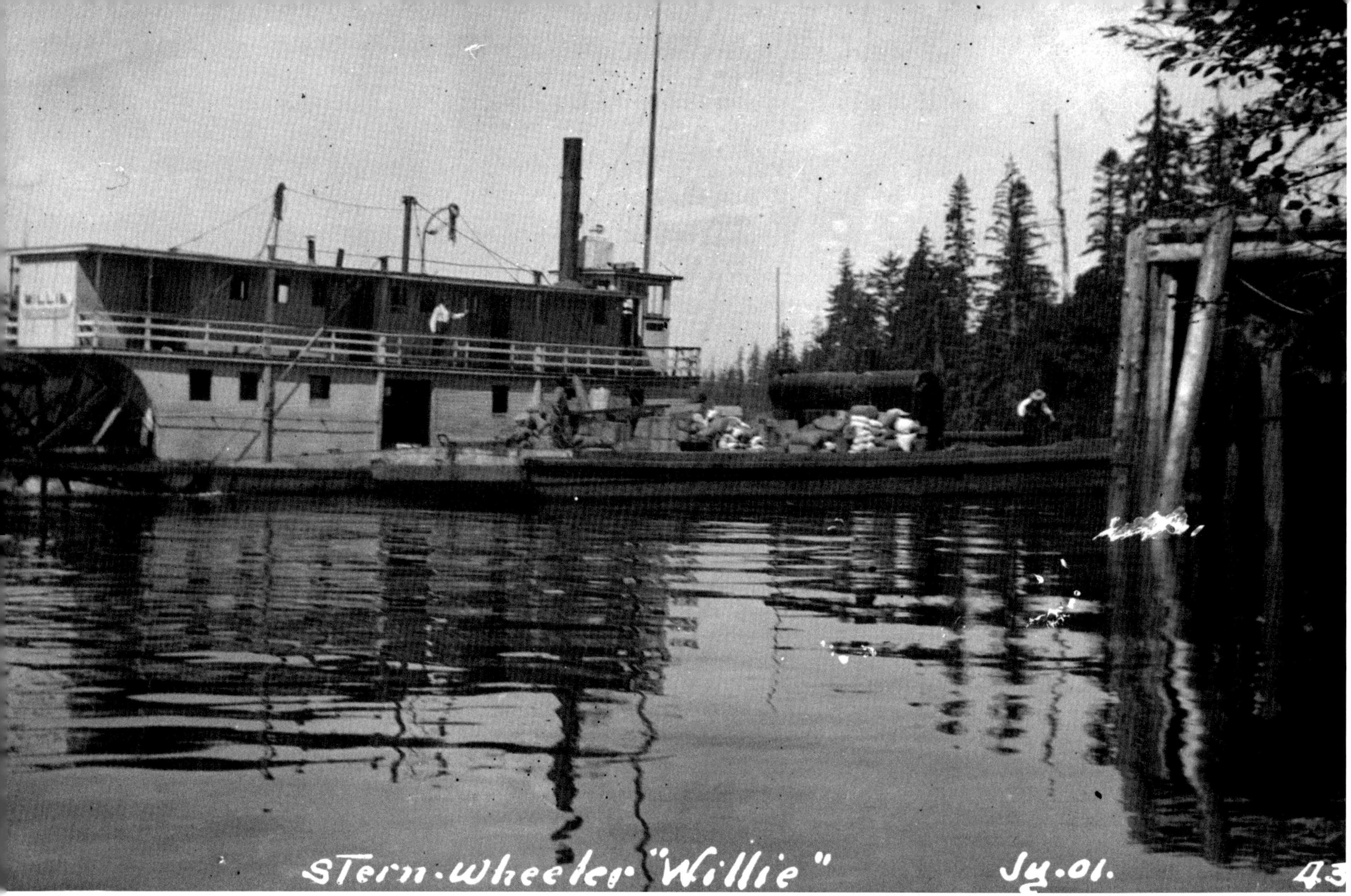

The sternwheeler *Willie*. The *Willie* was a small sternwheeler that was about 20 metres long. Built at Seattle in 1883, it operated on the Washington coast until the mid-1890s, when it was sold to a company in BC. The *Willie* was used mainly on the Fraser River. Image I-33641 courtesy of Royal BC Museum, BC Archives.

cub bear on board." The trip to Victoria was made on the SS *Yosemite*. At the office Swannell completed the field notes for the Texada survey.

During the first half of June, Swannell worked on surveys in the Victoria area. For the last two weeks of the month he and his friend, George Copley, who had also enlisted in the 5th Regiment of the Canadian Artillery, were at military camp, spending much of their time on infantry and gun drills. In June Swannell began mentioning Ada Driver in his diary along with their activities, which took place mainly on weekends. He, Ada, George and Mabel celebrated Dominion Day (Canada Day) by going on a picnic to Albert Head southwest of Victoria.

On July 23, Gore, Swannell and the survey crew left for Bamfield on the west coast of Vancouver Island to survey the site for the Pacific Cable project. They travelled by train to Nanaimo and then by stage for 45 kilometres before stopping for the night at a boarding house. The next day they reached Port Alberni. While there, Swannell went fishing on the Somass River. On July 25 they "went by sternwheeler *Willie* to Bamfield Creek and pitched camp. Mosquitos very bad."

They began their survey the following day by making a traverse along the ocean shore. In his journal Swannell noted seeing HMS *Egeria*, a Royal Navy

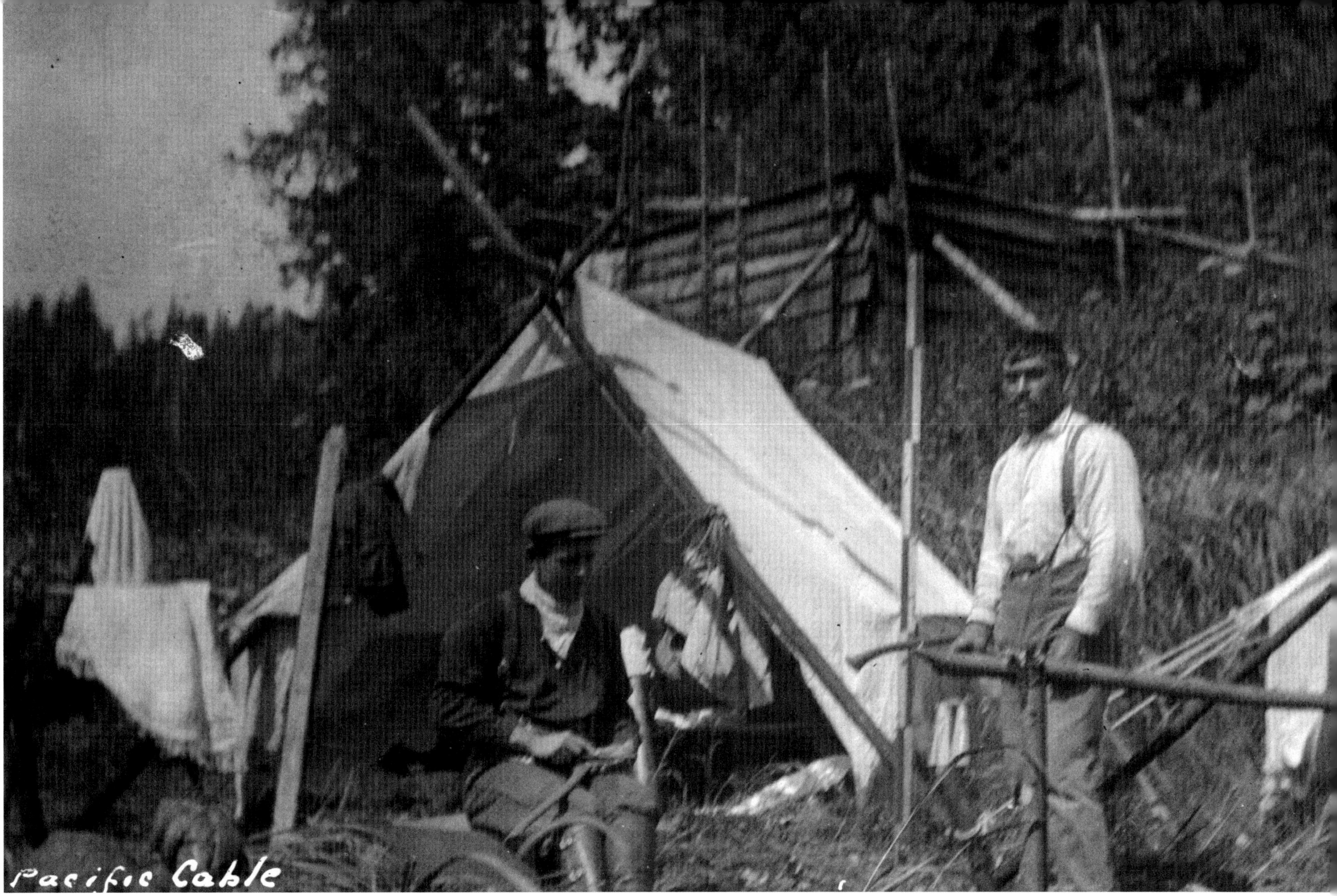

steam sloop that did hydrographic surveys on the British Columbia coast for several years and was involved in the Pacific Cable project. The men continued surveying the site around Bamfield on July 27. That evening, Swannell went with Johnny Klootisee to visit the nearby reserve at Dodgers Cove.

The next day they moved camp to Grappler Creek. After work Swannell tried, unsuccessfully, to obtain fresh meat for the crew by going deer hunting in the canoe. However, he was able to shoot a duck the following day. The survey for the Pacific Cable site continued until early August, but Swannell's diary for the beginning of the month is missing.

On August 21 Swannell, Gore and the survey crew left for Vancouver, taking an overnight voyage on the SS *Charmer*. He spent the next day in the city, walking around Stanley Park and staying overnight at the Hotel Badminton at the southwest corner of Howe and Dunsmuir streets. On the twenty-third the men left for Britannia on the SS *Defiance*. Swannell observed smoke from forest fires obscuring the mountains. They spent three days surveying the Kalispell mining claim. On the twenty-fifth Swannell briefly described a difficult day of surveying: "Ran two lines Kalispell. Thick timber, very bad line. Flies bad. Greslock slashed his knee." After completing the survey Swannell returned

❖ The Pacific cable station survey camp at Grappler Creek. Frank Swannell is in the middle of the picture. Image I-33643 courtesy of Royal BC Museum, BC Archives.

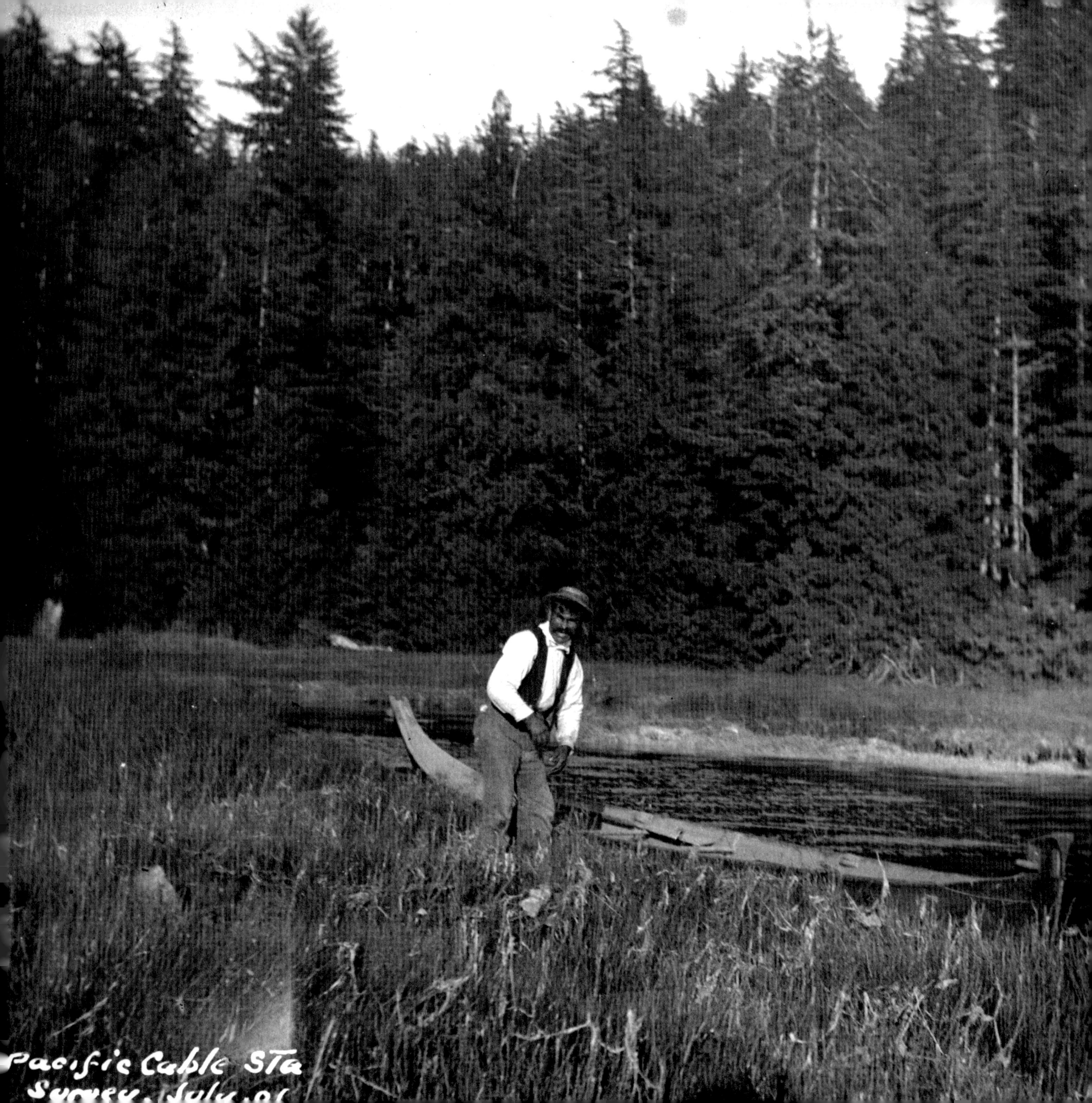
Pacific Cable STa

to Vancouver, where he spent a day at the Hotel Badminton copying his notes and drawing the plan for the survey. He wanted to complete the office work because they were heading north on a project instead of returning to Victoria. Swannell also wrote to Ada, describing his survey adventures that week.

Hotel Badminton, Vancouver
Mineral Claim Survey
Britannia Mines—Howe Sound
August 27, 1901

Have just got back from Howe Sound, am again a gentleman of leisure, accordingly will sling some more ink at you.

We had a splendid trip up of about 7 hours—calm except off Atkinson Point where we struck a tide rip, which made the little *Defiance* try all sorts of violent gymnastics. I shan't attempt to describe the scenery—narrow fjord-like stretches of water between towering mountains or among rocky fir-clad islands—here and there a narrow bay with a strip of sandy beach and perchance a log boom and a shake-roofed cabin or two, with a skid road zigzagging up the mountain slope. The south peninsula of Gambier Island was all afire—smoke for miles, through which the snow-capped peaks and jagged fir-crowned bluffs loomed indistinct, tremulous.

This fire made photography almost impossible, so I will have very few mementoes of Howe Sound for my girl in Victoria, albeit I think I have four good pictures.

Landing at the float at Britannia Beach we procured a pack cayuse and cinched our belongings thereupon. Perhaps you have never witnessed the process of cinching up? Picture to yourself the unfortunate cayuse as bundle after bundle is hoisted on the alpareja (pack saddle). He furtively eyes each additional one with an air of abject dejection. The diamond hitch thrown over everything the packer braces his foot against the saddle and tightens the rope till you'd swear the unfortunate horse's ribs will stave in. Not a bit of it, however, that cayuse wasn't born yesterday, he'll have that load shifted loose in short order and then will the outfit gently pour blessings in his ear.

Twas a nice climb we had to altitude 2300 feet [700 m] up a trail which would make Mabel's hair stand on end—wasn't exactly a turnpike road, seeing it is quite a respectable torrent bed in wet weather.

Well, we arrived finally above the clouds at dusk, built a fire and indulged in your favourite recreation. We didn't have a tent, so imagine us four scattered along the mountain, each peacefully reposing among the boulders in his own particular lair. Great fun watching the cloud rack sweep past the jagged tree tops while Jane Creek roared in the darkness beside us—and you must remember there's never a gleam of light in a fir forest at night—you can feel the darkness. I much prefer sleeping under the open sky in fine weather though tis passing cold work for a greenhorn, ere he learns by freezing experience how to make a cold-proof chrysalis out of one double blanket.

The work was pretty bad—thick brush and heavy timber, and the Strathcona lads about as adept at handling an axe as a certain Ada Driver would be. One of them hit flat. As a result he neatly sliced the side of his puttee off and slashed his knee, although not enough to disable him.

Our cook is a jewel, a most inventive genius with a penchant for weird culinary concoctions. For the benefit of Mdlle Anno Domini, against the time when aforesaid mademoiselle will have to make cake for some fellow or other, I give his latest evolved recipe:

Cake a la Survey: Musty flour—a sufficiency. Baking powder: a fistful. Raisins galore. Marmalade—as much as will stick to a clasp knife blade: syrup (with sandflies therein)—considerable. 'Tin cow'—half tin

❖ Opposite page: Johnny Klootisee was a First Nations person from the Port Alberni area who was a member of their crew on this survey. He is standing in front of a canoe. Image I-33133 courtesy of Royal BC Museum, BC Archives.

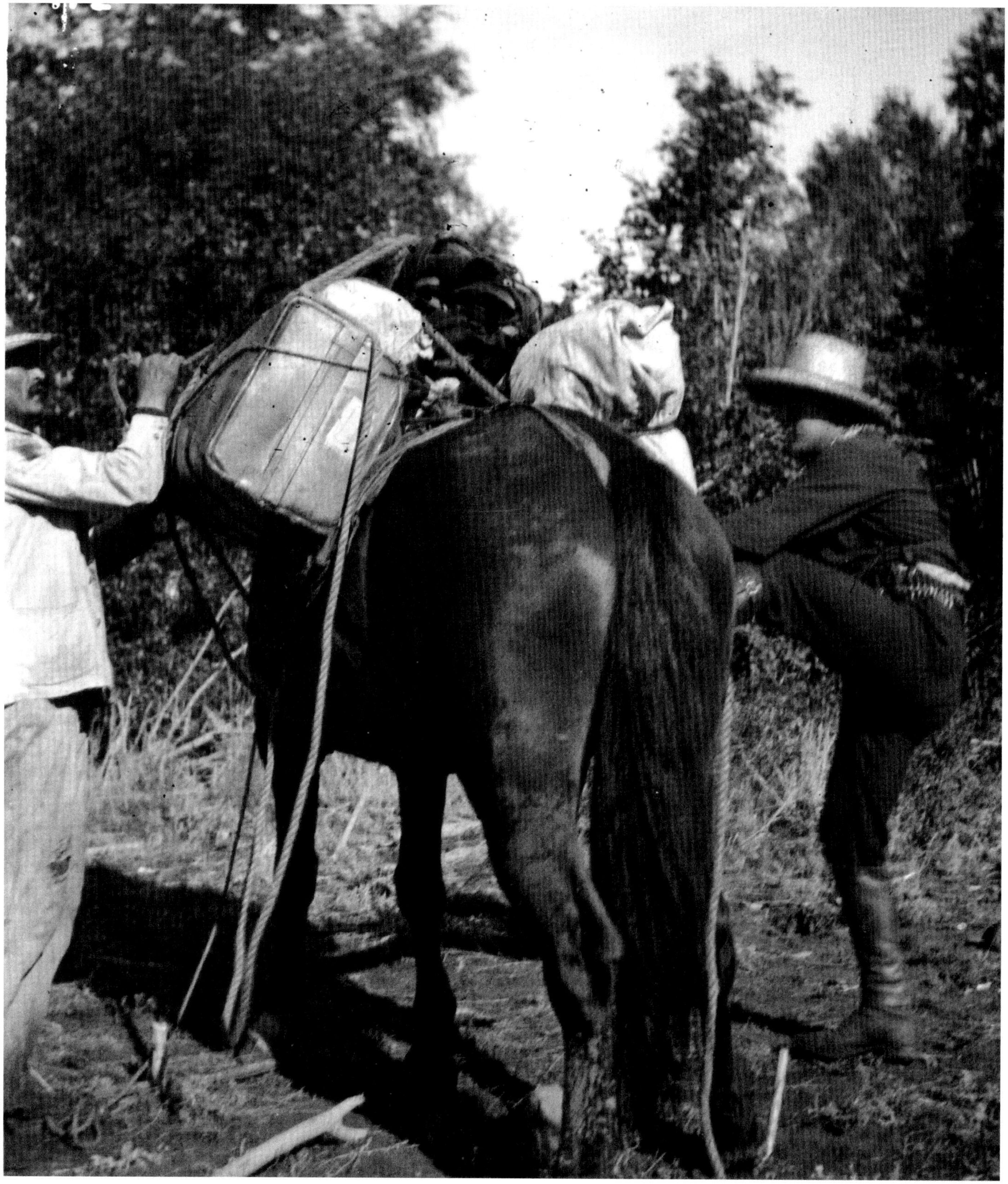

❖ Cinching up a pack horse. Image I-58357 courtesy of Royal BC Museum, BC Archives.

Vancouver, Westminster, Northern and Yukon Railway Company

The idea of building a railway from Vancouver to the Yukon in 1899 was one of the most fanciful in BC history. There was already boat service to Skagway, Alaska, and the White Pass and Yukon Railway was being constructed to Lake Bennett where there was boat access to Dawson City. Yet in 1899 the British Columbia government gave its approval for the incorporation of the Vancouver, Northern and Yukon Railway Company. The Company received no land subsidy but gained a virtually unlimited opportunity to construct a railway line through the northern BC Interior. In the company's petition to the government it stated that it wanted to build a line from Vancouver "running in a northerly direction by way of Seymour Creek, or the most feasible route, to the Squamish Valley; thence by the most feasible route through the Pemberton Meadows to Lillooet; thence northerly to Quesnelle [Quesnel]; thence northwesterly to Hazelton or some other point on the Skeena River; and thence northerly to the northerly boundary of the Province." The company also received permission to build any feasible branch lines. The British Columbia government was desperate for any economic development, and the fact that the company's petition was supported by several prominent people probably ensured that it received a charter.

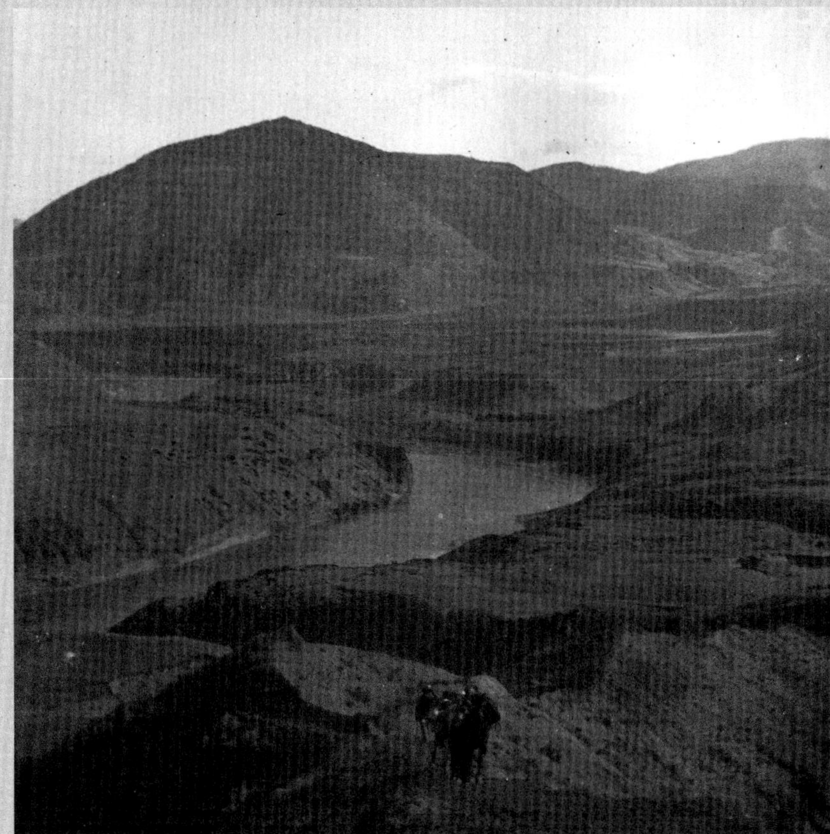

❖ "The trail to Quesnelle, 40 miles [64 km] above Lillooet." The Fraser River is in the middle of the picture. Big Bar Mountain is in the background. Image I-57558 courtesy of Royal BC Museum, BC Archives.

In 1900 the company again petitioned the government, this time to change the railway's name to Vancouver, Westminster, Northern and Yukon Railway. This small change was important because the railway now obtained permission to also construct a line between Vancouver and New Westminster. John Hendry, a prominent Vancouver lumberman and businessman and one of the main supporters of the Vancouver, Westminster, Northern and Yukon had close ties with J.J. Hill of the Great Northern Railway. Hill wanted to establish a railway line into the Vancouver area in addition to his ventures into the Kootenays, and he had constructed the New Westminster Southern Railway from the US border to the south bank of the Fraser River. Hendry anticipated that the two railways would connect once a railway bridge was constructed across the Fraser. The section from Vancouver to New Westminster was the only part actually built, and it was profitable. In 1904 a railway bridge was constructed across the Fraser enabling the Great Northern to operate a train service into Canada's main Pacific port.

The route north from Vancouver was divided into four sections with the government giving the railway company a fixed amount of time to complete each section. Despite the railway's impracticality, a portion of it was actually surveyed. Swannell was part of the crew that surveyed the second section, from Lillooet to the Blackwater River, north of Quesnel, in 1901.

In a letter to the premier in April 1901, T.S. Gore wrote that the railway had the potential to develop the economy of the Cariboo. "I know very well the part of the proposed route from Lillooet to the Chilcotin River ... From what I have seen and heard of the country I think it is a perfectly practicable and favourable one through which to build a railway. And until such a railway is built, the immense mineral and by no means unimportant agricultural resources of this large area of the best portion of British Columbia must remain undeveloped."

Although there was no work done on the railway north of Vancouver, the Pacific Great Eastern followed a large portion of the proposed route when it began construction in 1912.

cupful. Water: ad libitum. Mix: Place before a reflector with lots of fir bark on fire. Test for doneness with a hemlock sliver.

I grieve to state that our cook hadn't anything else to dump in except beans.

On August 28 Swannell and the survey crew left by train for Lytton, staying overnight at McMillan's Hotel. The next day they set up camp near Lytton, had a swim in the river, and went to a dance in the evening. A couple of ladies from Lytton visited their camp the next day, and once again they had a dance in the evening.

The crew left for Lillooet on the last day of August, travelling 30 kilometres to Maddock's ranch. By September 2 the men were camping near Bridge River. Swannell noted that there were no salmon in the Fraser. At this location Swannell received a letter from Ada and he replied before they left.

> Railway Reconnaissance Survey Lillooet to Blackwater Yukon Tel Trail
> Camp near Bridge River 14 miles from Lillooet, Sept. 3, 1901
>
> Our men were in Lillooet for the last time yesterday and woke Gore and me in the wee small hours by chucking mail at us, so we lit a candle and tackled the letters. Had you written a day later I wouldn't have got your letter for two weeks, the next post office being 64 miles [100 km] above us up the Fraser Canyon. Constant going over mountain roads has opened up that old axe wound and Gore wouldn't let me ride today, ordered me to stay and help move camp. Therefore I have a chance to write but have a very indifferent table and pen which I heartily anathemize [anathematize] ...
>
> We got to Lytton last Tuesday night after an 8 hour railway journey from Vancouver. Next day pitched camp "delate halo kloosh" [not very well—Chinook] near the town. My two companions got very chummy with the landlady at the hotel, principally because there was a rather pretty girl there I opine. Verily the ways of a man be strange in his youth! Greslock and I returning from a swim at the junction of the Fraser and Thompson found Orchard snugly ensconced in the parlour regaling himself on watermelon in company with the only two girls in the village possessively chaperoned by my host!—we joined the coterie to look after our cook.
>
> They got up a musicale in our honour this evening. We told everybody what a musical prodigy Orchard was, so he was removed from feminine temptation by being made to stick to the piano. We left the boss in camp to watch things.
>
> Next day the whole outfit visited us and took [us] driving. Greslock got into trouble going out—girl (Annie MacTaggart) at the bottom of the mischief as usual. He slipped his hand on the back of the seat going up a steep hill "to prevent the young lady from falling" as he explained. The driver in the front seat took a different view of the situation, scowled savagely at G and told him "quit hugging my Annie." At camp Lizzie Shackleton was rigged in bandolier & rifle, not to mention the cook's carving knife and a .44 Colt—a la desperado ... Had a dance that evening.
>
> We took over two days riding to Lillooet—the confounded pack cayuses were always trying to work their packs by rolling. The road was 1000 ft [304.8 m] above the Fraser most of the time. At Big Slide 28 miles [45 km] out the road is cut out of an almost vertical bluff, the muddy swirling Fraser far below, while above was a slide of loose rock rising up till the trees at top looked like bushes. Had trouble galore with our pack horse Calamity who tried to edge the other horses over into the river at the most ticklish place—there is no parapet. 'Twas a ticklish proposition.

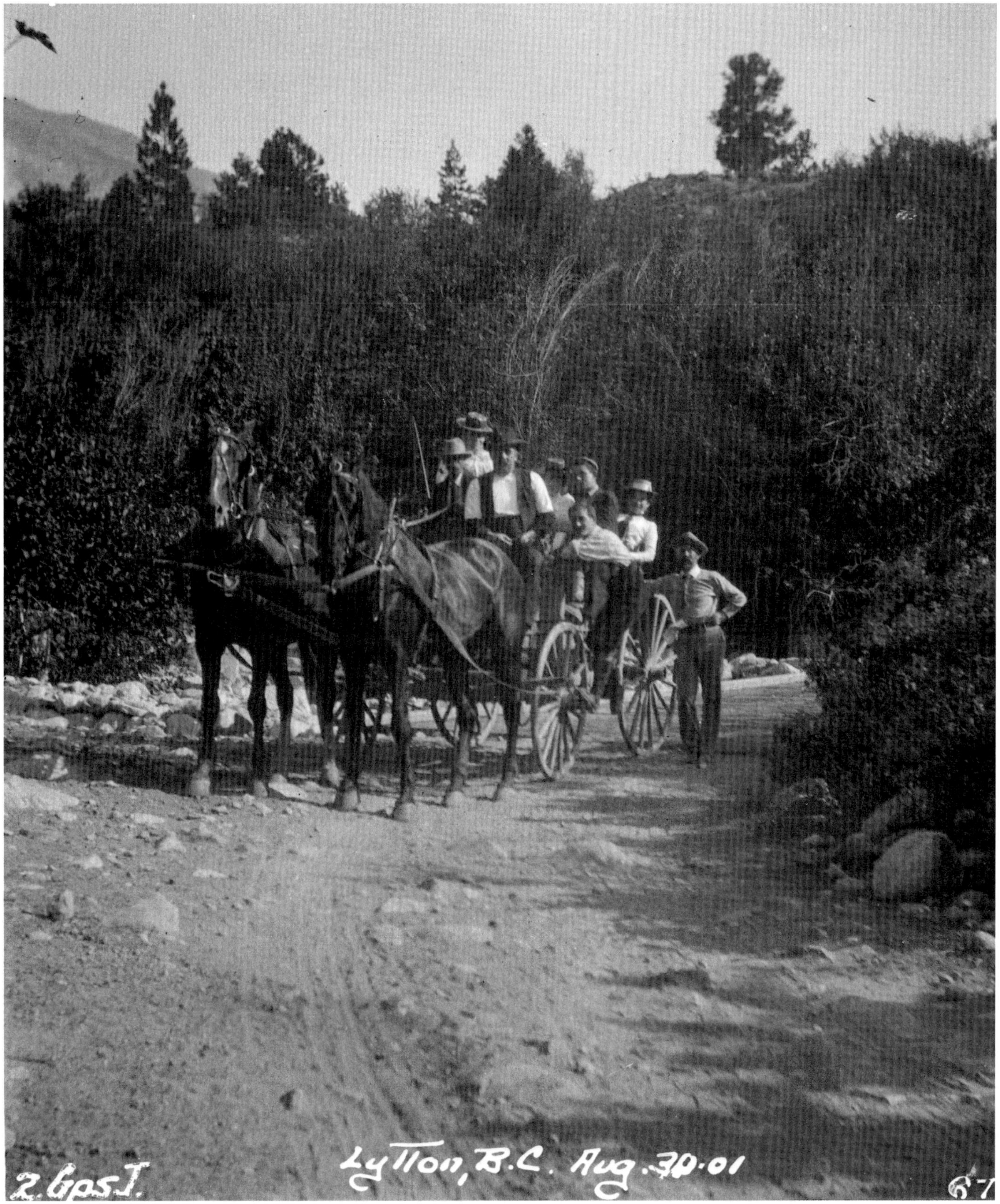

❖ The wagon ride in Lytton that Swannell described to Ada. Image I-58105 courtesy of Royal BC Museum, BC Archives.

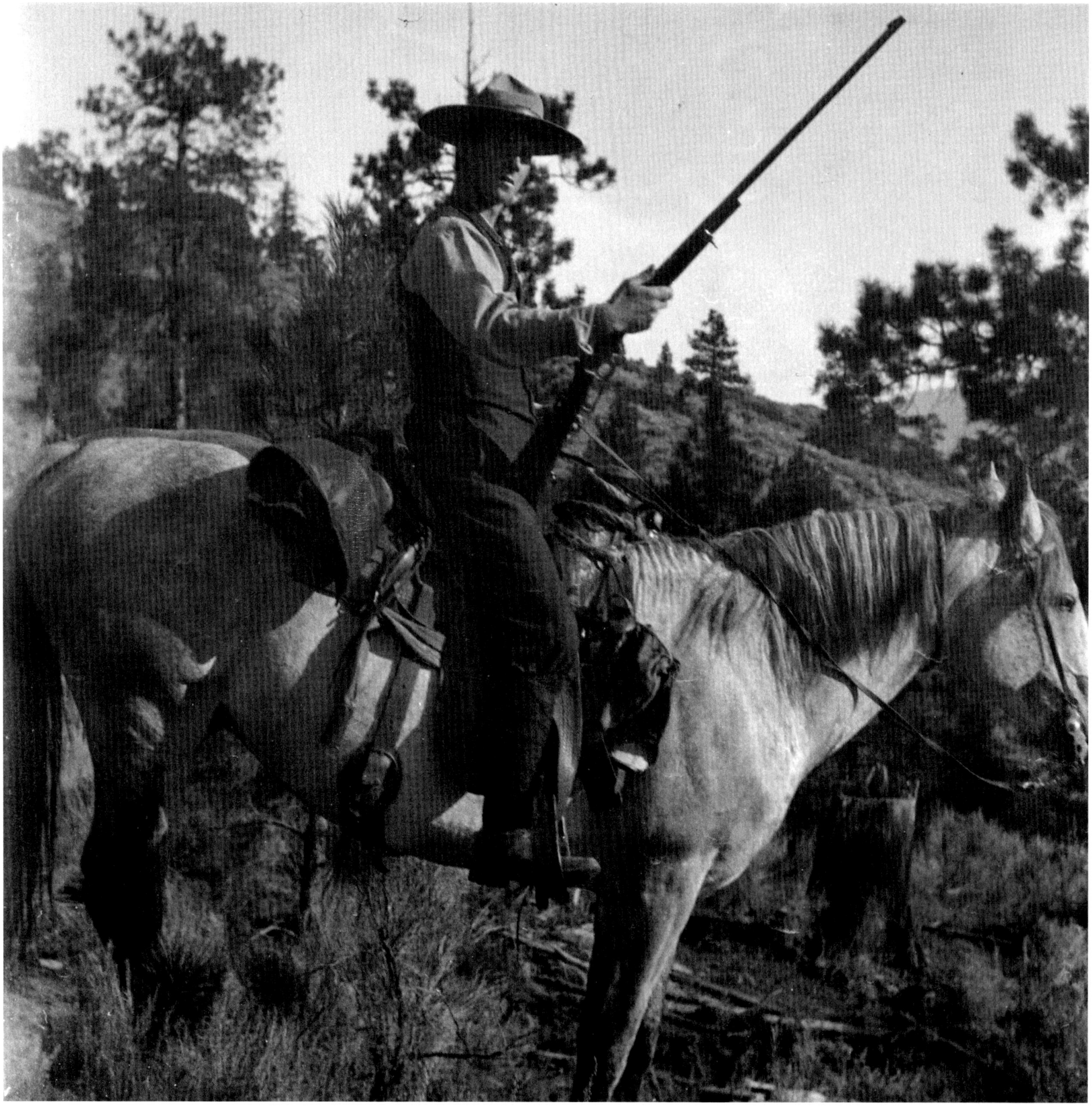

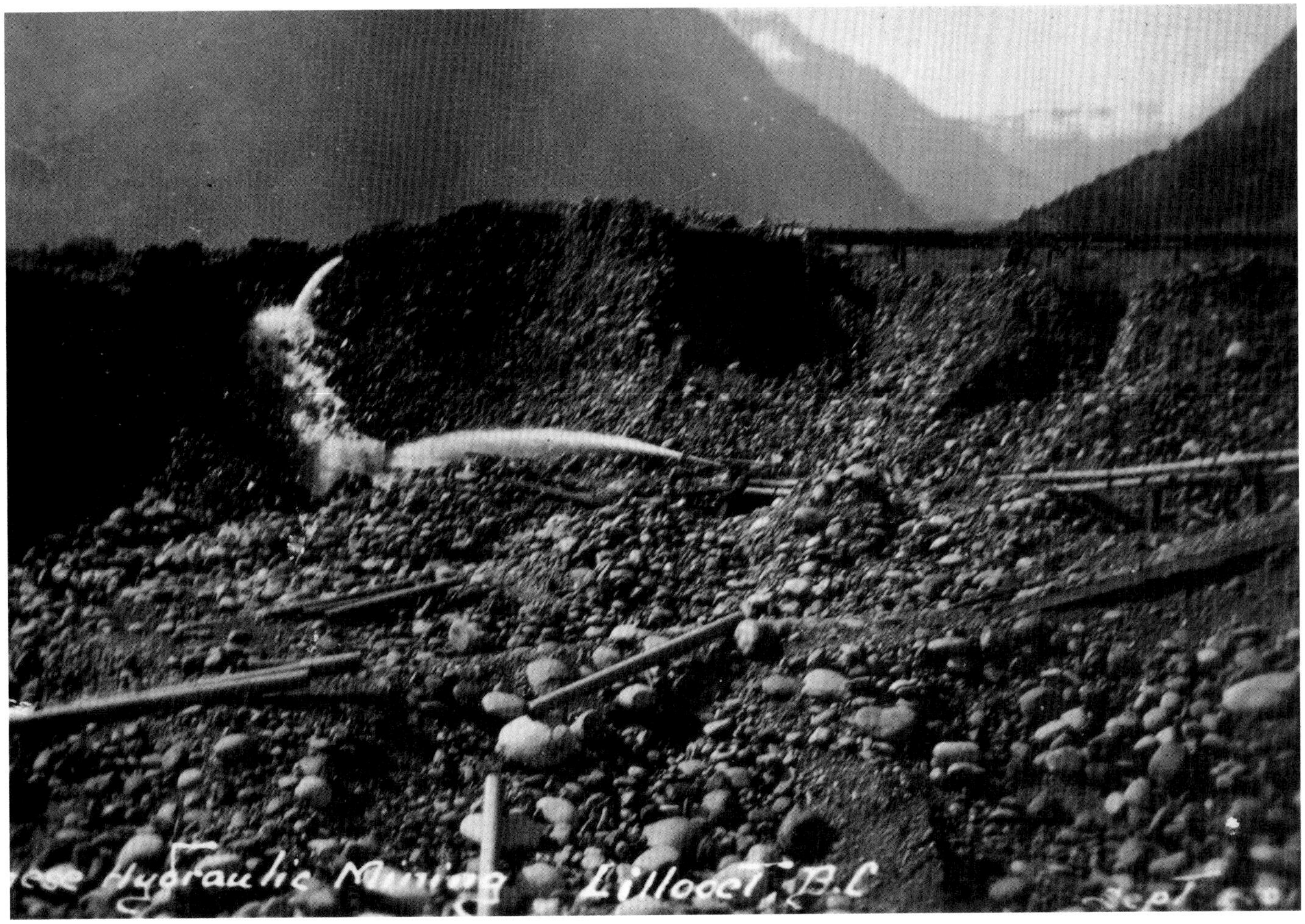

❖ Chinese hydraulic mining near the Fraser River. This was near the MacDonald ranch at Watson Bar. Image J-00315 courtesy of Royal BC Museum, BC Archives.

❖ Opposite page: Howard Greslock. In 1953 Swannell wrote this note about Greslock: "One of the most inexplicable things I have ever experienced occurred regarding Greslock—In 1915 at Ploegsteert in Flanders I heard we (116th) were relieving Strathcona Horse. I had completely lost track of Greslock since 1901 beyond hearing he had gone to the Klondike. I was standing at the mouth of a very deep communication trench—it flashed through my mind 'I wonder if that chap Greslock has rejoined his old South African war unit.' The Strathcona started straggling out. I stopped one of the first and asked, 'Have you a fellow named Greslock in your outfit?' He answered, 'I am Greslock.' I never saw him again, but he got a commission and was killed on the Somme. The trench was so deep I could not possibly have seen him before I spoke, indeed I did not recognize him then, for he had changed much in 14 years." Image I-58062 courtesy of Royal BC Museum, BC Archives.

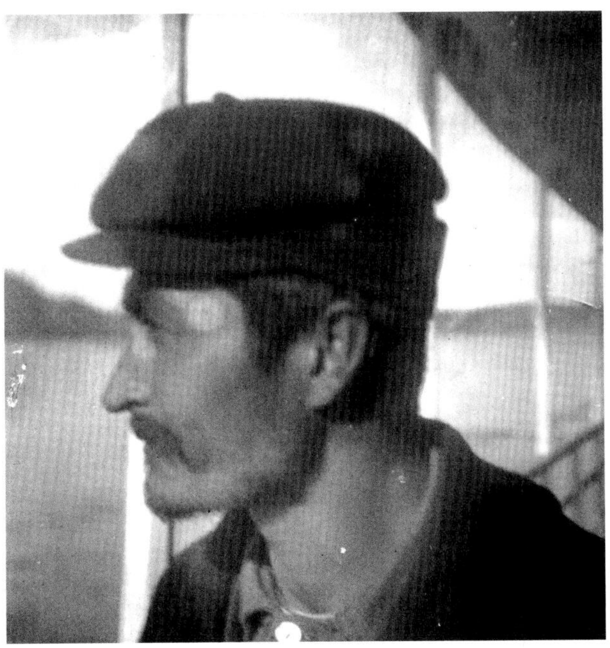

❖ Swannell noted that "Lytton Indian Jimmy (commonly known as Jimmy Lytton)" was their packer on this surveying project. After he returned to Victoria, Swannell sent Jimmy (also called James Dunstan) some photographs from the survey and received a thank-you note from him. Swannell wrote that he stopped at Lytton in July 1954 and tried to locate Jimmy, only to learn that he had died the previous year. It appears that Swannell took this photograph on the SS *Charlotte*. Image I-33199 courtesy of Royal BC Museum, BC Archives.

We had a bad time passing the stage coach—only two feet between the wheels and a jump off of nearly a quarter of a mile [400 m] …

Culinary? Well, my girl, here's our stove, very portable—bright tin, reflects heat on bread in pan inside … It folds up flat, so is easily packed. I'm afraid I'd better stop, or you'll accuse me of always thinking of the delectable but fearfully contrived chef d'oeuvres of Bert.

Now will wind up, trust to luck to get this posted. Will expect a longer letter than this, you needn't build it on the strictly stiff & conventional lines of a 4" collar—we're on pretty intimate terms.

I warrant I work as hard as "the wife" which is upcountry for Sweetheart which is

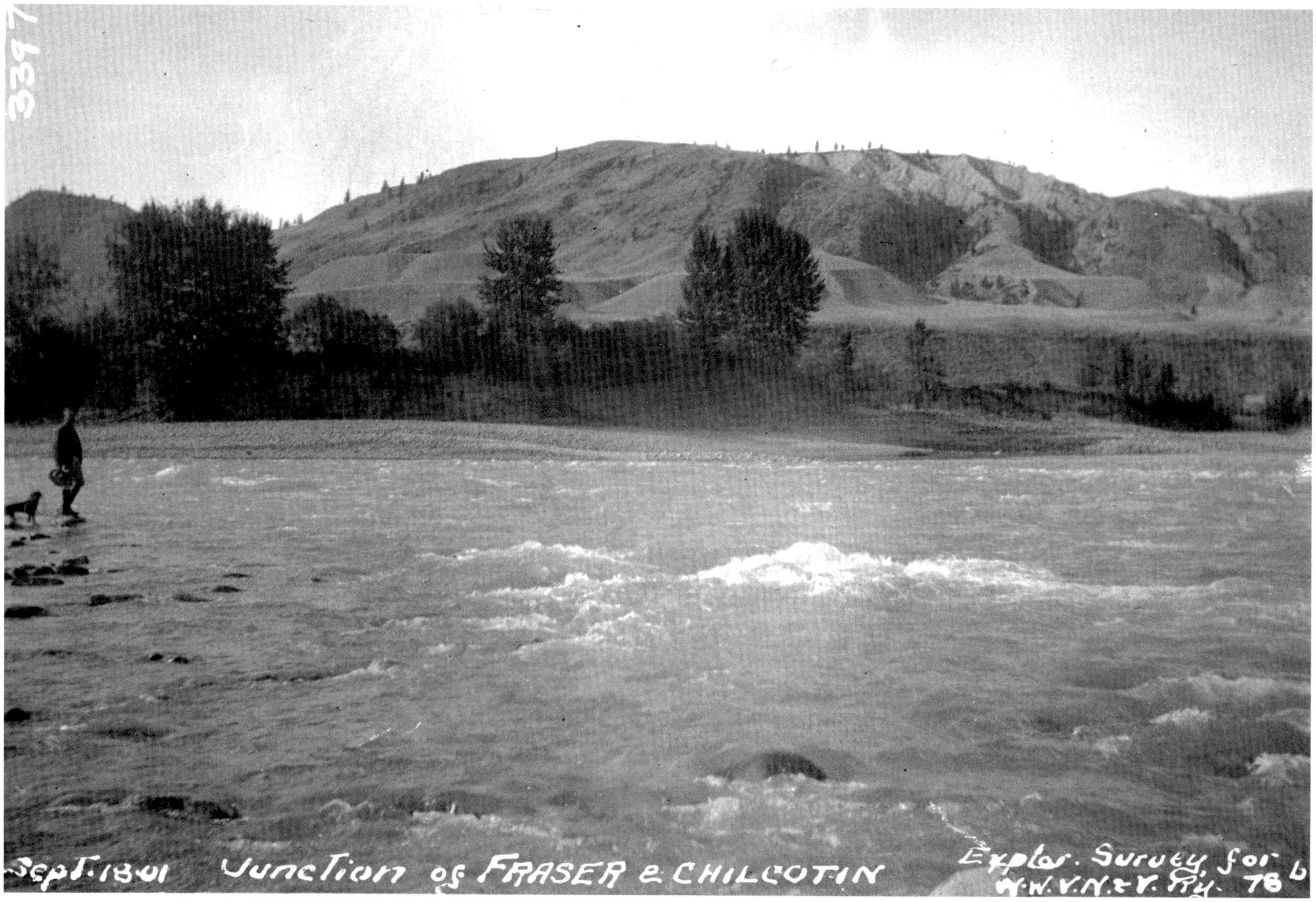

❖ The junction of the Fraser and Chilcotin rivers. Swannell took this photograph from the south bank of the Chilcotin River. The man and the dog are on the north bank with the Fraser River behind them. This is an important historic location along the Fraser River. In the immediate vicinity there are kekuli pits, indicating First Nations habitation of the area. As well, there is evidence of the Chinese miners working along the banks of the Fraser River during the Cariboo gold rush era. Image I-59684 courtesy of Royal BC Museum, BC Archives.

Ontario for "my girl Ada." Office hours 5:30 am–6:30 pm, 1½ hours for refreshments—2 hard tacks for lunch. Shall I bring you one?

Suppose I sign—Frank—His mark X?

P.S. Address—F.C. Swannell c/o T.S. Gore PLS, Empire Valley P.O., Lillooet Dist., BC. Will arrive there in about two weeks.

That night the crew camped on a "small bench 1000 ft. [300 m] above [Fraser] river. Too windy for tents—thunderstorm—made shelter but got soaking wet." During September Gore and his crew worked north along the Fraser. In his diary Swannell wrote about some of the people and places he came across. On September 4, they arrived at "Tinker's Ranche—walked all way, unable to ride. Mountains getting lower and benches wider." After a day in camp they travelled 16 kilometres to a Native reserve. On September 7, Orchard, the cook, left so Swannell "took his place as cook at $10 extra. Made to McDonald's by noon. Squaw speaks with strong Scotch accent. Very bad hill, 1400′ [425 m], bad trail & deep canon." While spending the next day in camp, Swannell observed some "Chinamen washing gold in river."

On September 9, the men "passed Irwin's ferry. Very bad hill 1100′ [330 m] to ranch. Camped 6 miles

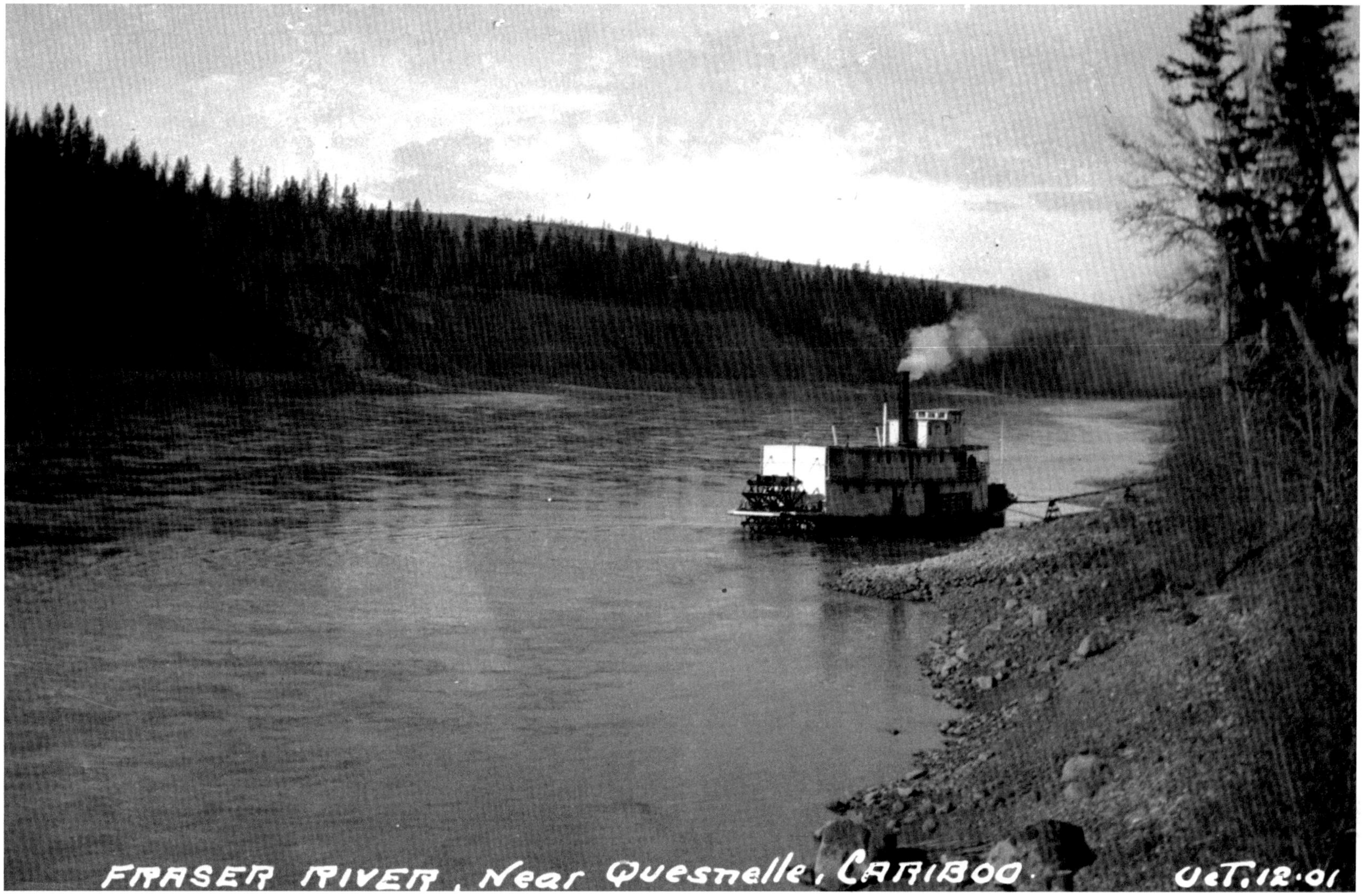

[10 km] past. Lost trail which runs high up." After a day examining the Big Creek area the crew "moved to above McKeowns. Trail washed out, horses scatter in timber. Park-like summit. Shot 2 grouse." There they spent a second day before moving on September 13 to Churn Creek over a wagon road. By September 16 they had moved to 16 kilometres past the Gang Ranch where they camped in timber at 900 metres elevation. After a day in camp they arrived at the Chilcotin River on September 18. "Near Chilcotin—splendid wooded upland. Stayed in empty cabin (bushy-tailed rats). Chilcotin 200′ [60 m] wide, very deep." On September 21 the crew "moved to Vedans [Louie Vedan ranch]. Swam horses over Chilcotin. Crossed in canon." The next day they "moved to Cotton's ranch on Riske Cr. Shot prairie chickens. Open plateau with bluffs and small lakes. Well wooded at Riske Cr."

They spent two days at this camp before proceeding to Chimney Creek. The next day they "went up canon in boat. Canon 2 miles [3 km] Fraser very deep. Moved to 3 [miles, 5 km] past Meldrum's on wooded plateau." By September 27 they were at the river opposite Soda Creek. The following day Swannell went hunting, took some photographs and wrote a letter to Ada. Only a fragment of this letter remains, and it describes a visit to a restaurant in Soda Creek that day.

SS *Charlotte* near Quesnel. Built in 1896, the *Charlotte* ran from Soda Creek to Quesnel and was the only boat operating on the upper Fraser River in 1901. Swannell travelled on this sternwheeler in 1908 and 1909 en route to surveying in the Nechako valley. Image I-33927 courtesy of Royal BC Museum, BC Archives.

A Visit to Soda Creek

❖ "Indian chief at Soda Creek Rancherie [Reserve]." The woman behind him is probably his wife. Both are dressed up for the picture. Image I-33193 courtesy of Royal BC Museum, BC Archives.

❖ Opposite page: "Soda Creek Rancherie." Swannell photographed these men while he was at the Xatsull Reserve. Image I-33176 courtesy of Royal BC Museum, BC Archives.

Following spread (pages 32-33)
❖ Left: In the village of Soda Creek, he took a picture of some young First Nations women on horseback. Image I-33177 courtesy of Royal BC Museum, BC Archives.

❖ Right: He also photographed a freight outfit travelling through the community. Image I-58366 courtesy of Royal BC Museum, BC Archives.

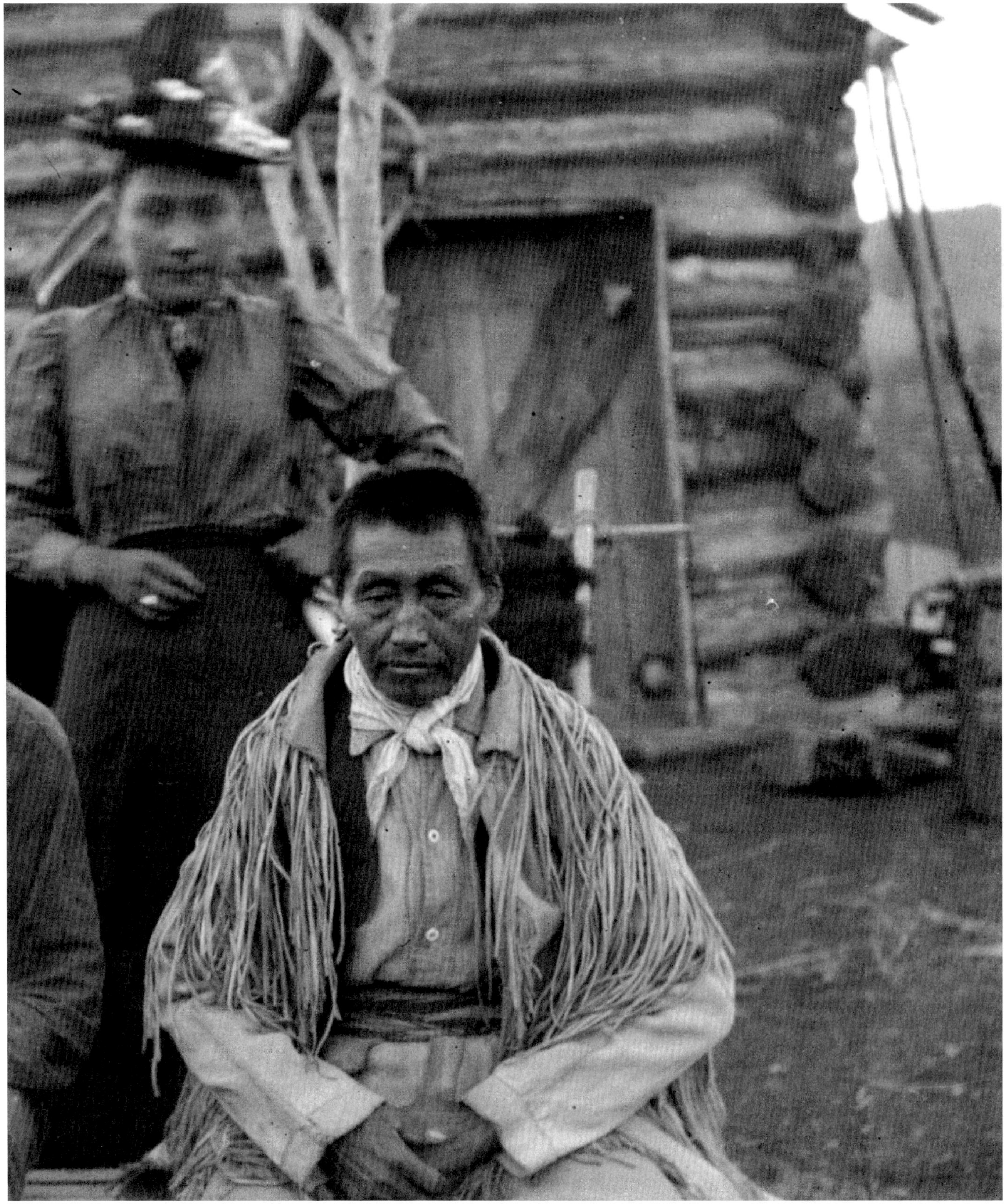

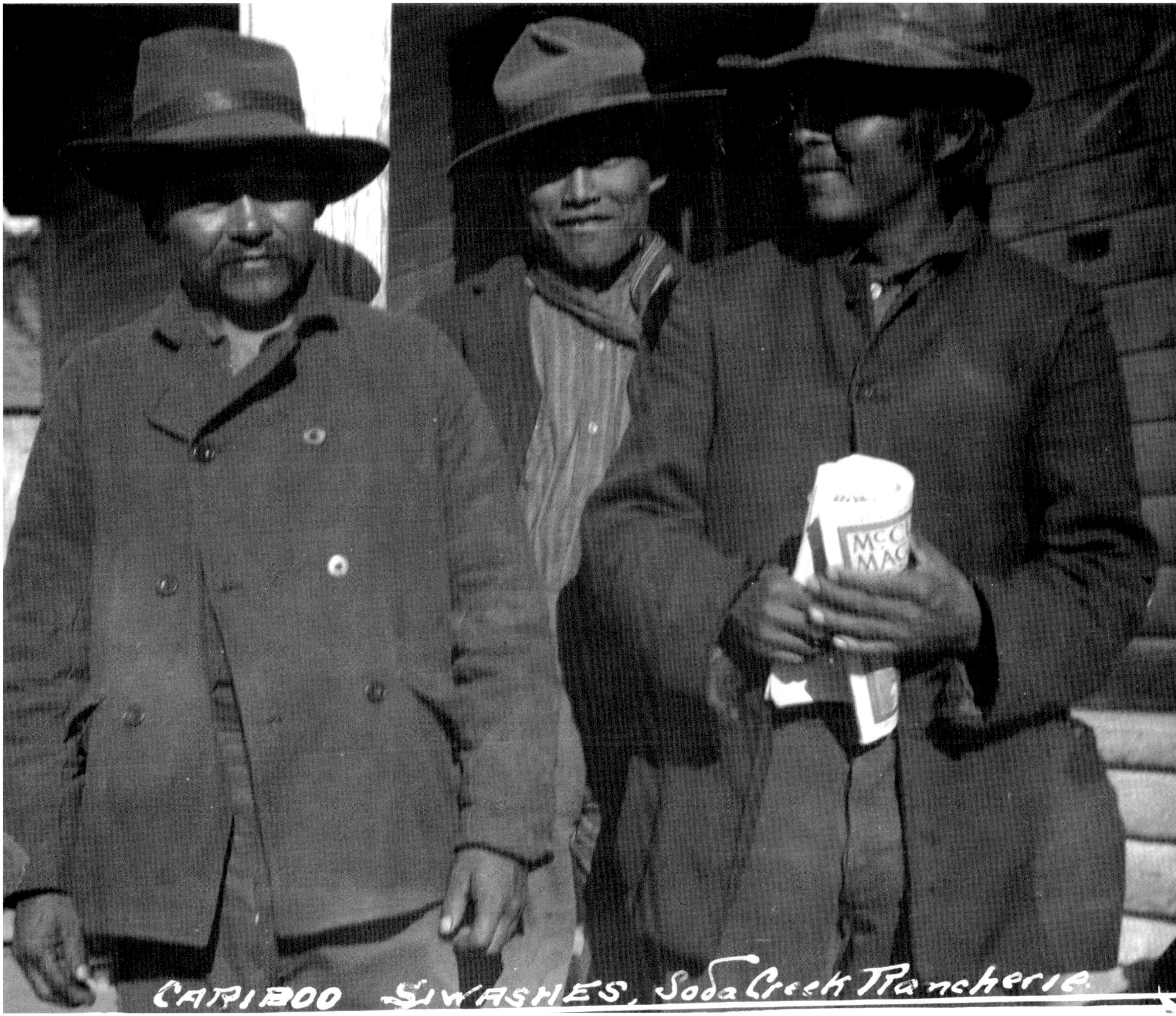
CARIBOO SIWASHES, Soda Creek Rancherie.

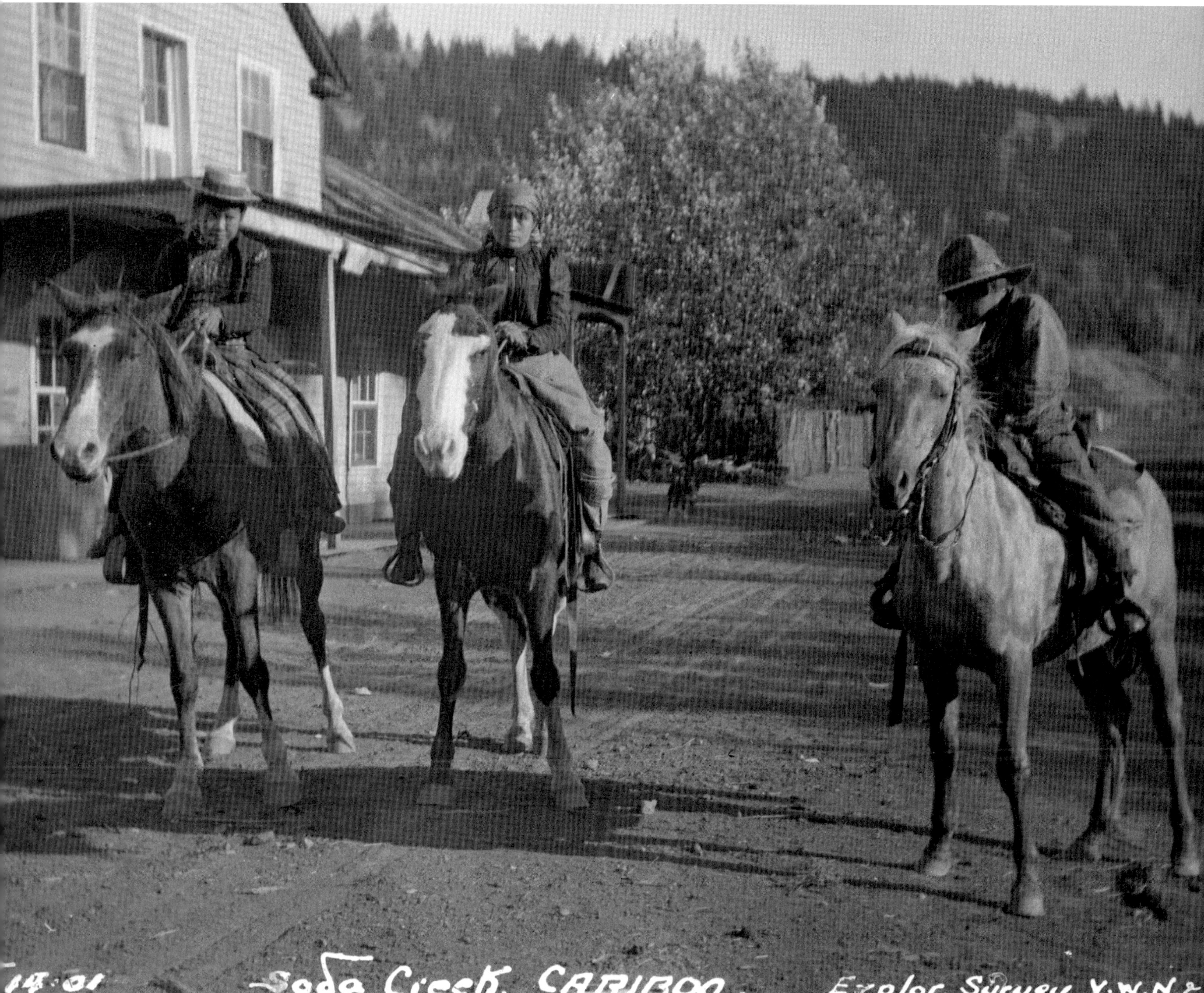
14·01
Soda Creek. CARIBOO.
Explor. Survey. V.W.N.

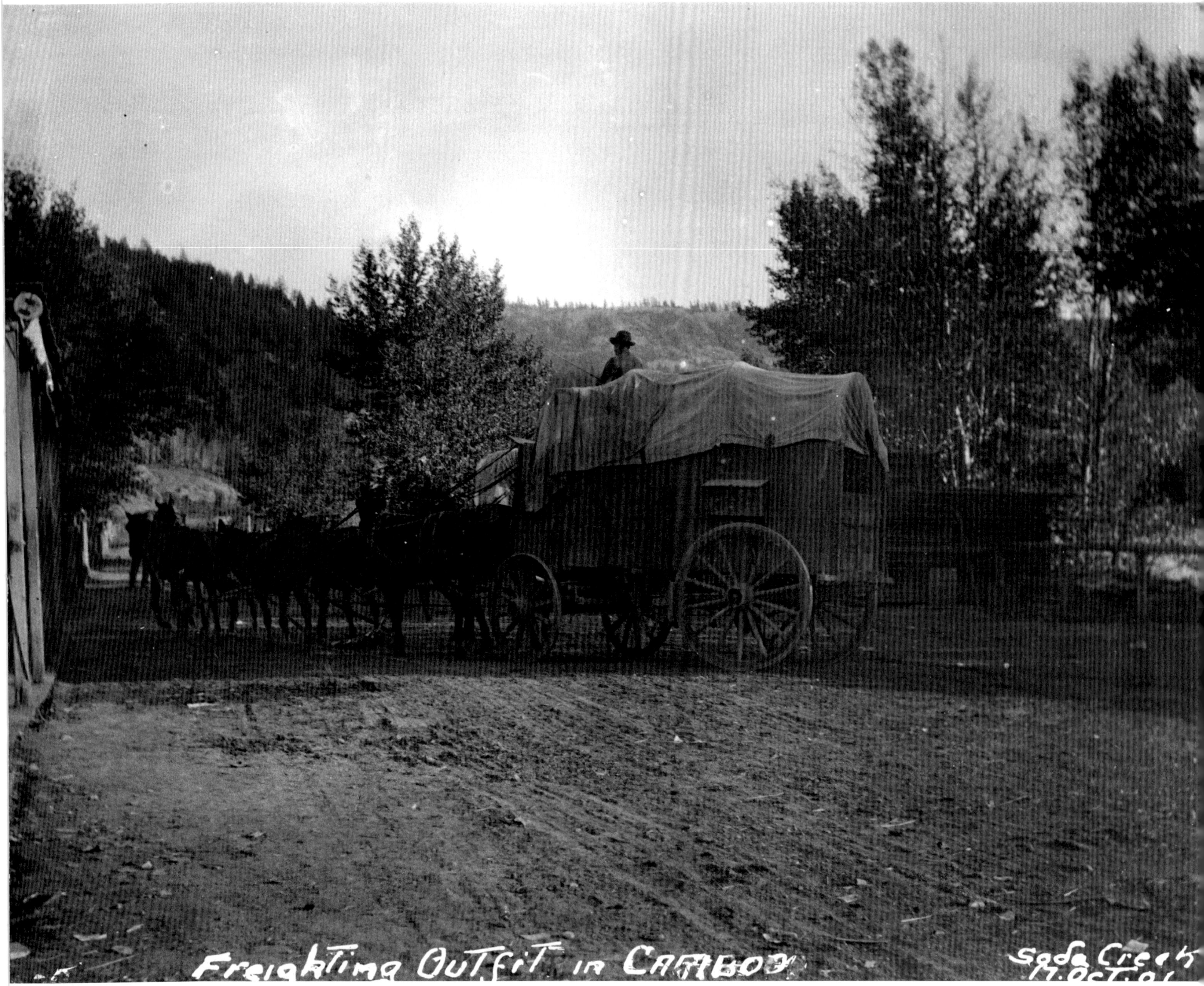
Freighting Outfit in Cariboo
Soda Creek

❖ Packing from Raven Mine to Cap Sheaf. Image I-33673 courtesy of Royal BC Museum, BC Archives.

❖ Opposite page: Prospector Tom Railson standing among Douglas fir trees that fell across the Gillies Bay trail during the wind storm. Image I-58237 courtesy of Royal BC Museum, BC Archives.

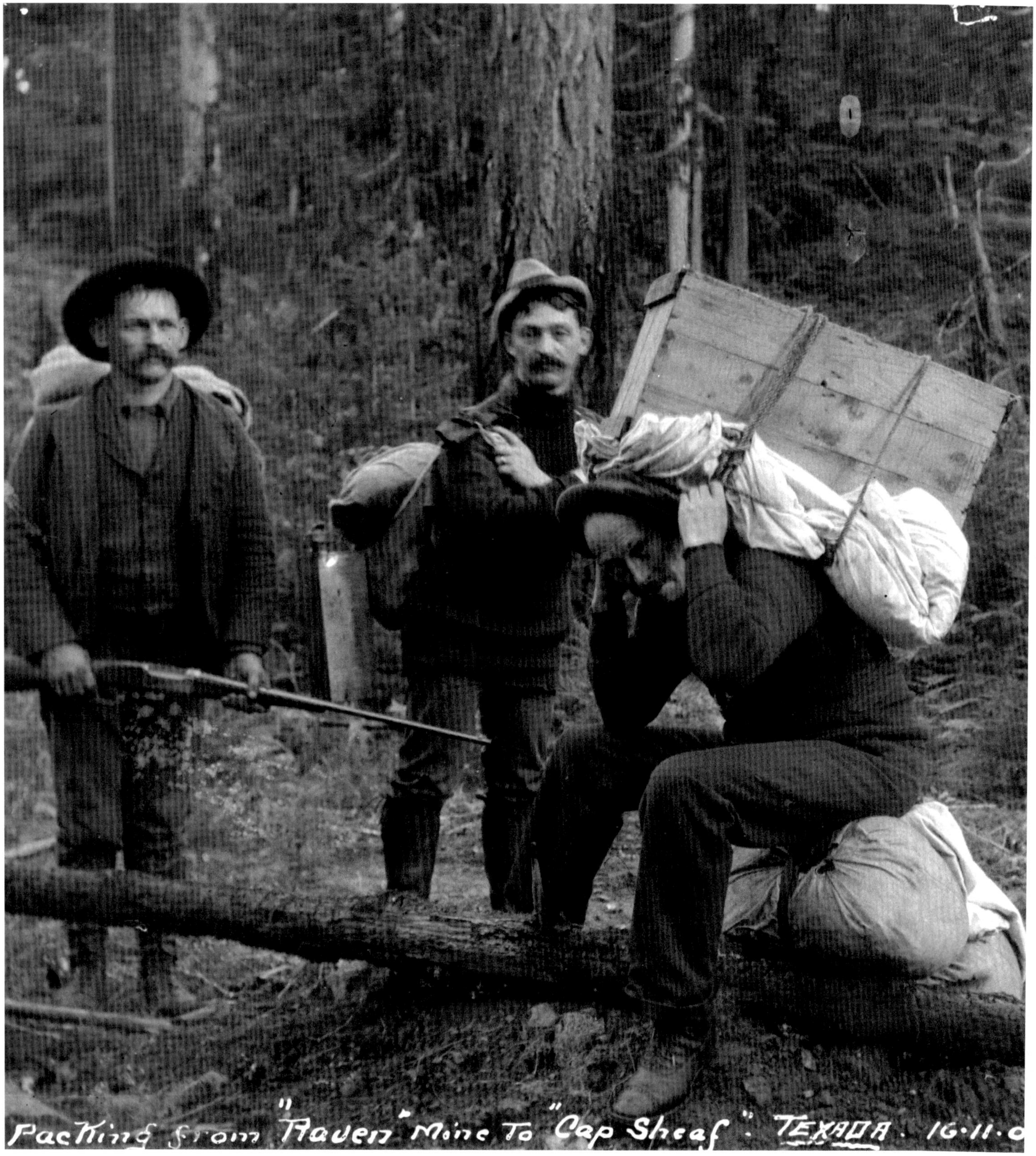

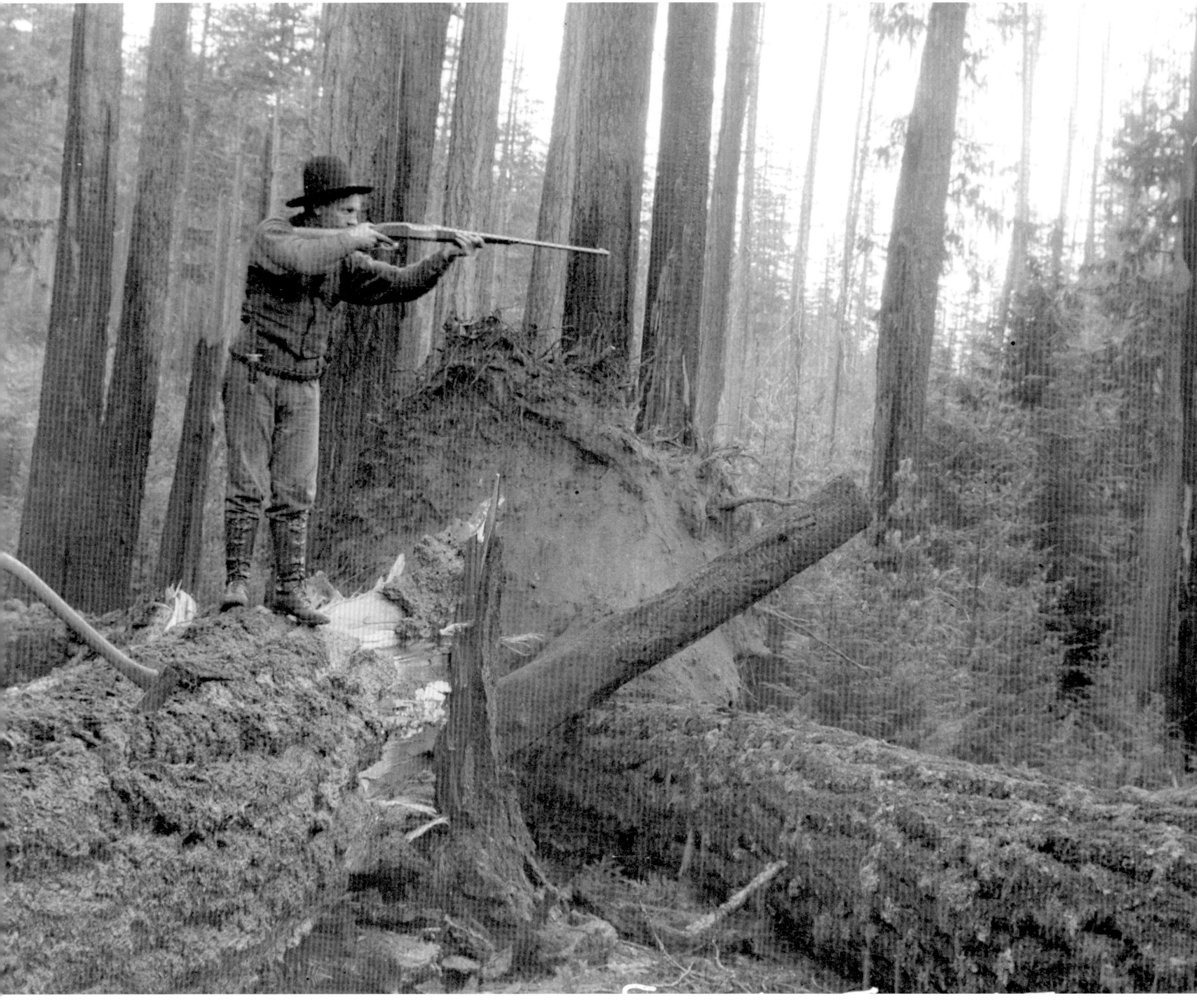

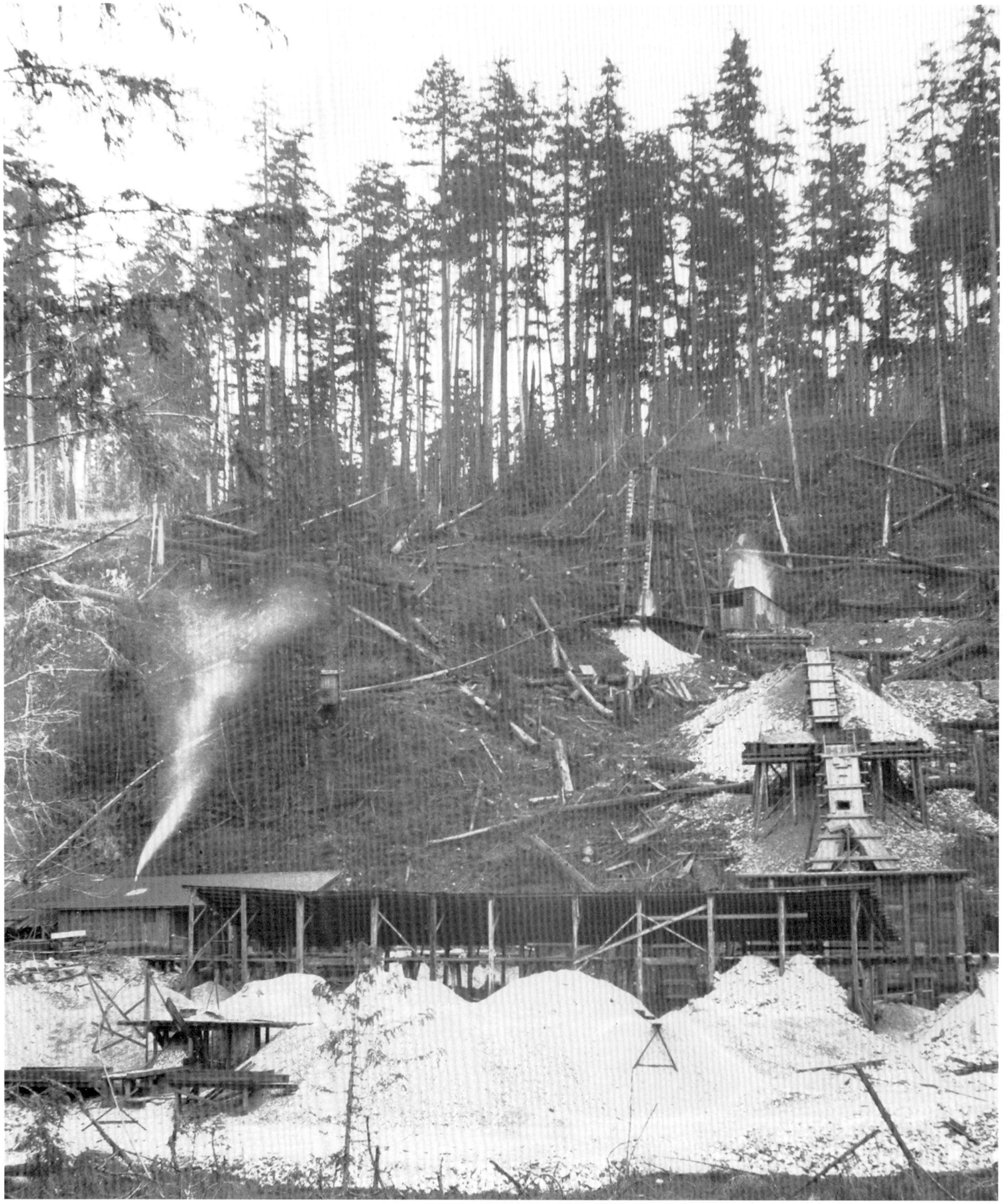

❖ The Cornell mine was one of the largest on Texada Island in 1901, and it operated for several years. Image I-57994 courtesy of Royal BC Museum, BC Archives.

> tine of a crotched stick. It took me an hour to fix that almost invisible thread onto the diaphragm by the aid of a magnifying glass and pine gum. A most delicate operation. We let the spider go on parole.
>
> Dinner 11:30, bread & butter and tea. 12 noon, off to work. Drizzling, woods soaking wet, every salal bush seems to shower a gallon of water over one. A leaden sky overhead, a bog underneath. That's Texada! One "sight" I took today I had to stand ankle deep in a creek for 15 minutes. You ought to see the fearful smudge I make of my field book trying to figure out angles & distances with the rain spattering on the page.
>
> At half past three it gets so dark I can't see to read the instrument without lighting matches. At four I gave up work in disgust & tramped home through the darkening forest, wading through pools and stumbling over rocks and logs tangled in the soaking hemlock scrub and salal. Made camp at dark—boots and leggings like soggy blotting paper. Well I tumbled into dry togs and tackled your letter.
>
> 5:30 Supper. Beans mixed with Australian rabbit. Bread and prune pie. (All side orders extra and unobtainable.) Read your letter I don't know how many times and got unmercifully chaffed by the boys therefor.
>
> We've had such a fine time in the interval between the first and this part of my letter that I can't compose myself to write. Sang every song they could remember from My Bonnie to Clementine, then had an impromptu prayer meeting. Kuechberg fervently prayed for the boss. "O Lord, grant that he be more liberal with both bacon and pocket book. May he see the error of his ways in working us so hard."

December started with a fine day as they ran the south boundary of the X-Ray mining claim. On the third Swannell saw both a cougar and a deer. That night they had a fierce storm with hurricane-force winds.

Parr and his crew also surveyed a few mining claims for the Duluth and Texada Mining and Exploration company. One of these bordered Paxton Lake, and when they finished surveying the claim they fished in the lake. A couple of times they visited the Cornell mine.

On December 13, Swannell recorded some snow. By December 18, the crew had finished surveying on Texada and left on the famous Union Steamship boat SS *Cassiar*, which had been built just a few months earlier. They arrived in Vancouver the next morning and took the SS *Charmer* back to Victoria.

Swannell finished 1901 by making a map of the railway exploration survey. It was a busy year for Swannell, working on several types of projects in different locations. During his second year of articling with Gore he learned more facets of surveying, particularly how to run a transit.

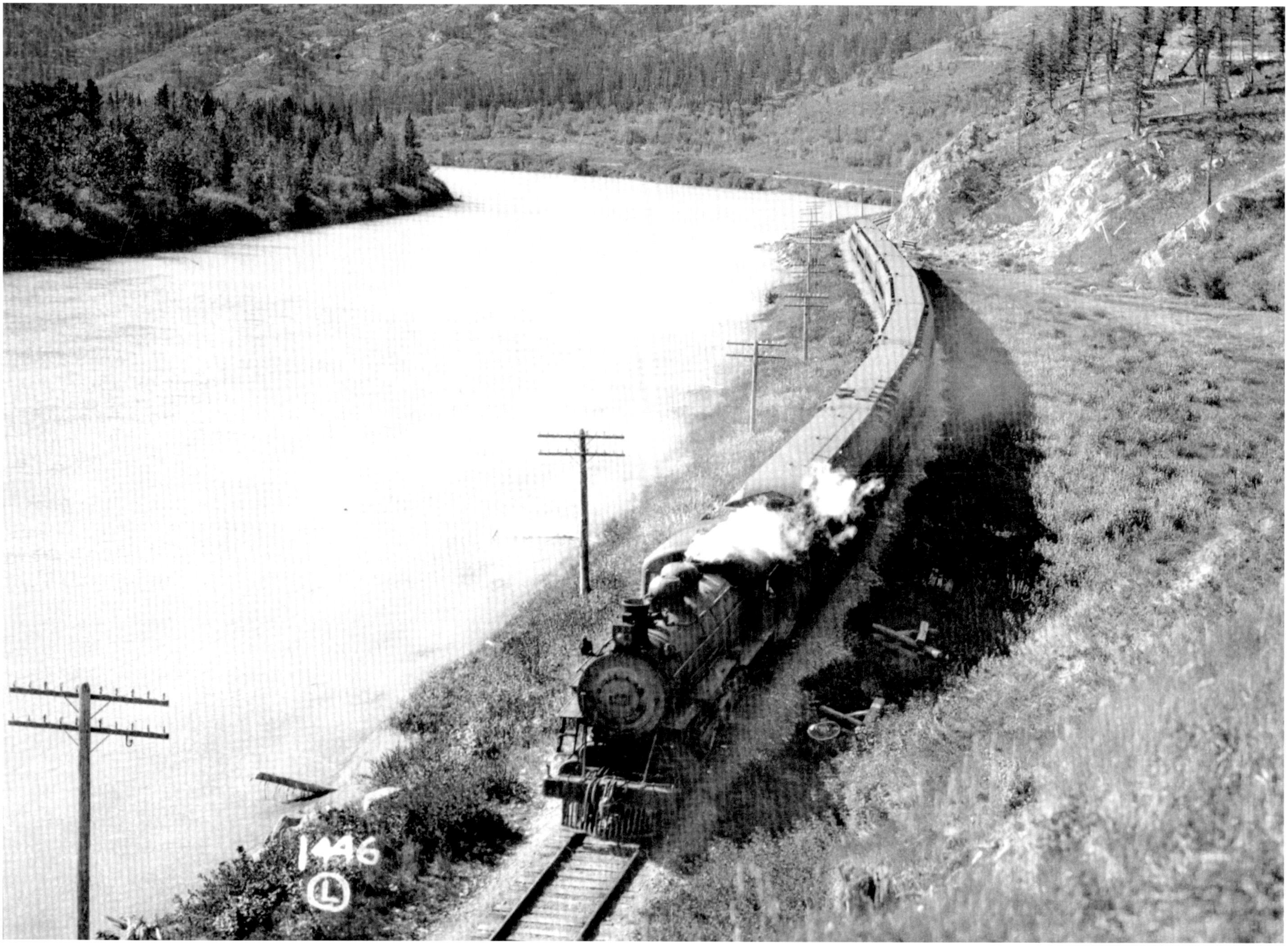
1446
L

The Canadian Pacific Railway in Southern British Columbia

From 1902 to 1905, a major portion of Swannell's surveying involved the large lots that were part of the CPR's land subsidies for constructing a railway line through southern British Columbia. These lots played an important role in the history of British Columbia at the beginning of the twentieth century, and Swannell's work brought him into two controversies.

When British Columbia joined Canada in 1871, the federal government promised to construct a railway that would connect the province with the rest of the country. The selection of a route through the mountainous province was based on both geography and politics. The federal government did not want a route that was too close to the American border, fearing that it could be taken over by the United States if there was a conflict between the two countries. At the same time it did not want the route so far from the boundary that the United States would be able to establish economic influence over the area near the border and drain revenue away.

The government felt that the route chosen—crossing the Rockies at Kicking Horse Pass, traversing the Selkirk Mountains at Rogers Pass and then following the Thompson and Fraser river waterways to Burrard Inlet—was the best compromise. The Canadian government fulfilled its mandate with the completion of the CPR in 1885. The railway was very expensive to build and unfortunately there was not a large amount of economic activity near the line.

In the 1890s mineral discoveries in the West Kootenay and Boundary regions of southern British Columbia created a mining boom in these locations. Geography made it easier to access this area from the United States, for the river valleys generally ran north and south, while east–west access across British Columbia involved traversing mountainous terrain.

By 1892 Daniel Corbin, an American businessman and railway builder, completed the Spokane Falls and Northern Railway from Spokane to Northport on the Columbia River near the Canadian border. Here the railway could connect with boats running up the river to the Arrow Lakes. The following year Corbin built the Nelson and Fort Sheppard Railway from Northport into the West Kootenay mining district. In 1896 he constructed the Red Mountain Railway from Northport to the bustling mining town of Rossland.

In 1893, the Great Northern Railway that ran from St. Paul, Minnesota, to Seattle was completed. It was the most northern of the transcontinental American railways, and its chief executive officer, J.J. Hill, wanted to access the economic wealth of the West Kootenay region. He completed the Kaslo and Slocan Railway in 1895. This railway ran from the booming mining area of Sandon, near Slocan Lake, to Kaslo on Kootenay Lake. There Hill established the International Navigation and Trading Company, which provided boat service from Kaslo to the foot of Kootenay Lake. He then constructed the Bedlington and Nelson Railway, which ran from the lake to the US border. There it joined the Kootenai Valley Railway, which went south through Idaho to the Great Northern at Bonners Ferry. In 1898 Hill purchased Corbin's railway lines.

The Canadian Pacific Railway had to respond to these attempts to divert the wealth of the Kootenays south of the border. In 1890 the BC government

❖ Opposite page: BC Southern train. Image G-01444 courtesy of Royal BC Museum, BC Archives. Photographer unknown.

❖ Following page: Map of southern BC land grants. The CPR produced this map to show their land grants in southern BC in 1902. The lots granted to the BC Southern Railway are east of Kootenay Lake, while the Columbia and Western Railway received the smaller blocks to the west. The status of lots 4593 and 4594, the two Columbia and Western parcels in the Crowsnest Pass area, was still undecided in 1902, so they were not included in the original map. However, these lots have been marked for reference. A comparison of this map and the one in the 1905 chapter shows the additional lots that the Columbia and Western received after the BC government made its final decision in 1903 to disallow lots 4593 and 4594. Map 7Atr3 old maps, courtesy of Land Title Survey Authority.

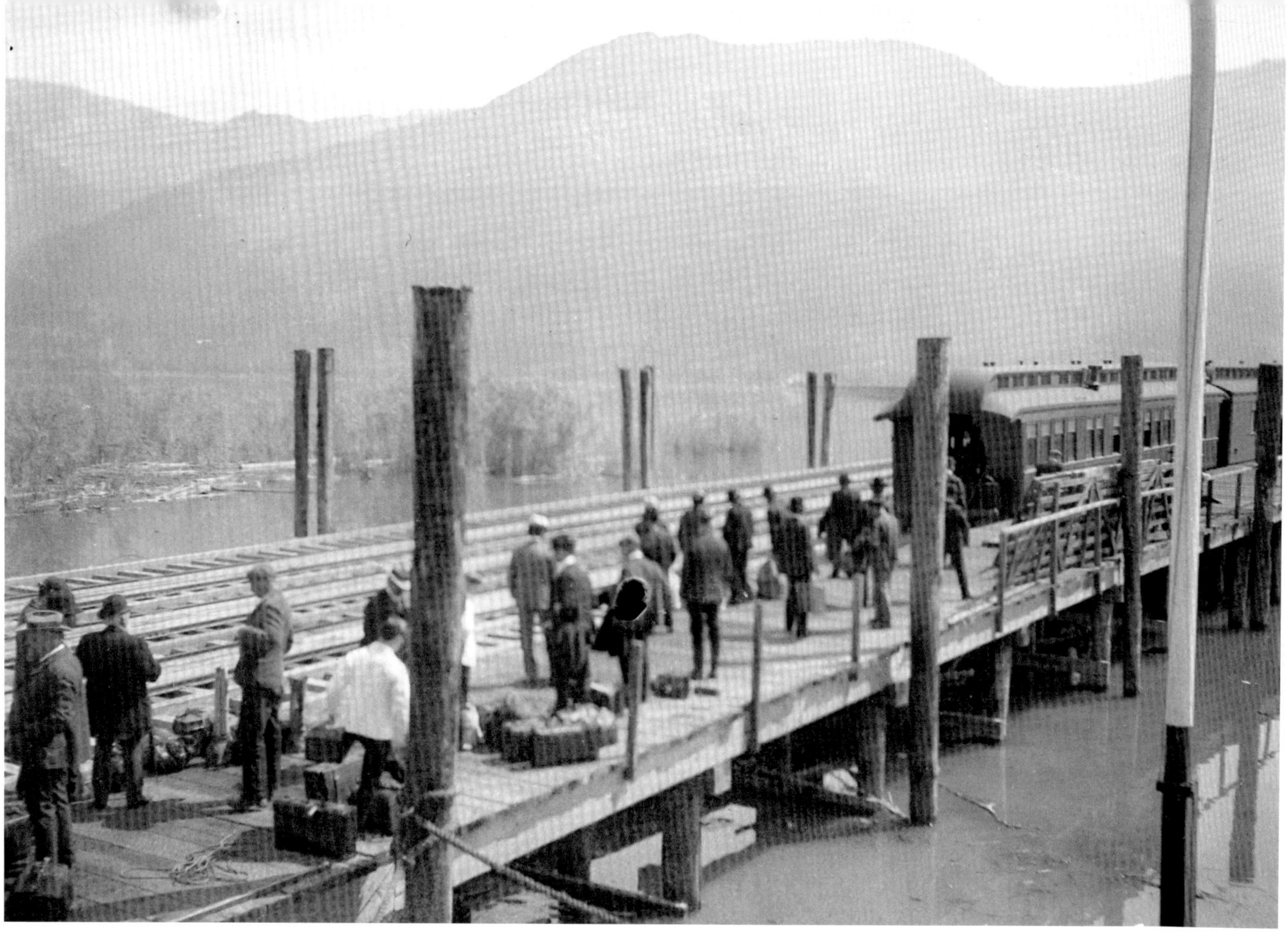

❖ CPR block survey crew at Kootenay Landing. The members of Gore & McGregor's crews are preparing to board a BC Southern train that will take them to the Yahk valley to begin their surveying. Image I-58080 courtesy of Royal BC Museum, BC Archives.

a rail line along Kootenay Lake from Procter to Kootenay Landing finally brought a continuous rail line across southern British Columbia.

The CPR's efforts to retain the economic wealth of the Kootenay and Boundary regions in the province were successful and it eventually purchased the Great Northern's railways and boat company. The BC Southern received $6,875 per kilometre ($11,000 per m) in cash from the federal government and a subsidy of 3,755,733 acres (1,519,891 ha) from the provincial government, consisting of seven large lots in the Kootenay region.

Lot 4588, the smallest one at 112,640 acres (45,584 ha), was located north of the railway line and encompassed the coal fields of the upper Elk River valley. Lot 4589, 498,893 acres (201, 895 ha), was in the Crowsnest Pass area to the south of the railway line. Lots 4590 (518,400 acres), 4592 (769,800) and 4595 (627,200) were each about 32 kilometres (20 m) wide from east to west on their northern boundary, the 50th parallel, while lots 4589 and 4591 (358,400 acres) were 32 kilometres wide on their southern boundary, the 49th parallel, which was the border with the United States. These two lots ran south from the BC Southern Railway line while 4590, 4592 and 4595 ran north. Lot 4596, at 870,400 acres the largest one, went north from the 50th parallel for 64 kilometres (40 m) and included much of the headwaters of the Columbia and Kootenay rivers. The subsidy encompassed a large portion of the remaining Crown land in the southern part of the East Kootenay district outside of the Kootenay River valley, and the CPR now owned vast tracts of coal and timber resources.

The Columbia and Western Railway received 1,348,225 acres (54,5607 ha) for their subsidy. The land that they received was between the Columbia River and Okanagan Lake in a checkerboard pattern of smaller lots. Although the railway had been constructed in the West Kootenay and Boundary regions, the Columbia and Western selected two blocks in the Crowsnest Pass area as part of their subsidy. Lot 4593, in the Flathead area in the southeast corner of BC, was

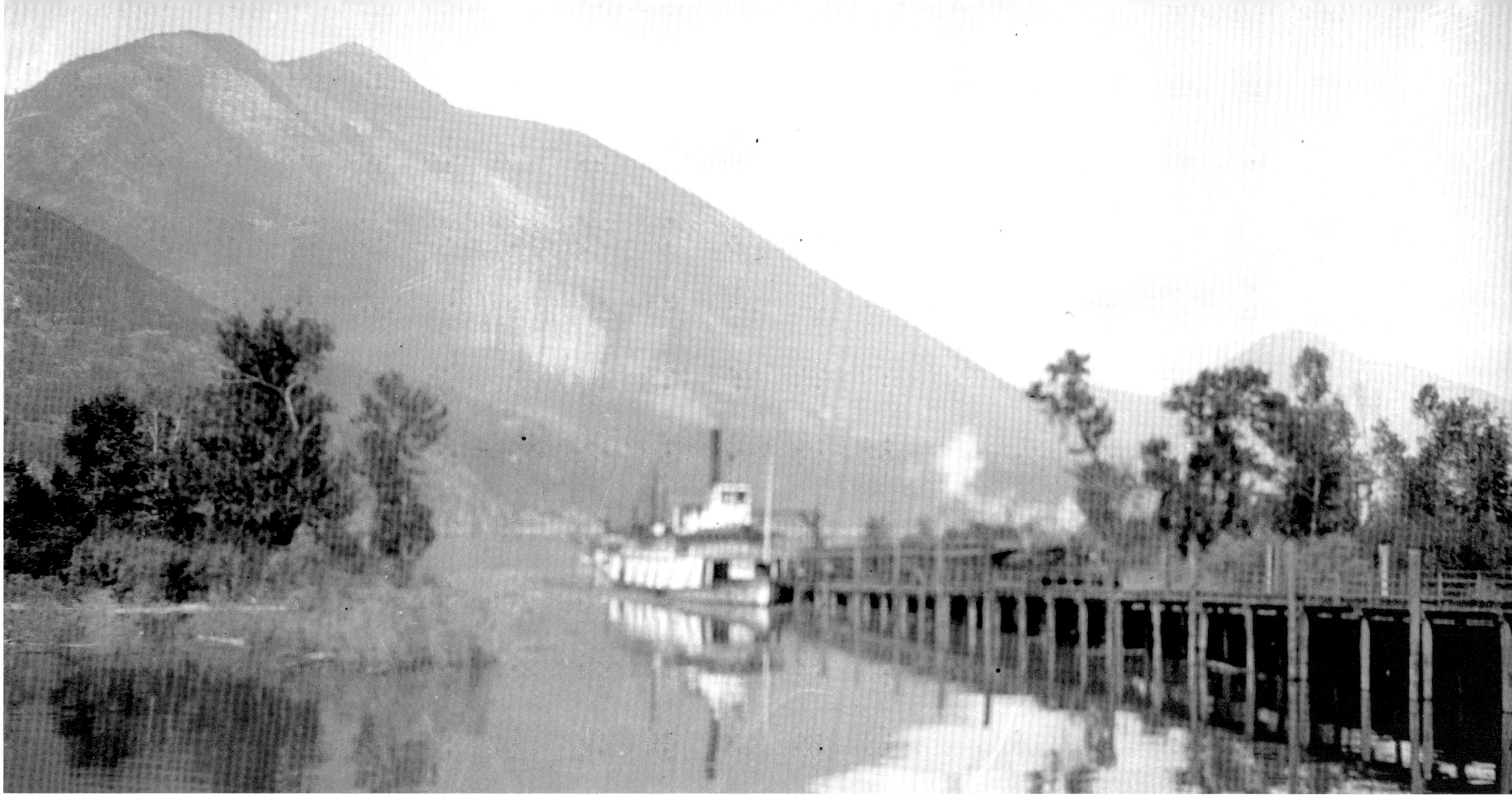

245,760 acres, while Lot 4594, in the Elk River valley, had 380,000 acres. Even though the railway was chartered as a separate company, the CPR claimed that since they were the owners they should be allowed to select their land anywhere along rail lines that they operated in southern BC. The two lots were both near the BC Southern Railway line in an area that the CPR thought had extensive coal and possible petroleum deposits.

The BC government, under Premier James Dunsmuir, initially accepted the request for these two lots. A few months later it cancelled the grant and this was ratified in the legislature in early 1903. Since the Columbia and Western received both a provincial and a federal subsidy for building the railway, there were rumours that the lieutenant-governor would not assent to these changes. In April a select committee was appointed to investigate the railway's land grant, and the issue received extensive coverage in British Columbia newspapers. The handling of the Columbia and Western Railway subsidy was one of the contributing factors that led to the fall of the government of Edward Prior and the election of Richard McBride as premier in 1903. The CPR protested to the federal government about the cancellation of the grant but Canada did not intervene. In June 1904 the BC government, which had placed a reserve on coal and petroleum leases within these lots, opened them to the public. This would later lead to another controversy within the BC government.

The Columbia and Kootenay Railway received a land subsidy of 188,593 acres (76,320 ha), mainly in the area between Castlegar and Nelson.

Once the British Columbia government confirmed the BC Southern's land grants in 1901, the boundaries for these lots needed to be surveyed. Gore & McGregor received contracts to survey most of these large blocks of land in 1902 and 1903. When the Columbia and Western Railway's land subsidy was finally approved in 1904, Gore & McGregor surveyed some of its lots the following year.

Sternwheeler at Kootenay Landing. This is the *Moyie*, the main CPR sternwheeler on Kootenay Lake in 1902. Image I-33983 courtesy of Royal BC Museum, BC Archives.

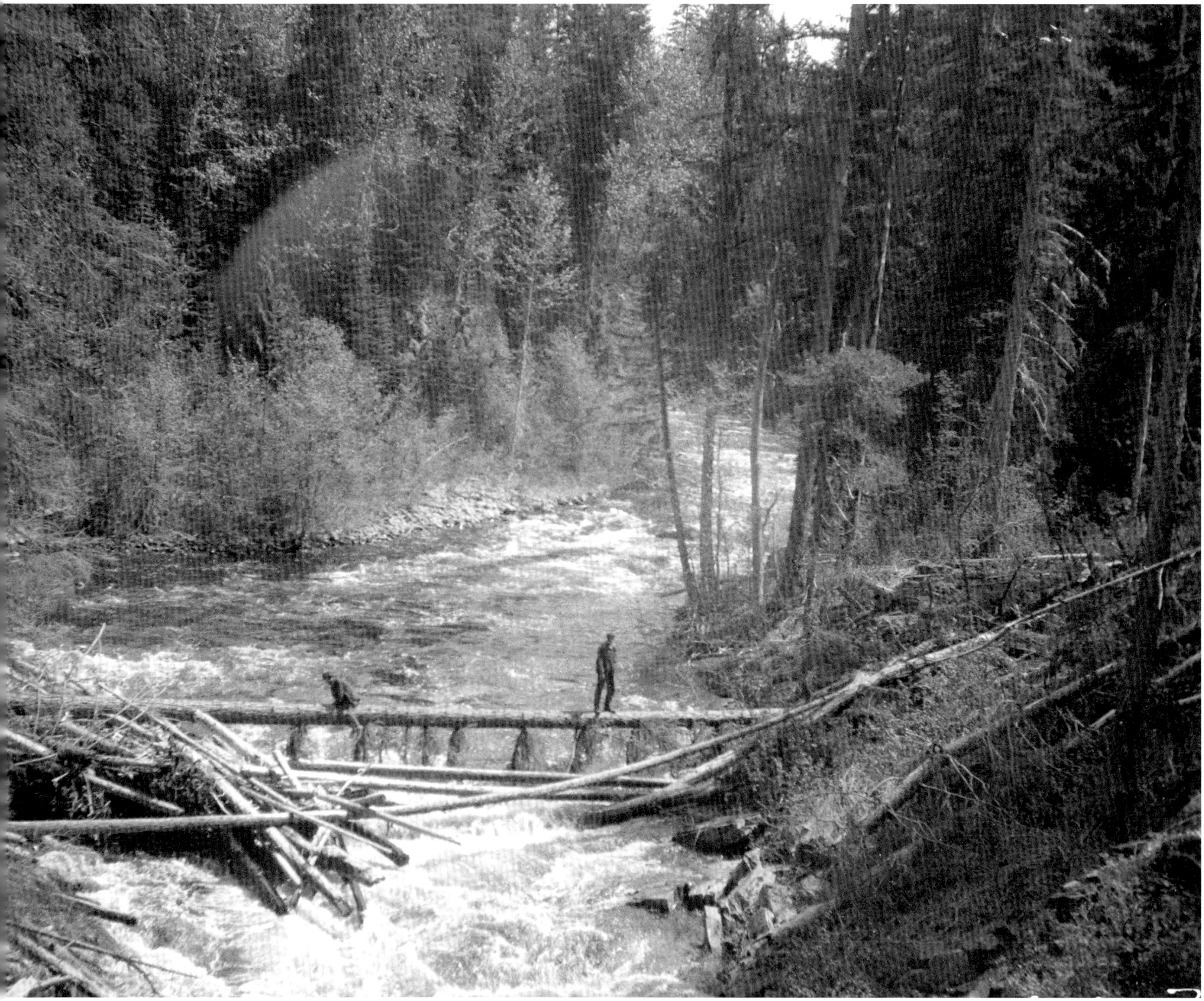

1902

During the winter and early spring Swannell worked in and around Victoria. He and Ada went on several outings with friends, and Swannell photographed some of the scenery and activities.

Swannell spent a lengthy 1902 field season in the Kootenay region working on the BC Southern Railway land grant surveys. Gore & McGregor received a contract from the CPR to survey portions of the five large lots of the BC Southern Railway land grant. Both men had crews for this project and they also hired J.G. Cummings (PLS #70), a surveyor from Fort Steele, to assist them. He brought his younger brother, Alfred, who was articling with him. Swannell was in his third and final year of articling to be a surveyor and Gore continued to give him more responsibility. Swannell was sometimes in charge of the crew with Gore supervising, and at times he was even left on his own when Gore became injured or had other duties to perform. Swannell didn't keep a diary but wrote several letters to Ada describing the work that he was doing. Swannell's photographs of the Kootenays from 1902 to 1904 are among the earliest ones taken in the region.

Swannell left Victoria with Gore, McGregor and other crew members on May 9, travelling to Revelstoke on the CPR railway. From there they took the train to Arrowhead, where they boarded a sternwheeler to Nakusp. At this community they took the Nakusp and Slocan Railway to Slocan Lake, where they boarded another boat to the foot of the lake. There the Columbia and Kootenay Railway took the crew to Nelson. Cold, inclement weather delayed Gore's departure for a few days so Swannell spent his time visiting his father and his stepbrother Fred and family. Then the crew went to Procter, where they boarded the SS *Moyie* for a trip to Kootenay Landing. At this locale at the southern end of Kootenay Lake the men once more travelled by train, arriving at their surveying area near Yahk on May 14.

At the beginning of the season, Gore, McGregor and Cummings all worked on the west boundary of Lot 4591. One group of men surveyed along the lot line from the railway line south to the international boundary, while another made a traverse along the railway track and down the west side of

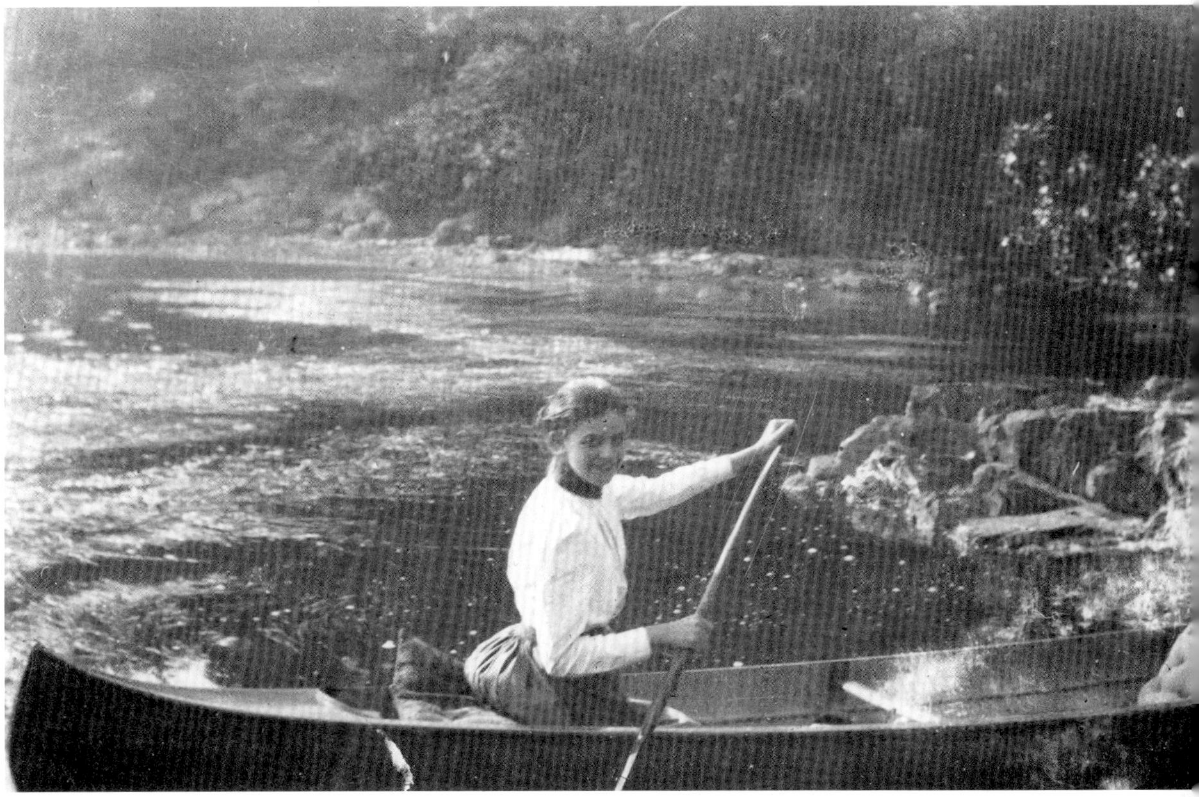

❖ Above: Many of Swannell's activities with Ada and friends included picnics and boat outings around the Victoria area. There are several photographs of these activities in his photograph albums. This picture of Ada was taken on an outing in the spring of 1902 before Swannell left to survey in the Kootenay region. Image MS-0392 vol 50, pg 36 courtesy of Royal BC Museum, BC Archives.

❖ Opposite page: Crossing Meadow Creek in flood. Creeks and rivers are often difficult and dangerous to cross in the high water of spring runoff. Image I-58077 courtesy of Royal BC Museum, BC Archives.

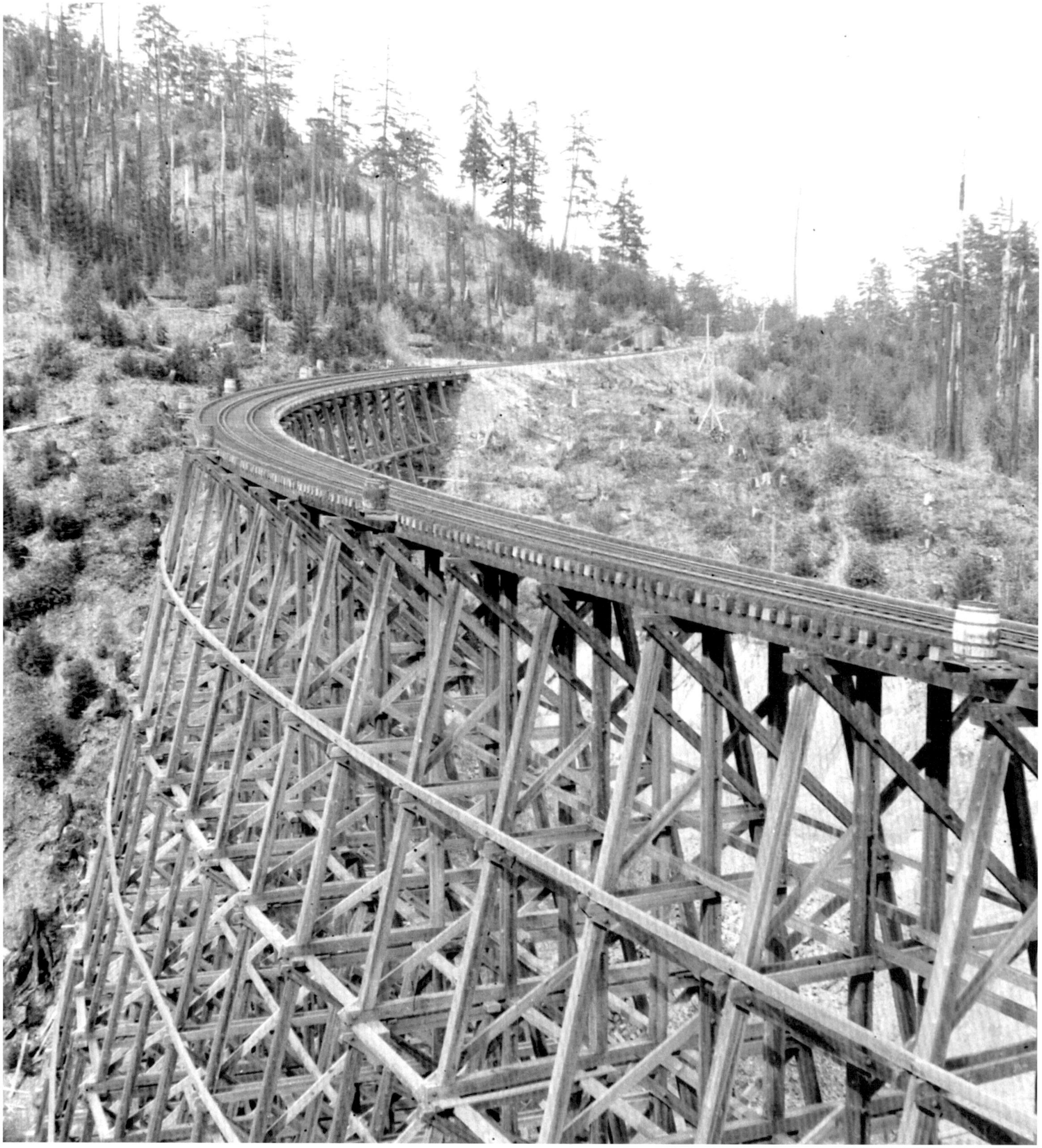

the Moyie River to the monument on the international boundary. The men had difficulty getting across the Moyie River, which was in flood, so the local CPR land agent lent Gore a canoe.

Swannell wrote to Ada a week later.

5 miles East of Yahk Station
Crowsnest Railway, May 21, 1902

We, us, Co. were dumped off between Yahk & Tochty last Wednesday. Made camp. That day ran 1/4 mile [400 m] of line.

Thursday. I was to start work running W. Bdy Lot 4591 to the International Boundary about 7 miles [11 km] = 15 days work. All hands spent day trying to build a raft to cross the Moyie without success. The river, which ordinarily is about 1/2 as big as Goldstream, is now a foaming torrent 200–250 feet [60–70 m] wide and running 7 miles [11 km] per hour. All snow water and icy cold. Has overflowed its banks everywhere. We had to altogether abandon hope & after working like beavers, chopping up to our waists in ice water for an hour at a spell. Peculiar sort of picnic surveying is!

Hickling, my chainman, tumbled off a log and went clean out of sight. Owen & I fished him out but he left his axe in the river. Next, Owen made a spring for shore from the butt of a rotten log, which broke under him. He hit the water on the broad of his back and made a most prodigious splash. We had a little excitement getting him back to terra firma.

Casualties up to date. Everyone outfit of 17 except 5 has tumbled in, all of us have been pleasantly damp continually. Gore (the chief) put his shoulder out of joint. We flagged a

❖ Below: Sealing schooners tied up at Victoria. In the late nineteenth century, there was a fleet of almost one hundred schooners based in Victoria that hunted fur seals for their pelts. Most of the schooners went to the Pribilof Islands in southwest Alaska to obtain their furs. By 1902 the number of sealing schooners at Victoria was declining because the number of fur seals was decreasing rapidly. Image I-58287 courtesy of Royal BC Museum, BC Archives.

❖ Opposite page: E & N trestle at Niagara Creek. The Esquimalt & Nanaimo Railway operated on Vancouver Island between Victoria and Nanaimo in 1902. The trestle bridge over Niagara Creek, in present-day Goldstream Park, was one of the notable features on the rail line. Swannell's photograph from 1902 shows the bridge that was in place between the original trestle and the cantilever bridge that was placed over the creek in 1910. Image I-58324 courtesy of Royal BC Museum, BC Archives.

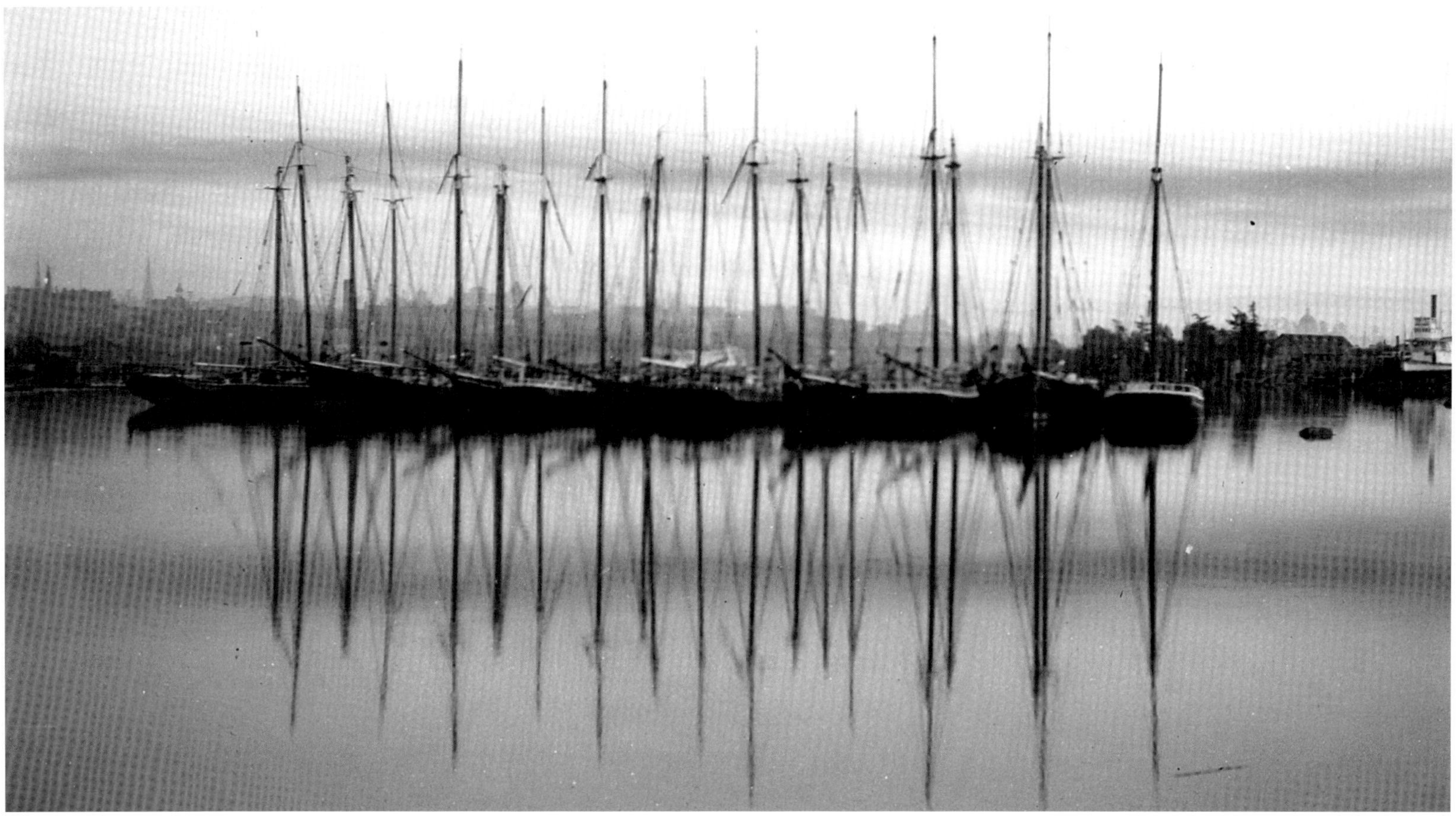

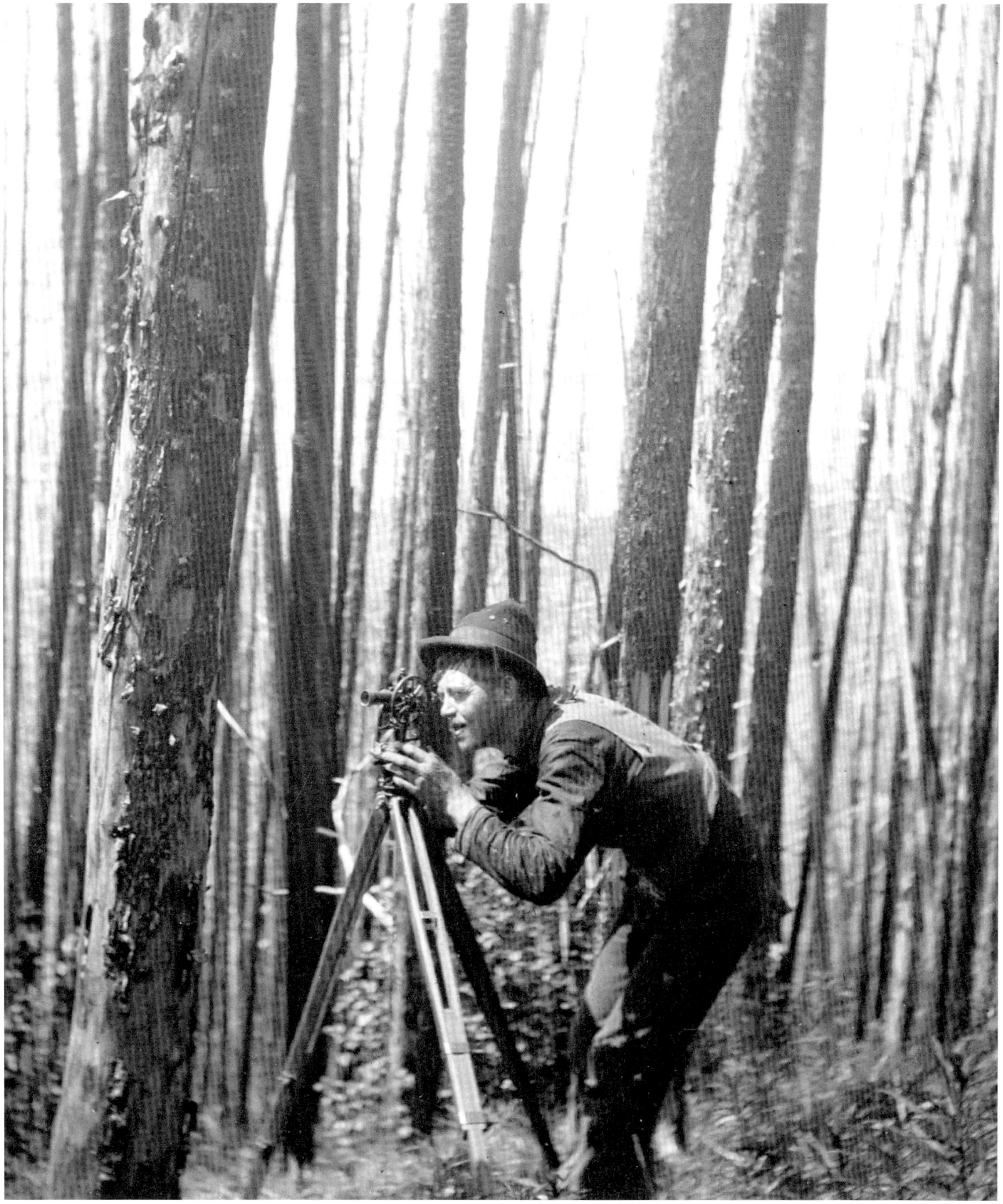

❖ Cummings at transit. Alfred Cummings became BCLS #10 in 1906. Image I-58072 courtesy of Royal BC Museum, BC Archives.

(no walls) 5′ x 7′ [1.5 x 2 m] ground area. Who wouldn't want to be a surveyor!

For the next two weeks Cummings & I (in dual charge, I as transitman) ran letter R of Moyie River on the map to Boundary due south. Have obtained a canoe for crossing the river.

Mac runs due north from same point on east boundary of Lot 4592 to Lat. 50°N.

Next, Cummings & I start at Fort Steele to run the W. boundary of L4590 and E. Bdy L4596 (to corner of yellow lot near Canadian on map title).

Cummings, Mac & I are the only Canucks on the gang—every man jack of the others is a blasted Englishman, don't cher know …

Excuse pencil—delate halo ink.

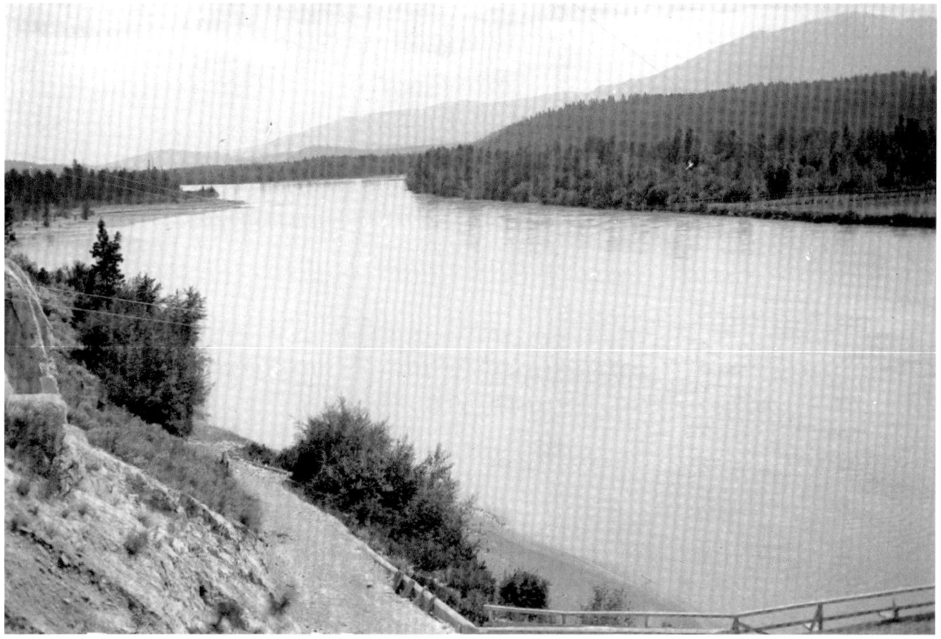

❖ Kootenay River at Fort Steele. The bridge across the Kootenay River is visible in the lower right of the picture. Image I-33937 courtesy of Royal BC Museum, BC Archives. For 2014 image see page 153.

Swannell's next letter to Ada was written less than two weeks later. In it he explains why he was able to write her again so soon.

East Kootenay Surveys
Cranbrook, June 2, 1902

Rather peculiar circumstances led to my getting blue envelope No. 1 today. To wit, we were at mile post three, south of Yahk, no prospect of seeing a soul for a fortnight. Along comes a terrific toothache, lasted two days and nights without let up. As I was seriously afraid Mc Recording Seraph would need a stenographer to take down the explosive language I was making the scenery lurid with I resolved to go to Cranbrook. I tried hitching one end of a piece of tackle to the offending bicuspid and the other to a bent sapling and then letting the tree spring back—it did and incidentally my jaw (to judge from the sensation), but the tooth remained.

Had a nice walk over six miles [10 km] of windfall and 5 [8 km] of railway track to catch the train. Doctor had a regular tug-of-war, tooth apparently had roots clinched through my boot soles. He did it in 5 hauls, cracking one of Hall's fillings, ulcerating, between amalgam and nerve. Awful botchwork, I thought he filled 'em in a hurry.

We have a fine lot of men in our party—exceptionally nice fellows. Cummings (Alfred) and I have a 5 x 7 tent to ourselves. The rest of the outfit are together in one tent and a nice racket they kick up singing in the evenings when Cummings & I are wrestling with logarithms and computations.

Cummings I like very much, a nice quiet fellow with a girl in Peterborough who probably would inform you there isn't another such boy in the world—He made an awful fuss over my toothache, insisted on making a whopping big hard-tack poultice—whew! It was hot … He looks after the commissariat, moving campsite. I do all the transit work. After 4 months in a 5 x 7 tent we will be intimate. I think I have already heard a great deal about the best girl in the world (as if forsooth they had half as good girls down east as my Ada).

Our cook, Virden, a very important in-

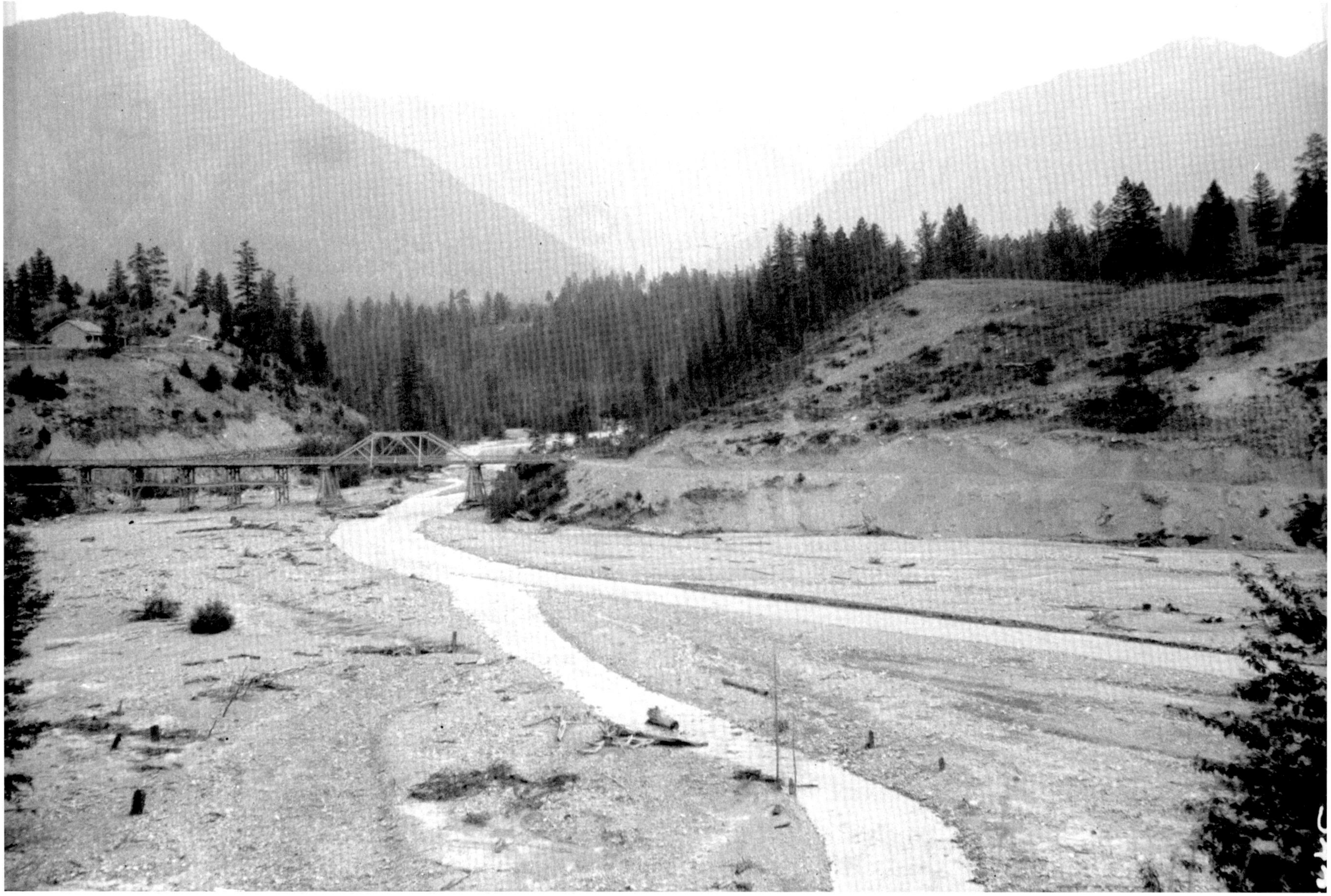

❖ Bridge on Wild Horse Creek. The present road crosses Wild Horse Creek at almost the same location. Image I-33946 courtesy of Royal BC Museum, BC Archives. For 2014 image see page 153.

dividual, has relieved me from a steady and arduous job of tossing flapjacks. He wears his socks over his trousers, in Ontario lumberman fashion, and seemingly has been wanting a shave for many moons. Yet he is all right, for did he not advance me $5 for my molar extracting trip.

Benson I like well—he never grumbles and packs a ridiculous load with undue groaning of spirit. Also wears an enormous .44 cal. revolver and cartridge belt. All of us practice on grouse with enough bird to flavour the beans.

Have had one crack at a deer, after dusk. Benson says I hit—I think I missed. In the touching religious sentiment embodied in a spirited Methodist hymn: "When I can point my rifle clear/ At red deer on the fly/ We'll bid farewell to pork and beans/ And feast on venison pie."

Swannell next wrote a brief note to Ada from Cranbrook after he finished surveying near Yahk and was on his way to Fort Steele.

Mining Operations on Wild Horse Creek

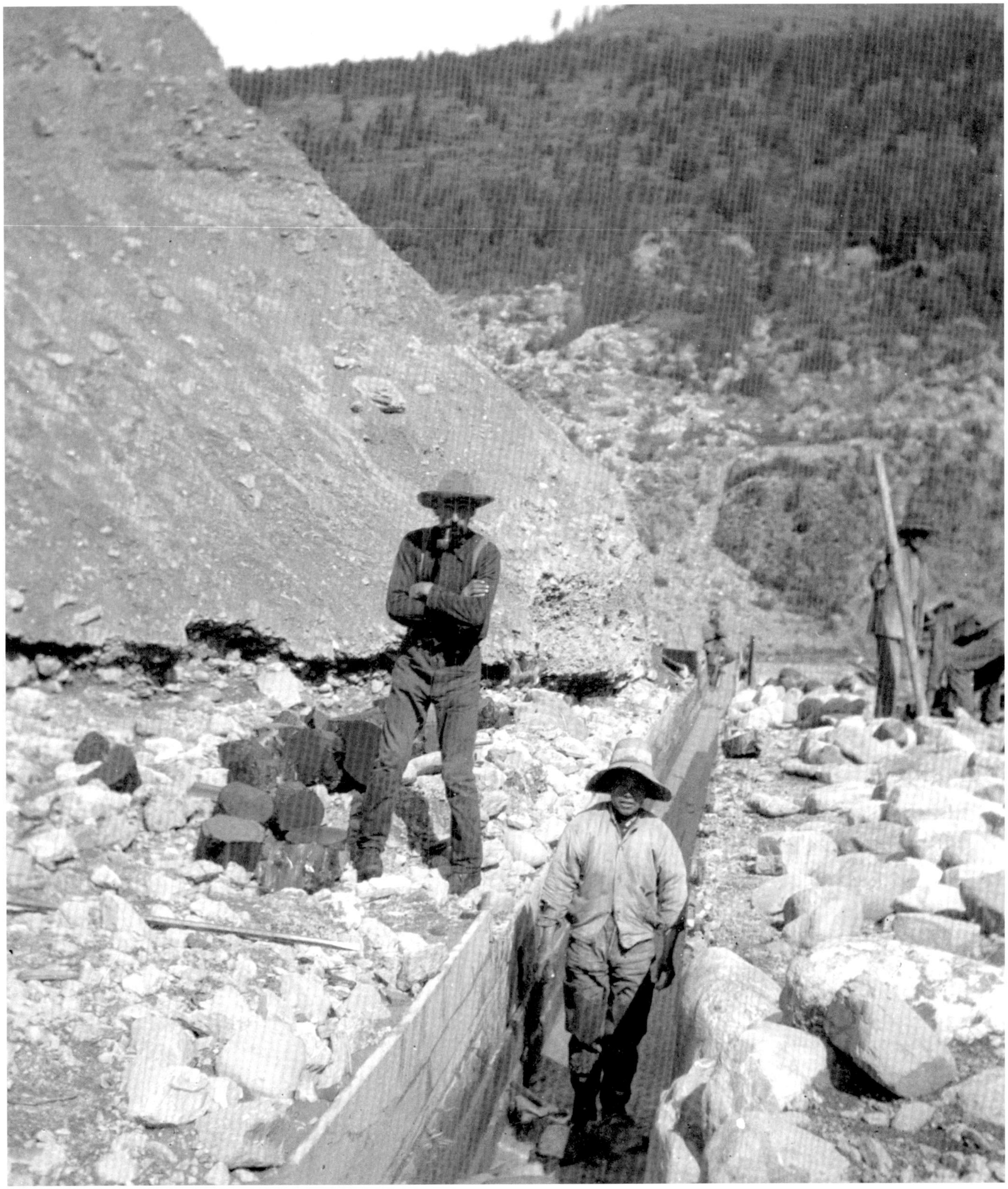

In the early 1860s there was a sizeable gold rush in Wild Horse Creek, a tributary of the Kootenay River, near Fort Steele. This occurred at about the same time as the Cariboo gold rush in the Barkerville area. From 1885 to the beginning of the twentieth century there was a second era of gold mining activity along Wild Horse Creek. Many Chinese miners reworked the claims, extracting additional gold. There was also more extensive use of equipment to remove rocks and dirt and to channel the water. Swannell took these pictures near the end of the second mining rush to Wild Horse Creek.

A monitor directed a powerful force of water that removed the rock and dirt from the hillside, sending it down to the creek where it could be processed. Removing this material made it easier to reach the bedrock where most of the gold was found. If there was sufficient water, several monitors could be used. Hydraulic mining quickly changed the landscape of a creek valley.

Gold-bearing material would be shoveled into a sluice. The water would separate the heavier gold, which would sink to the bottom of the sluice and become trapped in riffles.

Frank Paterson getting ready to remove the riffle bars and extract the gold. The large pile of dirt behind him is some of the material removed by the monitors. Image I-33950 courtesy of Royal BC Museum, BC Archives.

❖ Hydraulic mining on Wild Horse Creek. Image I-33521 courtesy of Royal BC Museum, BC Archives.

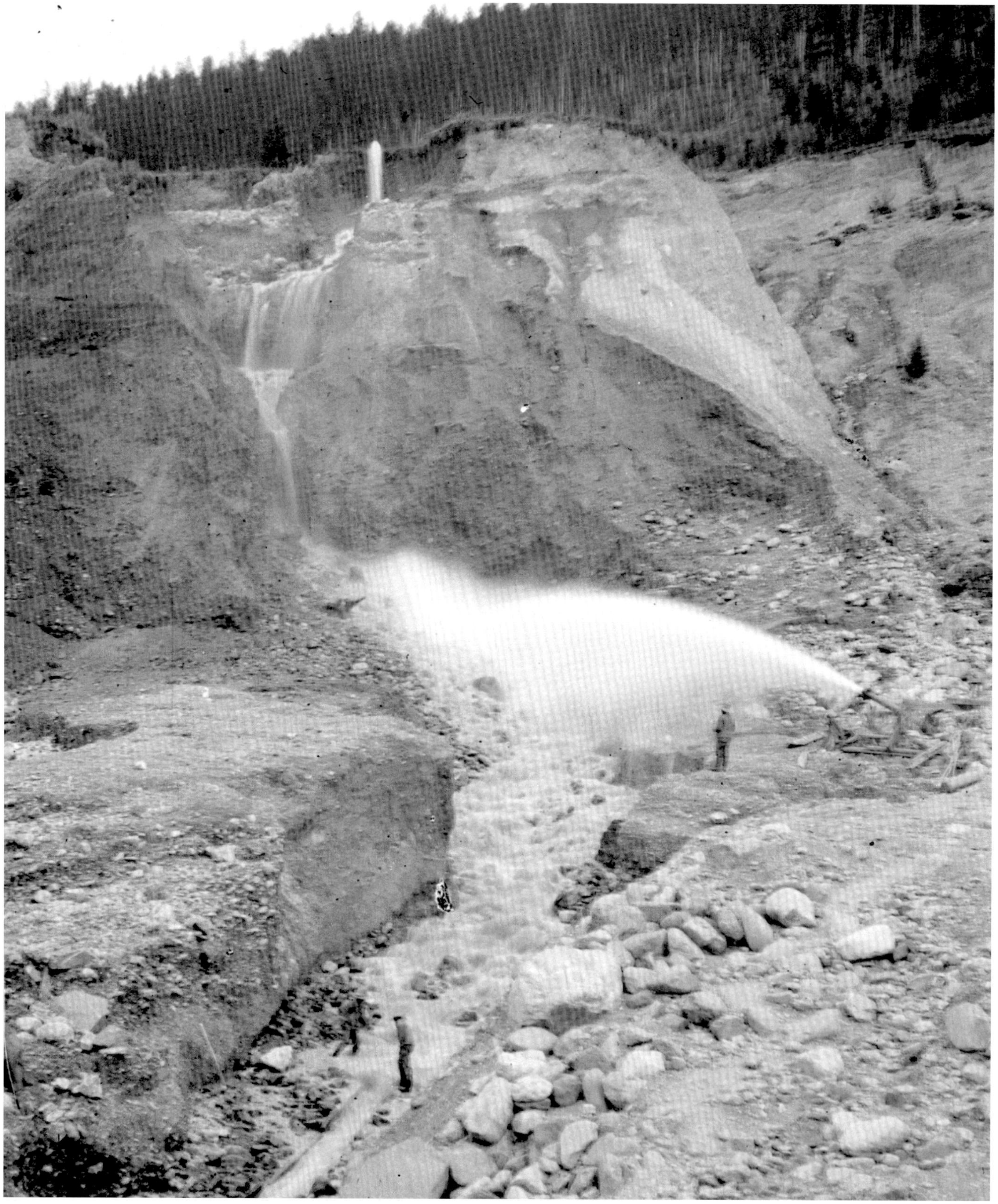

Opposite page

❖ Top: Chinese prospectors sluicing on Wild Horse Creek. In the foreground two of the miners are filling a wheelbarrow with material below the level of the creek bed. This would then be dumped into the sluice. Image I-33945 courtesy of Royal BC Museum, BC Archives.

❖ Bottom: A monitor used on Wild Horse Creek. Image I-33949 courtesy of Royal BC Museum, BC Archives.

East Kootenay Survey
Letter to Ada Driver
June 11, 1902
Cranbrook

First, a little summing up [of] our doings to date. Lest I become too lengthy will adopt Caesarian diction, Veni, Vidi, Vici, formula. After trip to doctor landed at Yahk 2am, walked until daybreak, ran gauntlet of the river 5am, packed five miles over heavy windfalls. Reached camp 6pm fagged out. Since packing and whooping it up on the line "grub" running out twice—lived on hard tack, one per man, two day, on scenery last day. Reached here tonight, after an exhilarating time on the

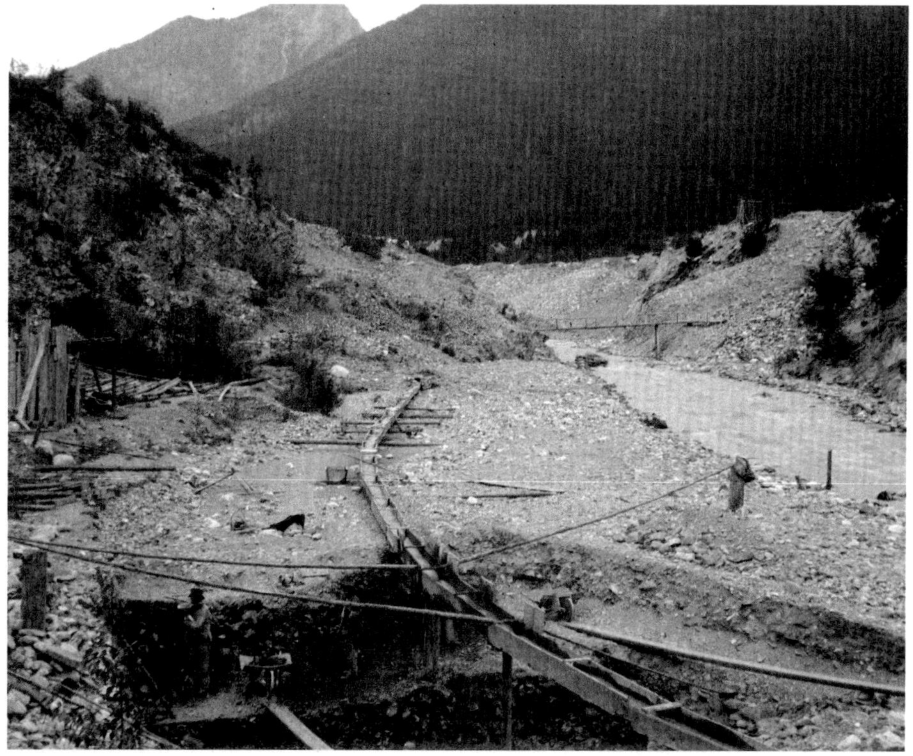

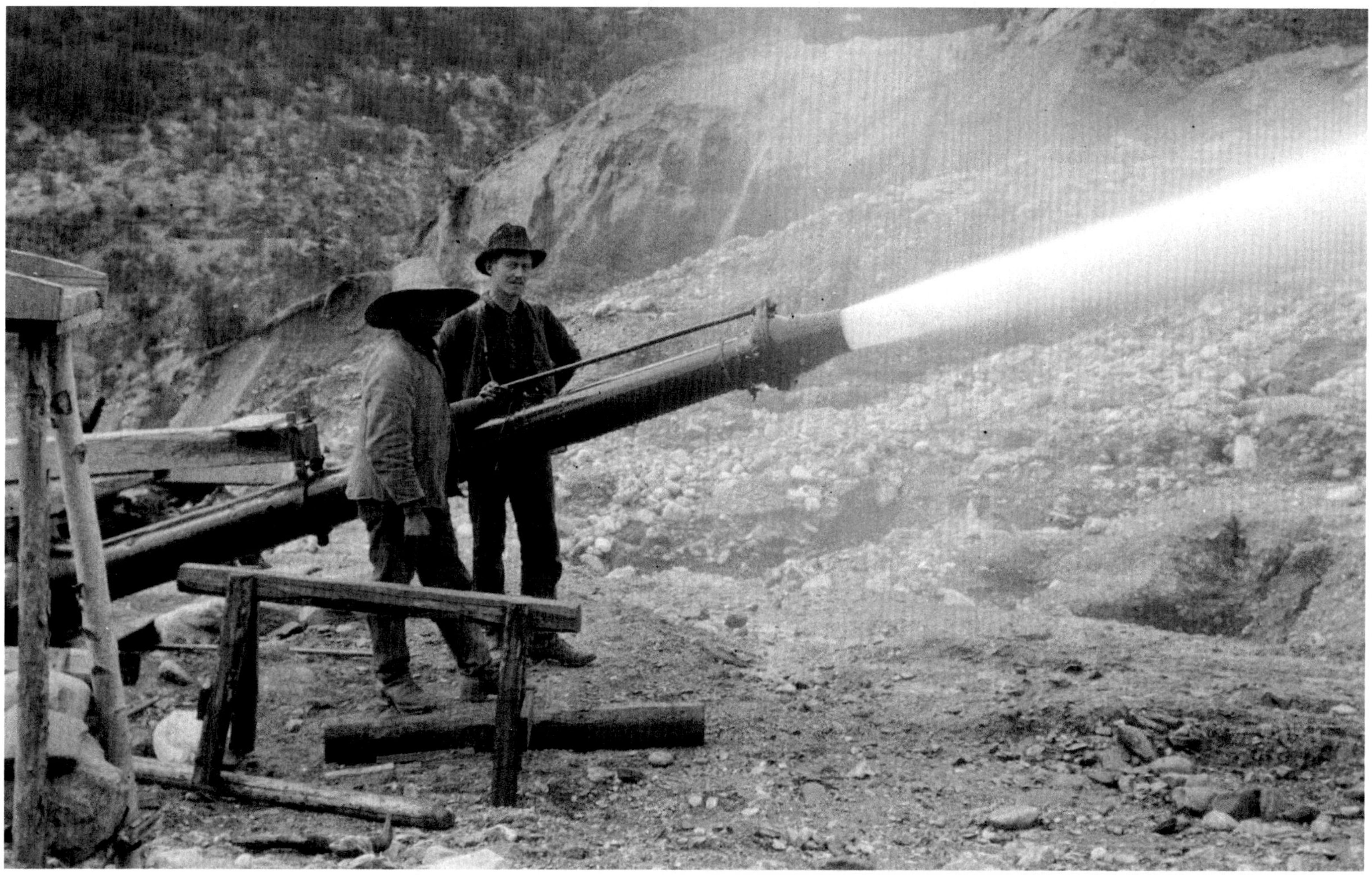

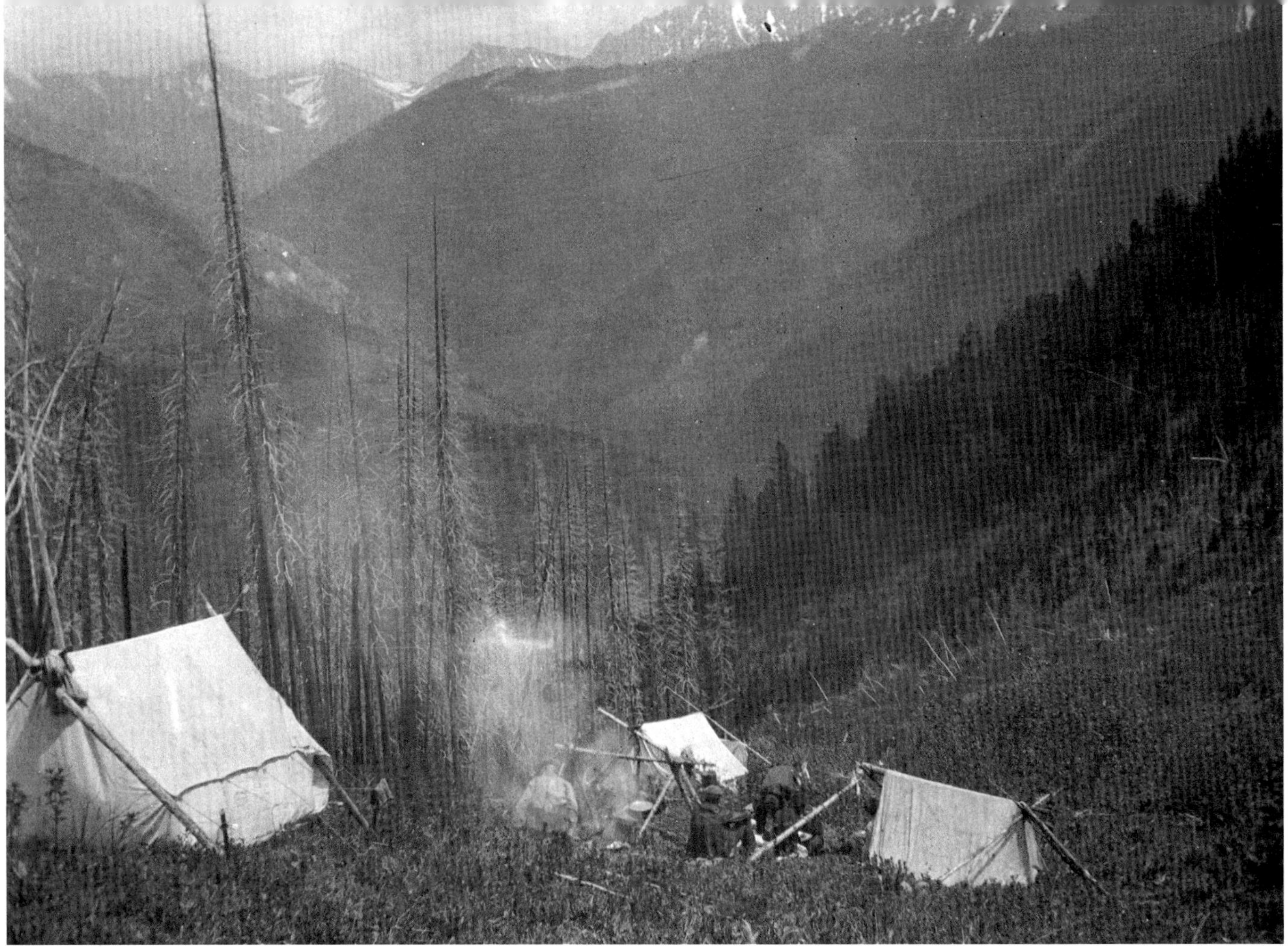

❖ Slide Creek camp. Image I-58069 courtesy of Royal BC Museum, BC Archives.

river (Moyie). None of the other lads would venture. Tomorrow start for 2 miles [3.2 km] below Fort Steele on the Kootenay.

While their survey crew was in Cranbrook, Gore wrote to the CPR land commissioner. He said that they had completed the west boundary of Lot 4591 and that McGregor had started north on the east boundary of Lot 4592 which ran directly north from Lot 4591. He commented: "Owing to the flooded condition of the Moyie River, we were unable at first to cross it. So in order to locate the International Boundary at the S.W. corner of Lot 4591 I made a very careful traverse from the monument on boundary at Moyie River up the valley to railway and along the railway to starting point."

Gore's crew then began surveying the west boundary of Lot 4590, spending about a week working near Fort Steele.

During that time Swannell visited the mining operations at nearby Wild Horse Creek where he took several pictures.

Wild Horse Creek, near Boulder Creek
19 June, 1902

(Written on paper torn from field book)
My dearest Ada,

Only about a week since I last wrote you. We move camp tomorrow. Frank Donavan (horse packer) goes to Fort Steele for grub. Horses with alpareras. I think I will receive mail every fortnight or so by our supply train (if I may use so grandiloquent a term for 3 pack cayuses). Don't imagine we are in touch with civilization however—I don't expect to see the flutter of a skirt for months …

Cummings has just come in from an exploration ahead. "Lots of snow ahead, trail blocked" his first words. Cheerful, eh? We've

been lucky so far but are going to catch it now—run slam over mountains twice as high as the Olympians. But snow (eternal snow I mean) doesn't lie below 8000 feet [2,400 m] here (McGregor party were in six feet of snow a fortnight below. Imagine me perched up on a pinnacle half as high again as the hills across the straits. Pardon for this smudge, am writing on a tinplate.

We have a splendid lot of men—all old hands except Benson. Just think how important an individual I am: "Take down that tree. Yes sir." Great fun being boss, only one has the entire responsibility and I figure I work an average of 11¾ hrs a day. Put in an hour after supper calculating. Best job in the world to knock a little of the inherent laziness out of me.

Trust pencil writing is legible. The landscape up the Wild Horse doesn't teem with stationery and clean envelopes.

Beyond Fort Steele the surveying of the lot boundary took the crew north for over 50 kilometres through rugged terrain and high into the mountains. The men had to clear the line wherever it went. At Lewis Creek northeast of Fort Steele, Swannell was able to send out a letter to Ada. It was the height of mosquito season and they were at high elevation where snow still remained, so Swannell's letter describes some of the difficulties surveyors faced in remote regions. The last part of this letter is missing.

10th July – North Fork Lewis Creek –
E. Kootenay

My dearest Ada,

Awful mixture this letter will be I'm afraid. The mosquitoes are rampant and considerably disturbing the even tenor of my thoughts. Don't be horrified by a few squashed samples ornament this letter.

Well, little girl, between skeeters and brule and a July sun on a scorched side hill I'm feeling most awfully cantankerous and so will explode right here into expletives of strength modified and softened by your ear. Here goes!

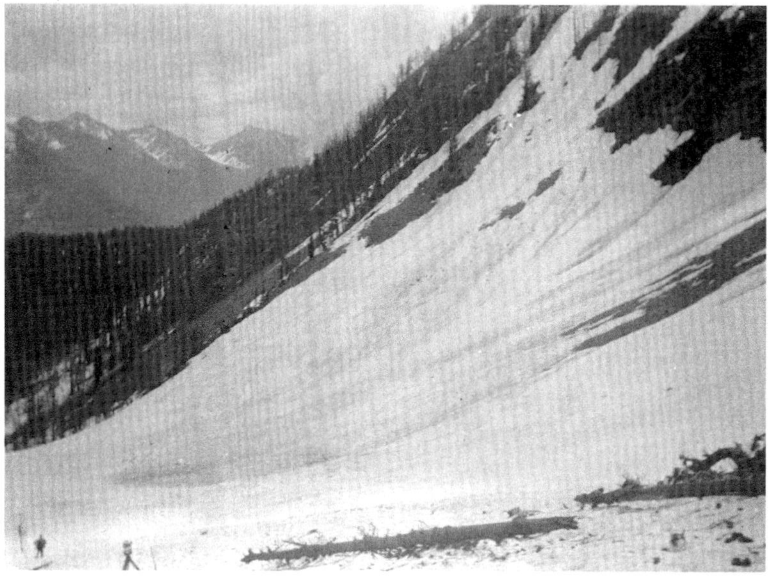

❖ Snowslide in the Rocky Mountains. This was typical of much of the terrain that Gore and Swannell surveyed on the west boundary of Lot 4590 during the early summer. The picture was taken in late June. Image I-59890 courtesy of Royal BC Museum, BC Archives.

Confound this blankety-blank E. Kootenay country root & branch. A malison upon its brules and fallen timber, may its rock and snow slides be anathemized. At the present moment more especially do I beseech the powers that be to blow the mosquitoes unto the 3rd and 4th generation. Amen!

(Please wait a bit, Ada, the tent's positively uninhabitable, must make a smudge.) All right now, skeeters disgusted with smoke and resting on the ridge pole. Here below the smoke is hard on the eyes, a regular inferno.

We've at last got over that confounded mountain, are now far down the northern slope and are finally out of the snow, clouds and winter cold. Today is the first time in two weeks we've had dry clothing on. It's the second day in that time we've had sunshine and no rain!

The glorious 4th was a terror, over 8" [20 cm] snow fell. It was hail & sleet all night, splattered clear through the tent so that famous eiderdown quilt of mine got pretty soggy.

Shall I become personal? I am arranged not like unto the Lilies of the Field. Kindly regard my boots, just put 1/2 lb hobnails into them, for we wear out a pair of boots a month ($4.50) and a set of nails a fortnight. One's life wouldn't be worth an hour's purchase in these mountains but for the same nails. As for the remainder of my attire I will merely remark that canvas patches are more conspicuous [than] the original material of my overalls, while my blue flannel shirt owes much to a flour back, collar absent, tie not conspicuous.

What did you do on the 1st? I was thinking of Albert Head [where they had had a picnic on the previous Dominion Day] all the time we were packing over the snow, floundering near waist deep sometime.

Noticing our bread of late had a rather more weird taste than usual I enquired of the cook and obtained the recipe used, he explained, when flour is scarce. Flour is the backbone, as it were, of the composition, then oatmeal and a generous percentage of hardtack biscuit. Next comes cheese pounded to a paste, a little tea and finally baking powder and salt. As the proportions are varied according to the supply, it being a case of cutting your coat according to your cloth, I can't give exact amounts of each ingredient. Therefore I am afraid the like of this bread will never be obtained.

In Swannell's next letter he discusses an important item in a field camp, particularly after they have been away from any luxuries for a while.

Diorite Creek—E. Kootenay, 22 July, 1902

Nice disappointment today, expected some mail, and got left owing to Cummings mislaying it. We have been completely and totally isolated since June 27.

Ever since Dominion Day a series of ridges and peaks 6–8000′ [1,800–2,400 m] alt[itude] have separated us from our supply depots in the Kootenay Valley. Consequently it has been packing and running the line on short rations. A fortnight ago all luxuries disappeared, not but flour, bacon, beans, rice and apricots left. A week ago bacon, rice, baking powder & salt became remarkable by their absence. Then we lived on bannock made with snow water, no salt, no baking powder. Very iron clad affairs they were. When did I last taste sugar or salt? Echo answers, when? Yesterday matches and beans became exhausted, we even boiled the small number the boys used as poker chips. Really, Ada, our own idea of heaven is very narrow—a little sugar, no mosquitos and no limit as to the size of a plateful of beans. What more could one desire?

The line runs slam over the mountain peaks—snow, ice, clouds and sleet ad libitum. Last Friday I had a bad time, crossed the range and managed somehow to get away down a gorge the other side (I was after grub, but the packer wasn't at the rendezvous and the lads nearly starving across the summit). Pleasant mood I was in!

I started back dead tired and had just pulled myself up a projecting crag—a rustle in the jack pine—and then three deer in full run only 180 yards [165 m] away. Hadn't time to elevate the rear sight. Bang. Down with the lever and bang again. My first shot told, a big buck rolled over and over down the steep slope. I thought he'd never stop and dolefully calculated how I'd like packing him up again. Down I scrambled to where he was caught be a projecting rock. Shot clean through the spinal cord at the back of the neck and stone dead—a three pronged buck too.

I had an awful time getting the hindquarter packed over the mountain—had to pull myself up, hands and knees for it was nearly a precipice and the snow wasn't safe. Fifty feet

[15 m] at a spell was the utmost I could do. Then I had to climb down 3000 feet [900 m] the other side looking for camp (it had been moved that day). Ridge after ridge, rock slide and bluff and fallen timber. At 9:45 I heard a faint shout, a turn round a ridge and there was the faint glimmer of the camp fire. Don't know how I reached it—it seemed an eternity before I staggered into the circle of firelight, and fell on my face, venison, rifle and all, never was so played out in my life before.

So, Ada, that's how the situation was saved, and we lived on venison & saltless bannock for 4 days more. Why we haven't seen any outsiders except Gore since June 20th. We don't even use table knives nor forks.

After completing the survey of the west boundary of Lot 4590, Gore's crew travelled to the northeast corner of Lot 4592 which was at the southern boundary of Lot 4596. Here they met McGregor's crew. From this corner Gore ran a line along the fiftieth parallel east along the south boundary of Lot 4596 back to the northwest corner of Lot 4590, while McGregor surveyed west on the fiftieth parallel to Kootenay Lake.

3 Miles north of Skookum Chuck Cr.
8 August 1902

This is a bad country although the mountains here don't compare with the Rockies. Fearfully thick tangle of spruce, pine and hardhack with pleasant (sic) intervals of brule and springing windfalls all on slopes of as steep a pitch as a house roof.

We have for 10 days been on a fly camp (no large tent and abridged rations) … Rain all last night and today, turning to snow now. Cheerful prospect—can't get blankets dried.

A week ago Gore and I connected with Mac (J.H. McGregor, PLS) … Heard Mac yelling stadia [a type of surveying] from afar off. Whereupon Gore shouted out, "Don't make such a racket." Then we clambered down and shook hands. First Mac, "Hello Gore." Then, "Oh, here's the young stranger." Mac, in the joy of his heart gave us each a hunk of cheese to nibble (unwanted luxury).

Saw his men next day—all bearded and patched plenteous altho' they insisted our party is the wildest and woolliest. Mac runs the north boundary L4592 to Kootenay Lake, 32 miles [50 km]. Expects to make it by October 1st. We run south boundary L4596 20 miles [32 km] away due east to where we quit our previous line, then east to Elk River 32 miles (52–55 miles [83–88 km] in all). The country to the Elk is a sea of mountains—the Rockies, no trails and utterly unknown. We will pack all our grub with us and trust to luck to get through by Oct. 15.

Little girl, you ought to see our camps. A fire for the cook is all. We each select a spot in the steep sidehill and dig out a level spot with the back of an axe, placing a log on the downhill side to prevent us rolling down in the night. Next, down comes a balsam, whose brush we use as a mattress—of course if we strike camp late we just lie on the ground without going to all that bother. So rolled up in the blankets we lie under the open sky all over the woods within a hundred yards of the fire.

When the survey line crossed the Kootenay River, Gore sent Swannell and Benson to get more supplies. From Wasa, Swannell wrote a brief note to Ada on September 1. "Regular picnic I'm having—Have today run a boat 28 miles [45 km] down the river to here. Made it in 7 hours all alone. Last Sunday I, with another man, trailed, pulled and poled that boat to 8 miles [12 km] above Skookum Chuck. Have taken nearly a roll of photos of this trip."

Swannell wrote Ada a letter at the end of the field season describing their adventures during the last few weeks and telling her why the survey was not completed during 1902.

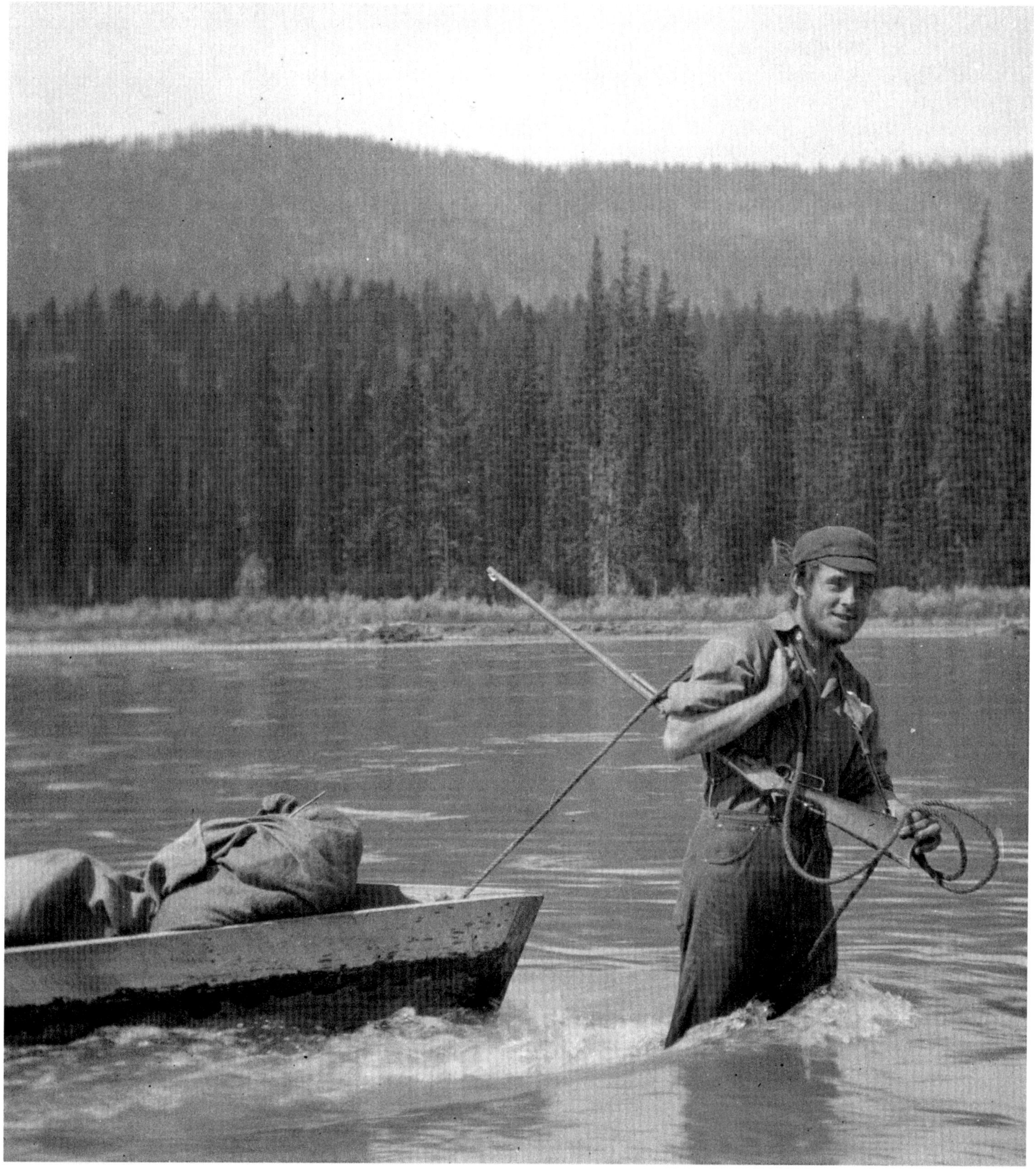

❖ Swannell lining a boat up the Kootenay River. This is a well-known photograph of Swannell as a young man. Image I-58076 courtesy of Royal BC Museum, BC Archives.

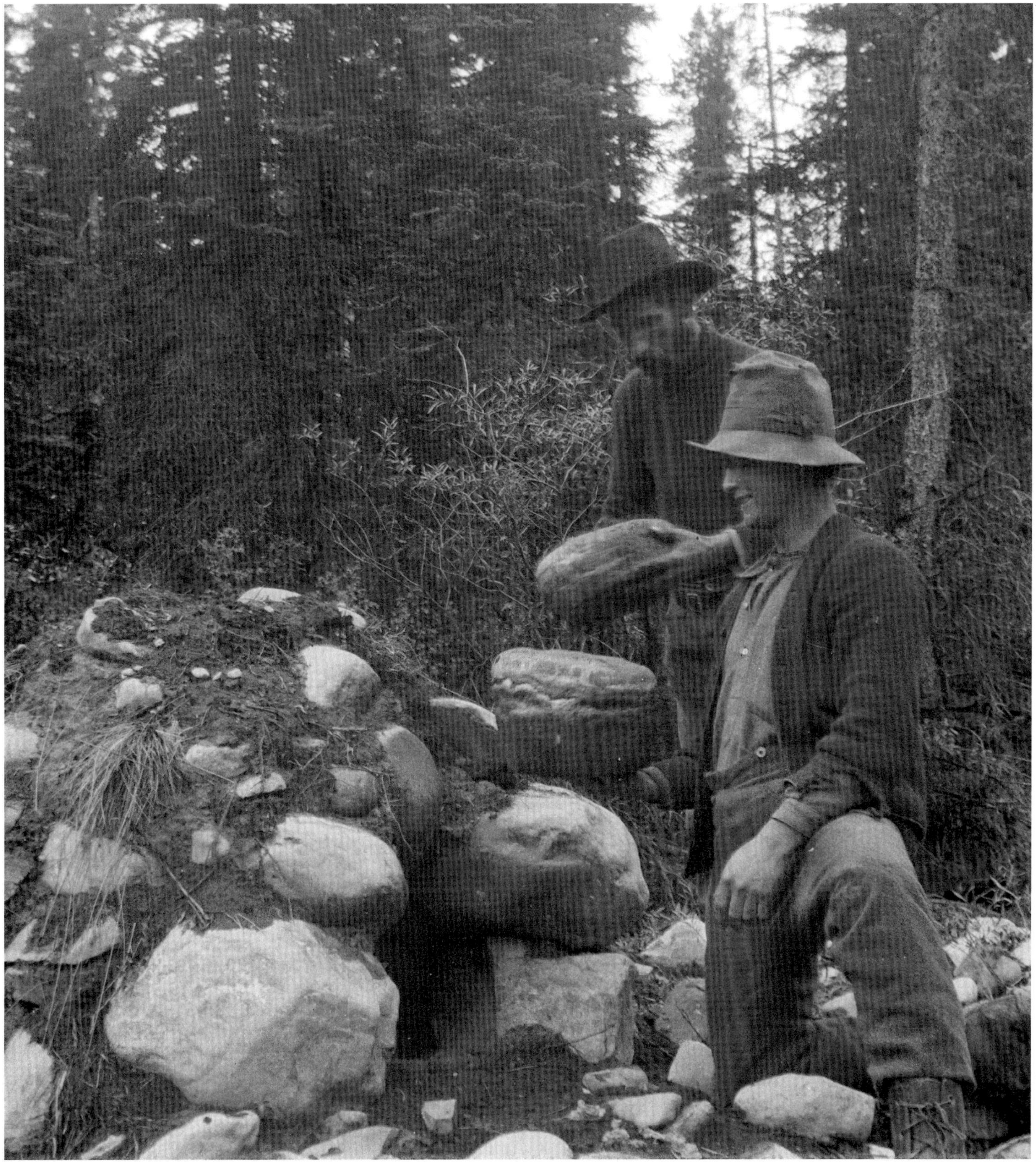

❖ Taking bread out of a stone oven. Making bread was one of the important activities for the cook on a crew. A stone oven would only be made if the men were at a camp location for a few days. Image I-58074 courtesy of Royal BC Museum, BC Archives.

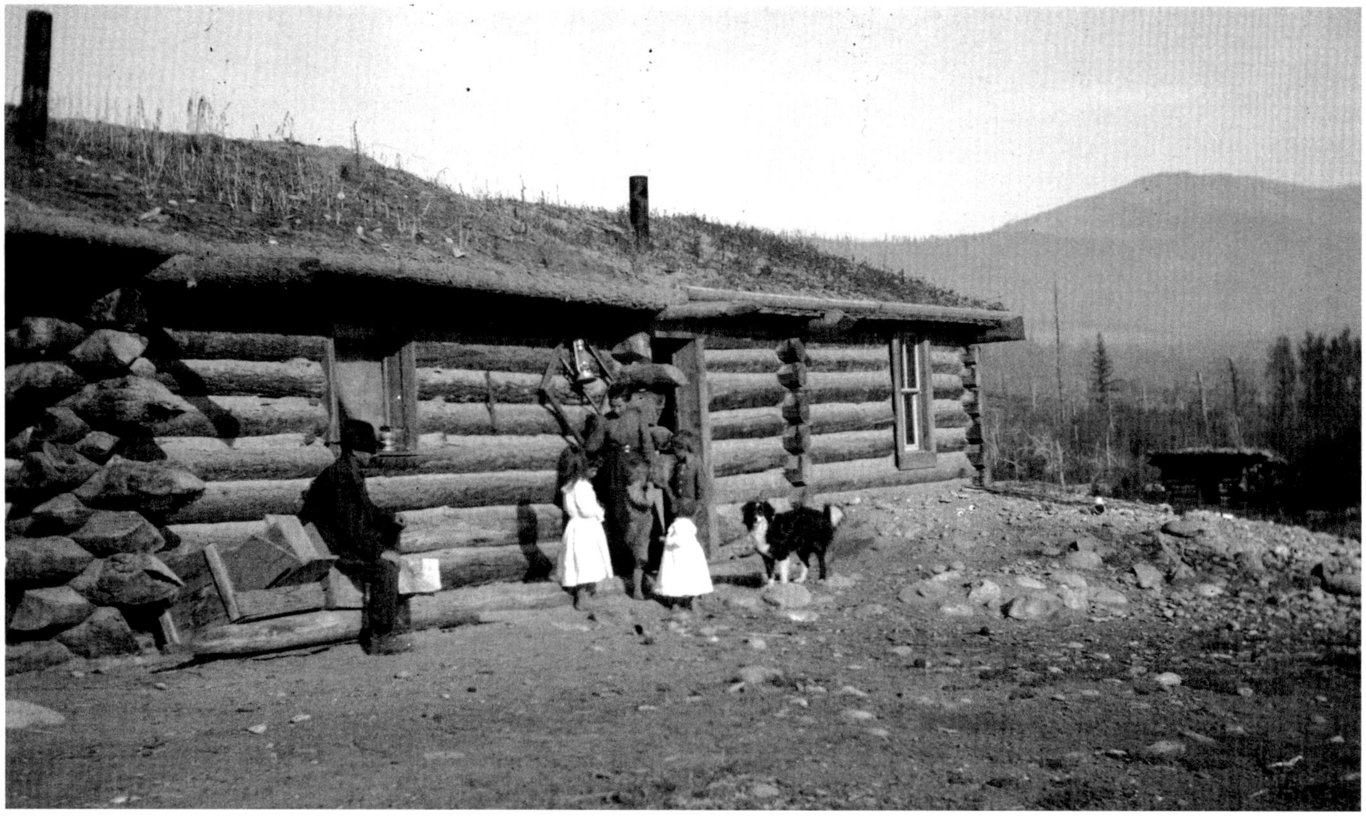

❖ Miller's cabin, Skookumchuck: a family's pioneer cabin with sod roof near the Kootenay River. Image I-58255 courtesy of Royal BC Museum, BC Archives.

October 1902
Imperial Hotel, Fort Steele

Don't bother your head about me, my dear, I'm faring sumptuously every day & am attired in brand new overalls ...

A true & veracious narrative now follows entitled:

Altitude 9000′ [2,700 m] and Halo Muck-a-Muck

Or the wicked axemen who waxed
Fat on scenery plus beans & bacon
& did presume to kick because
They did desire even such
Luxuries as flour & sugar
For which & also china—food
We did gently make
Search; but behold
It was not

To which is appended a sequel

Y-Clept [to call or name]

Ye Pursuite of ye Grubbe of ye same, 700 lbs

"Bouche va Toujours" [literally: mouth always giving]—especially on surveys

Sept. 6—sugar ran out, 10th rice and oatmeal, so down the list. 16th men mutinied & Gore had to let them go after 700 lbs [320 kg] supposed to have been left at a cabin at the summit between Sheep Creek & Wildhorse. Cut out 15 miles [24 km] of execrable trail through a windfall, but couldn't find it—came back at 2 a.m. after two days search.

Meanwhile, I & Fred Vedette (French-Canadian) had made 6 miles [9.65 km] of trail to Coyote Ck. A week ago Tuesday Gore, 12 Frenchmen, Benson & myself started, chopped trail through, reached (K) at dusk (15 miles) [24 km]. Rain & snow all day, no food. Spent night without shelter or blankets around a fire. At daybreak found empty cabin (F) on divide. Fred & Gore stayed there. (Benson had dropped behind previous day.) Tom [Gore] & I pushed on 5 miles [8 km], where the former deserted & hit for Steele (20 miles [32 km]).

I returned, left Gore who was hardly able to hobble, at our fire at K. Fred & I went 4 miles [6.4 km] up trail to A, found no cabin. Returned, left Gore, & pushed back, met horses & rest of men at forks of Sheep Creek. Got a little bannock & handful of hardtack (first for 36 hrs). Frank Donavan & I (leaving Fred) then got a saddle horse & another horse with a little cornmeal, beans & bacon & managed to get to Gore that night (awful job, I couldn't walk, had already done 28 miles [45 km] that day) so I stuck in the saddle somehow over a regular nightmare of a trail!

Spent another night by the fire—cornmeal mush & beans cooked in the teapot for breakfast at 6am. Gore & Donavan started for our old camp at B. Gore couldn't stand on his feet, had a dislocated shoulder. I tried the trail to A again, finally found a cabin in a rock-walled basin—but no grub! Cut across the mountains, waded Sheep Creek 25 times, got to camp by dark. (I had passed the right trail at I as one couldn't notice it because of the snow.)

Next day Benson & I went up the mountain to bring down the transit & picket—nasty climb, snow & mist & cold, terribly cold. Benson played out on the top, swooned a couple of times & couldn't see anything for a while. However, I packed the chain, transit & tripod & he got all right as soon as we got below the snow & into the rain.

Next morning we got up at 3:45, started with the whole outfit at daylight. Gore & I went ahead, Gore riding most of the time.

Reached a ranch at 1 p.m., nobody home. So we broke in and helped ourselves—14 miles [23 km] from camp. Pushed on 16 miles [26 km] to Wasa that afternoon, had a fine time. Next day the rest of the gang arrived & we all went to Steele (14 miles [23 km]).

Tuesday I got directions, took Gore's horse & rode over the summit & found right cabin at E—at dark (26 miles [42 km]). Broke open a case of hardtack, a can of corned beef & a tin of butter—had a sumptuous meal. Took a lot of photos & rode back next day. Imagine the trail when it took me 6 hours to ride the 1st 12 miles [20 km].

Such is a true & unvarnished narrative of why the job was abandoned. Everybody left, but Gore & I—the former is now at Marysville. I have been making a plan at the Government Office today.

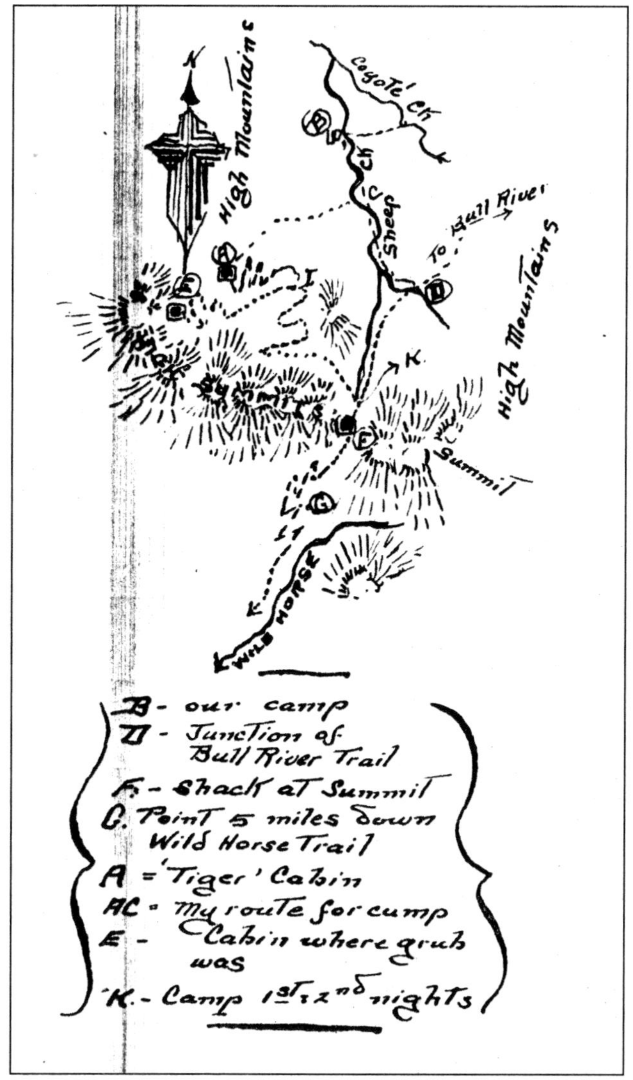

❖ Swannell included this sketch in his October letter to Ada.

Gore and his crew finished surveying the south boundary of Lot 4596 but only covered about 6 kilometres of the north boundary of Lot 4590. After they crossed Sheep Creek, the survey ended on September 15 on a mountain near an old snow slide, where Gore left a survey post to mark the end of their work for that season. In a letter to J.L. Doupe, Assistant Land Commissioner of the CPR, Gore wrote, "owing to various adverse circumstances I have been obliged to abandon, for this season, the survey of the north boundary of Lot 4590." He noted:

The 20 miles [32 km] east from the N.E. corner of Lot 4592 proved very bad country to get through and took much longer than I anticipated. Then when I at last got through to upper Sheep Creek, three of my men left me, and to add to my misfortunes, after cutting out a trail up Sheep Creek I spent several days looking for supplies I ordered sent in via Wild Horse Creek and failed to find them. We were out of food and obliged to come in. If I had found the supplies as expected I do not think I could have run more than five miles [8 km] further this year, as after that the line crosses a succession of limestone peaks and ridges that already have more than a foot [30 cm] of new snow on them and would be almost impossible to climb.

Although Gore and McGregor had not completed all of the surveying they had planned, the season was a good experience for Swannell. He had been given more responsibility for handling most aspects of the surveying and, along with Alfred Cummings (also articling to be a surveyor), had run the operation when Gore was injured and absent.

On his way back to Victoria Swannell stopped at Nelson, where he spent a day visiting his family again and took some photographs of the area.

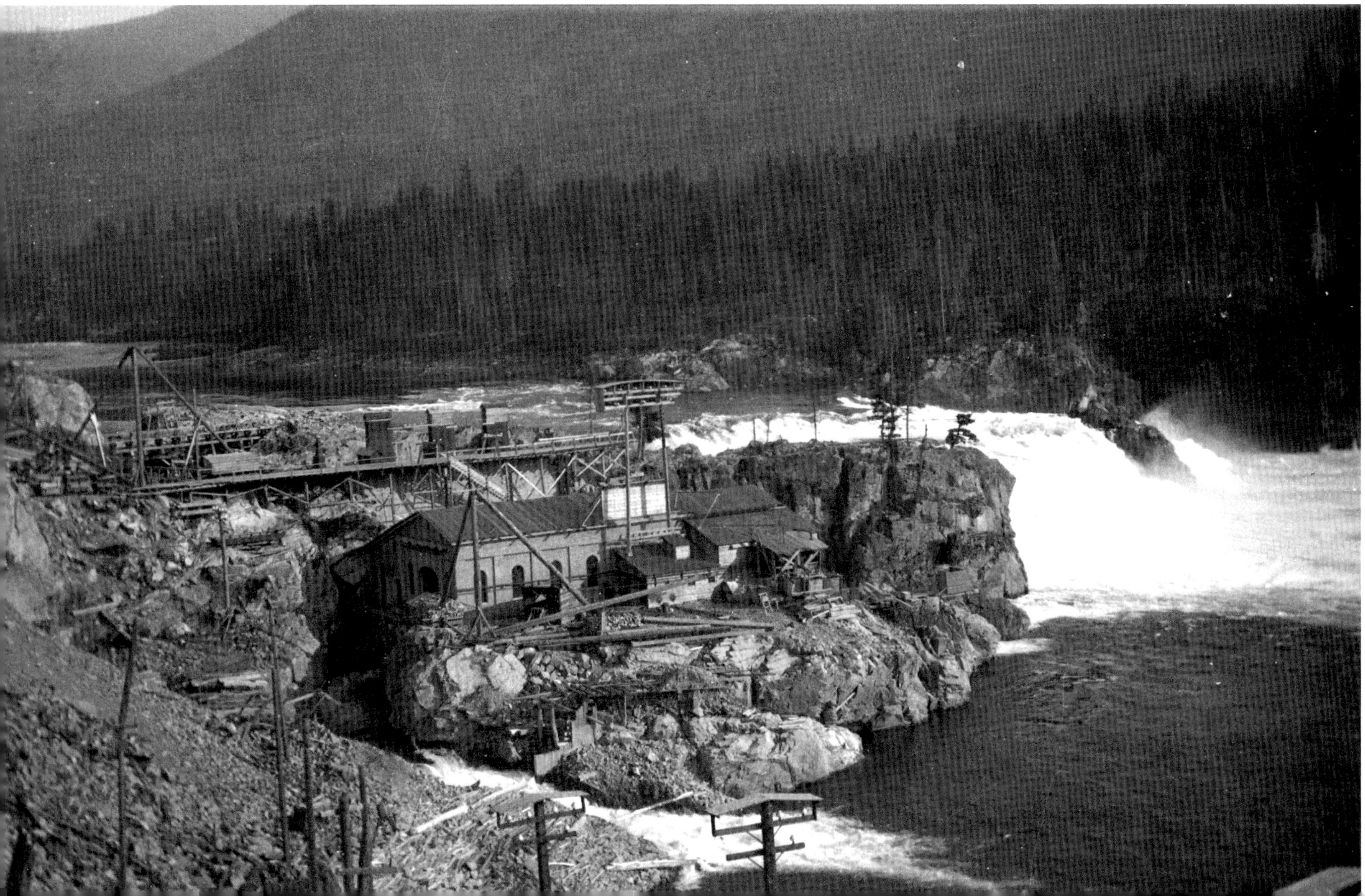

❖ Bonnington Falls power plant. Without constructing a dam, Kootenay Power and Light built a plant in 1897 to generate hydroelectricity at Lower Bonnington Falls on the Kootenay River below Nelson. Image I-33939 courtesy of Royal BC Museum, BC Archives.

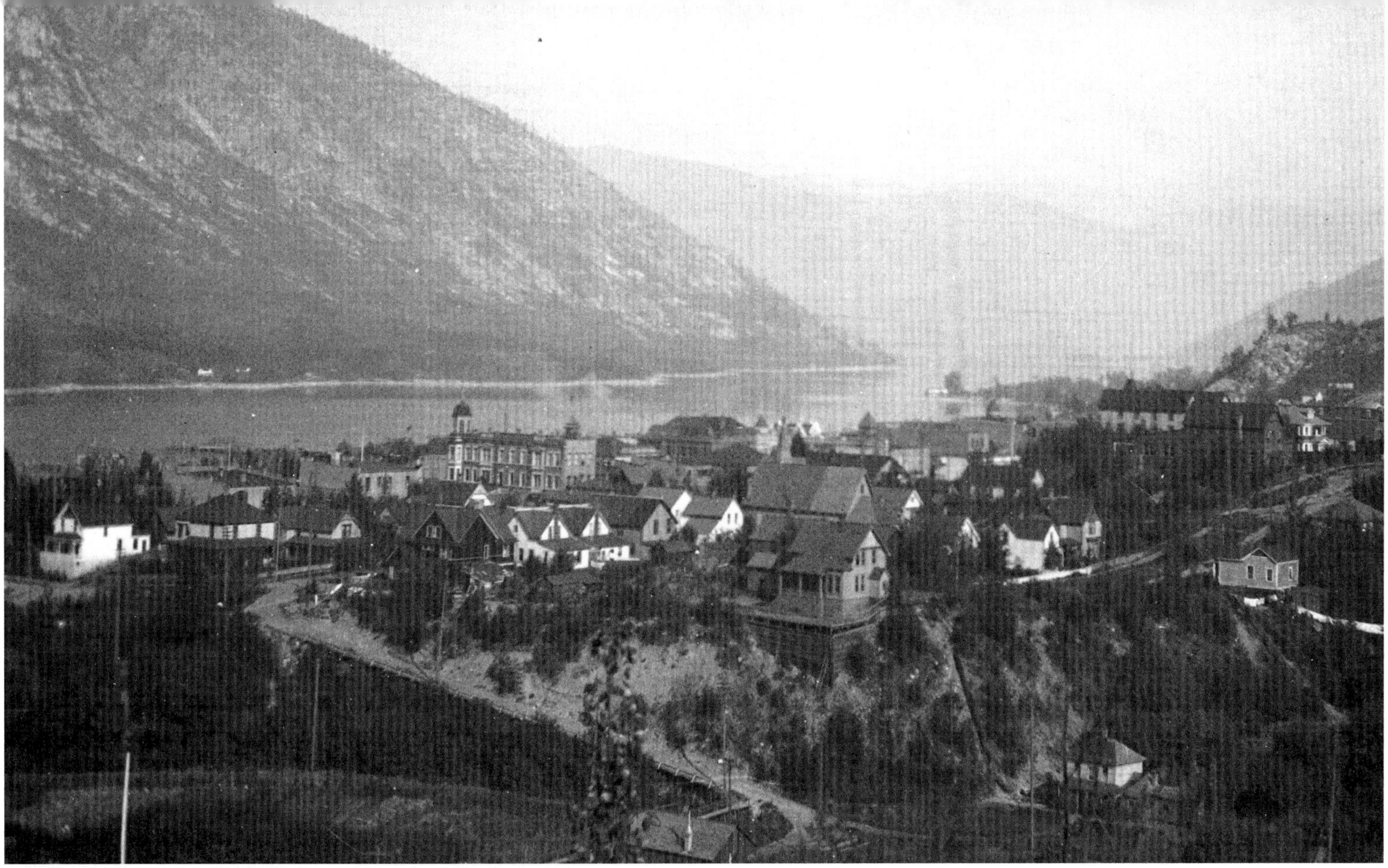

❖ Nelson. Swannell took this photograph from the Rosemont district. In the foreground is the Cottonwood Creek drainage. Highway 6 traverses this area today. Behind the drainage there are some residences while buildings of the downtown area can be seen in the centre. The large building with the turret is probably the KWC Block. Built in 1901 it was the biggest mercantile block in Nelson. Image I-58030 courtesy of Royal BC Museum, BC Archives.

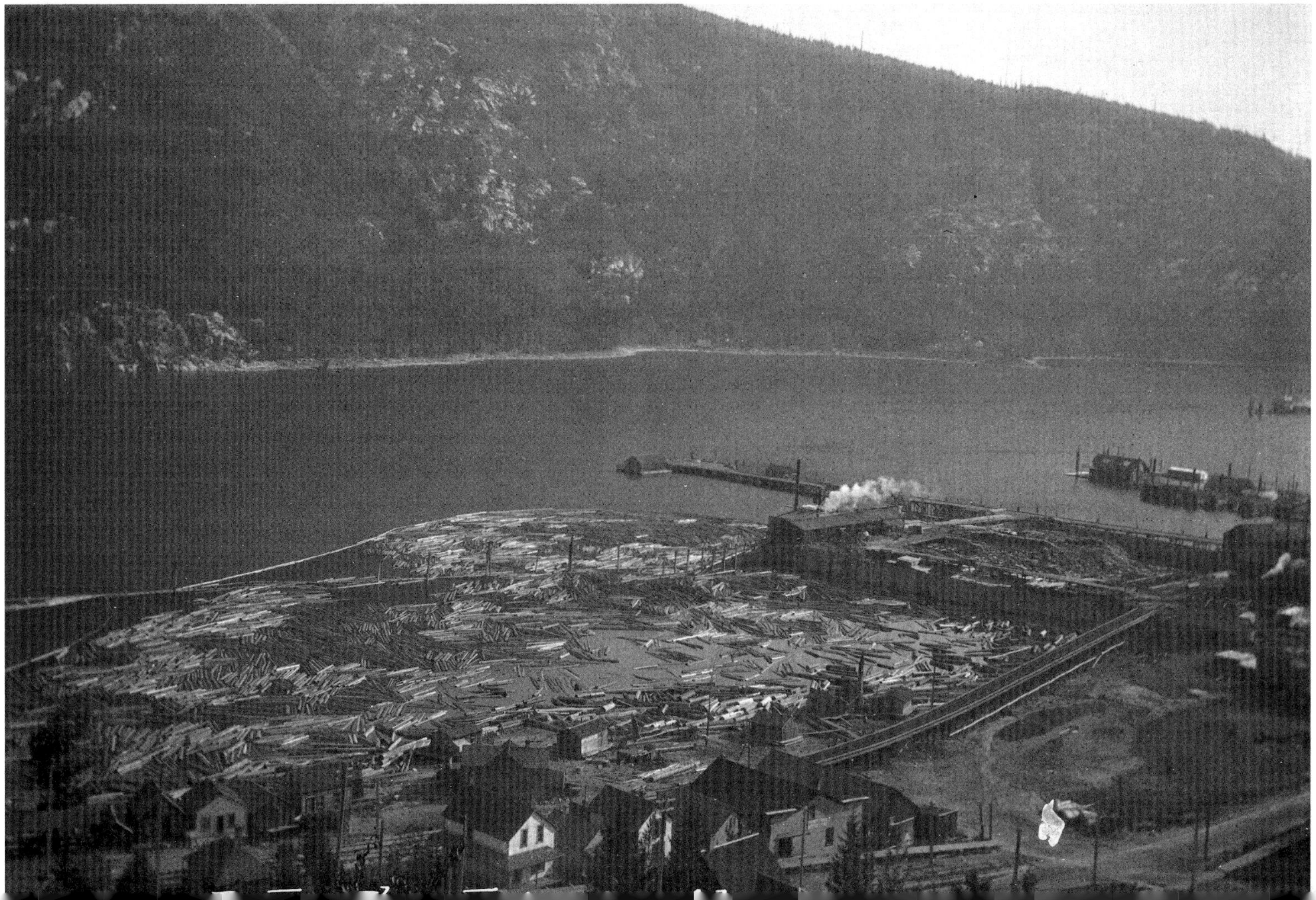

❖ Part of the waterfront of Nelson. This picture, taken from the upper part of one of the downtown buildings, is one of the earliest photographs of the Kootenay Lake waterfront at Nelson. It shows a sawmill with a log boom and a pier. Today much of the waterfront near the mill has been filled in and is the site of a large mall and shopping area. Image I-33543 courtesy of Royal BC Museum, BC Archives.

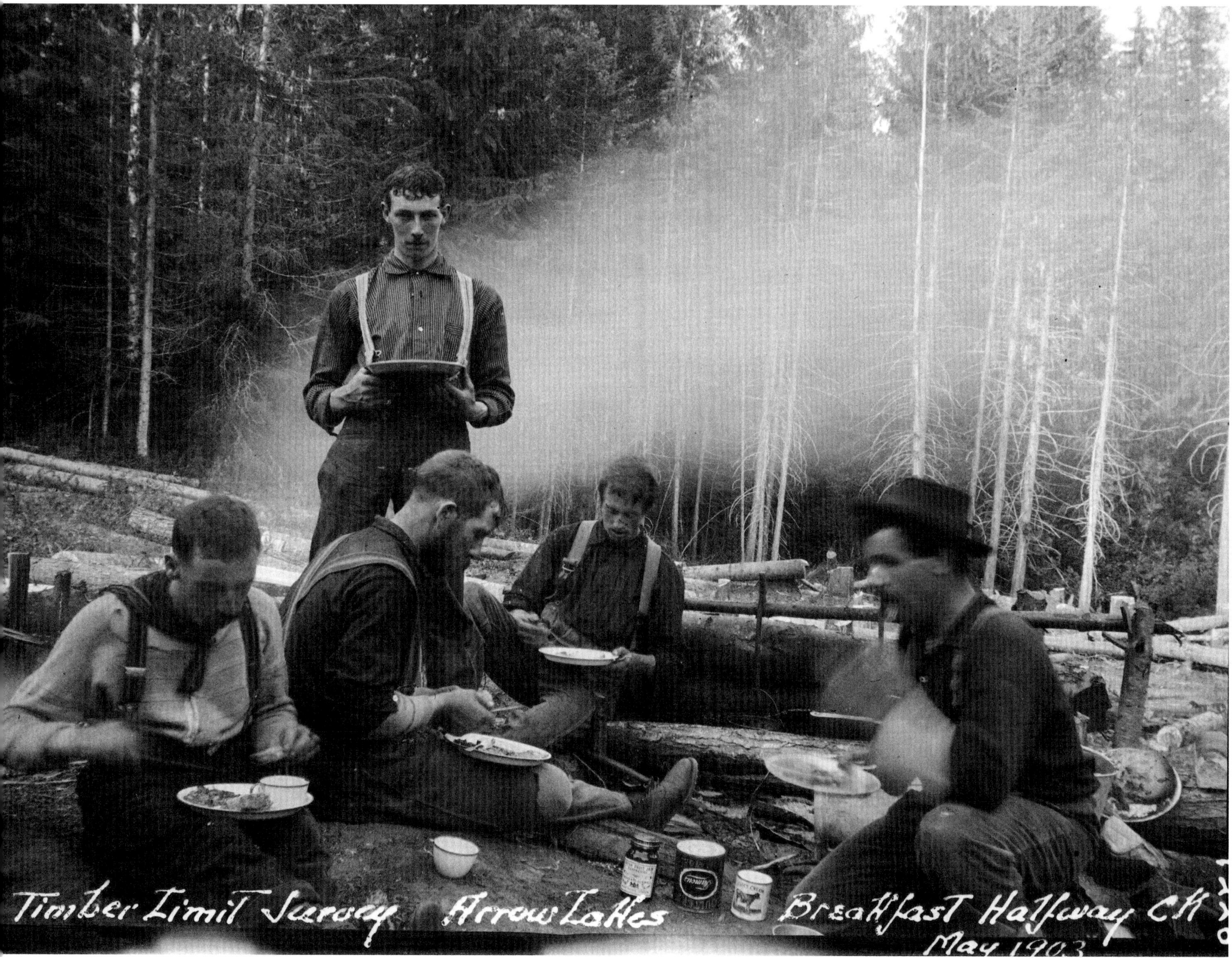

❖ Arrow Lakes timber survey. Image I-58222 courtesy of Royal BC Museum, BC Archives.

1903

By the end of 1902, Swannell had successfully finished his three years of articling to T.S. Gore. Gore & McGregor had only a small amount of surveying scheduled for the winter of 1902–3 and there was very little office work, only the few large lots that they surveyed in 1902. Swannell was laid off for the winter, so he decided to use the time to take a trip back to Ontario and to study for his upcoming examination, the final step to becoming a provincial land surveyor.

On January 10, 1903, acting surveyor general W.S. Gore wrote to Gore & McGregor informing them that their field notes and survey plans had been verified and accepted. The surveyor general noted the remaining boundary lines that needed to be run for the BC Southern railway land grant and stated, "The whole of these surveys are required at the earliest possible moment, except the boundaries of Block 4596 and the north boundary of Block 4590, which can be surveyed at a later date." Two days later T.S. Gore wrote to J.S. Dennis, assistant land commissioner for the CPR, and included W.S. Gore's letter.

> You will see this letter makes a reference to the lines the Government wish to have surveyed as soon as possible. The reason for this is that they are receiving numerous applications for the purchase and leasing of land in the vicinity and are unable to deal with it until the boundaries of the land grant are defined on the ground. I shall be very pleased to again enter into arrangements with the Company to continue the survey of the land grant—and I should like to do so at as early a date as possible to suit your convenience, as I am sure you know by experience, that as early as he can in the winter is the time for the surveyor to plan the coming summer's campaign.

In a letter that Gore wrote to Swannell in Toronto in late January, he stated, "Though it is too soon yet for prospects to be quite settled, still I feel confident that we will have plenty of work ... if you come back here and take your examination in April I think I can give you a job for six months or so, at $100 a month. In any case, if you fail to pass, which I think very unlikely, I can find work for you that will at least keep the wolf from the door."

Swannell spent most of January in Ontario. On his return trip he wrote to Ada from western Ontario on January 29.

> Ever try writing in a train. If so you will appreciate the difficulty of the undertaking. Left Toronto on Tuesday [January 27]. At Nipigon last night 7 cars of the fast freight ahead ran off the track on this part of the line. Small wonder considering the way they swing trains around on this part of the line. Well, we stopped until they got the wrecking train from Schreiber past us. We're just 13 hours behind time now. No hope of getting to Winnipeg today as the line is pretty badly blocked with snow. Fun galore, sleeper just crammed. Lady and youngster going to Michel [a small coal mining community in the Crowsnest Pass region of BC] in same compartment with me. Nearly all the others have four occupants so we are lucky.

Crow's Nest Pass Lumber Company

Although the main purpose for the railway line through southern BC was to access the mines being developed in the region, timber was also important. In September 1902, the Crow's Nest Pass Lumber Company was incorporated, with John Breckenridge as president and Peter Lund as secretary-treasurer. The company's mill at Wardner was at an ideal location where the BC Southern Railway crossed the Kootenay River. Not only did the company have ready access to the railway (which ran east–west) for bringing in and shipping timber, the Kootenay River provided transportation north and south through the Rocky Mountain Trench. The river current was normally slow at this location and there was a backwater channel where logs could be stored. The Crow's Nest Pass Lumber Company was the largest producer of timber in the East Kootenay region for many years and operated until 1957.

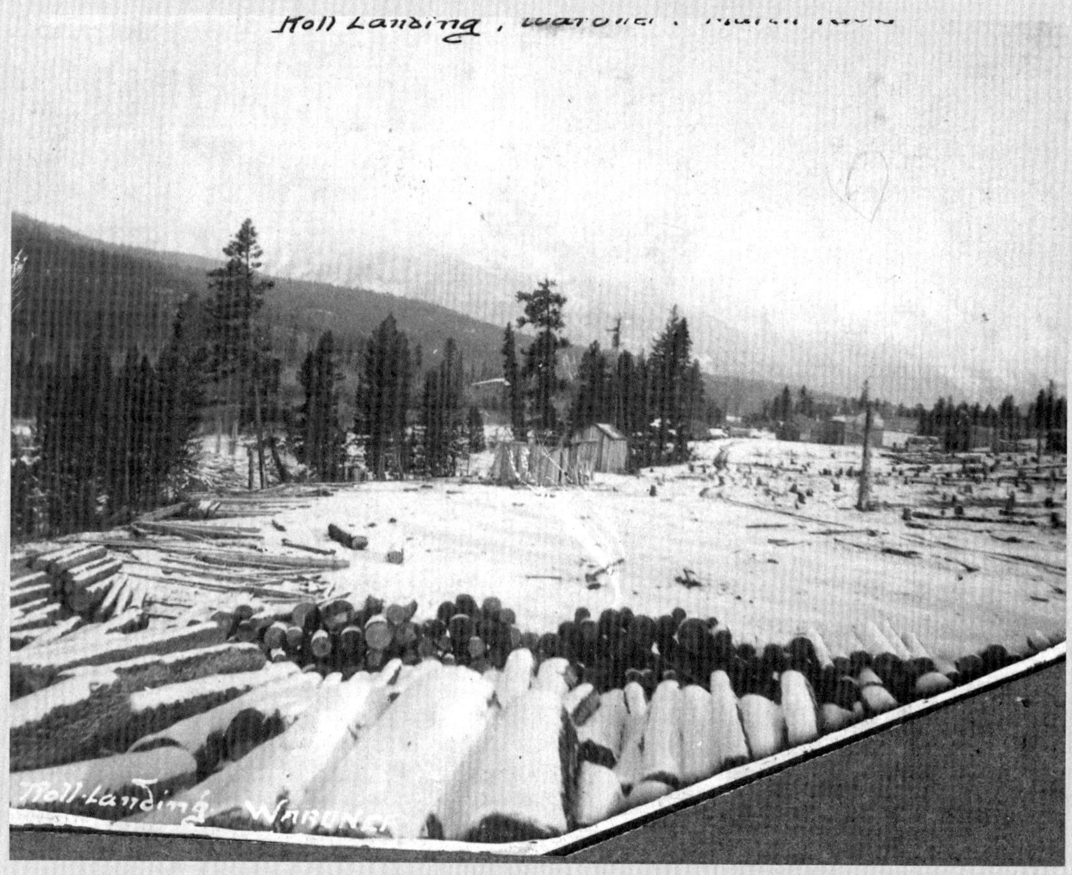

Roll landing at Wardner. Image J-00317 courtesy of Royal BC Museum, BC Archives.

He arrived in Winnipeg the next day at 3 a.m. with the temperature at -31C.

Instead of returning to Victoria, Swannell spent a few weeks at Wardner, where he worked for the Crow's Nest Pass Lumber Company. Swannell provided a brief description of his situation.

> Wardner, BC
> 8 February, 1903
> Breckenridge & Lund, Railway Contractors
>
> My dearest Ada:
>
> You will see by this letter that I am right down to business again. Arrived here Tuesday 11:30 a.m. Started work 1 p.m. Quit for the day at 9:45 p.m. So you see I haven't had a very long and tedious search after a situation. Work in fact is running after me open-mouthed.
>
> About 2½ ft of snow. Thermometer 10° below 0 [-23°C]. None of your Victorian weather here. Wednesday you ought to have seen me cavorting around the tall timber on a pair of clumsy bearpaw snowshoes—took a couple of headers into the beautiful. You'd have seen a pair of moccasins frantically thrashing around, I floundering somewhere out of sight below …
>
> I am quite comfortable here, have a bunk in Lund's office—a log cabin. He, the foreman

of construction and I eat together at a small cabin in the cook house. The rest of the outfit, 40 in number, sit at a big table …

After receiving another letter from Gore, Swannell wrote again to Ada with some exciting news.

14 February
With Breckenridge and Lund
Wardner, BC

Read the enclosed letter from Gore. $100 a month (expenses included) is not bad for a start is it? Another reason for my being pleased is that next autumn I think I may be taken into the firm as partner and will be able to marry you on about your 20th birthday. However, I must quit building castles in the air and state a certainty, namely I must run no chance in regard to passing my examination. Am working most evenings here, so can only study Sundays, therefore shall be in Victoria early in March and devote a month entirely to study.

Swannell realized that if he passed his examination and became a licensed provincial land surveyor it would provide certainty for his working career. With his secure employment he and Ada, who he had now been courting for almost two years, could plan to get married. He wrote one more letter to Ada from Wardner.

22nd February
Crowsnest Pass Lumber

My working hours have been irregular. Yesterday from 7:15 to 6 p.m. I was out on the log roll landing scaling. I was not very busy and could have taken a peavey and helped the men but my head was too bad. After supper I draughted till 8:30, had then to give up and go to bed. This morning I went 5 miles [8 km] up the river to superintend the sheer-boom piling … Expect to leave here Wednesday next. Consequently, allowing for a day's stop-over at Nelson would arrive at Victoria Saturday night. Breckenridge wanted me to go to Alberta with him to locate a railway spur to their coal mine, but I shan't be able to manage it.

Swannell passed his examination in April and became PLS #75. As a registered surveyor he could now run all operations of a survey crew and sign the completed surveys.

In early May the CPR's BC land commissioner, J.L. Doupe, wrote to William Whyte, assistant to the company's president. "As you know, the Government have been pressing for the completion of the surveys of this land grant, and I assume it would now be well that we should take up the matter of putting them in hand and having them carried to completion in accordance with the system which has been approved. The work, however, will be somewhat of an undertaking, and will cost a considerable sum of money." The company appropriated $25,000 for surveying the lot boundaries in 1903.

At the same time they were able to get the BC government to accept a change in the method of surveying. The BC land commissioner for the CPR explained this in a letter to T.S. Gore. He wrote, "Instead of attempting to run the boundaries throughout it is proposed to carry instrumental traverse lines up the valleys situated in these blocks, and when the intersection with the boundaries of the blocks is determined, to run such boundaries each way across the valleys and sufficiently high on the mountains on each side to provide a boundary line when it is likely to be wanted. The traverse lines will be run along any roads or trails through the valleys, and properly marked with iron posts and mounds."

During the previous field season, Gore & McGregor had found that it was impossible to survey the exact boundary of some of the lot lines in the mountainous terrain of the East Kootenay region. The main demand for much of the land would be for the tim-

ber that could be easily accessed in the valleys of the area. Delineating the boundary line through the main valleys would be an easier way to survey these large lots and would mark the lot in the locations where the boundary line would most likely be needed. It was also a cheaper way for the CPR to survey these lots. Traverse surveys were easier to do. Unlike the lot lines, which had to follow a fixed direction, a traverse could follow the easiest and most open route until it reached the lot boundary. It also could follow terrain that was less steep. Since a traverse survey was quicker and less difficult, the CPR paid a lesser rate for this work.

However, this method of surveying also presented a problem. Since there was no continuous survey line along the boundaries of the lots, it would be more difficult to find a mistake if there was an error in the surveying. There was no provision in the CPR's contract to ensure that these surveys made a loop back to a known surveying post to check that the work was done accurately.

During the 1903 season, Gore remained in Victoria while Swannell and McGregor continued the railway land grant surveys, each man having a survey crew.

On their way to the East Kootenay area, both men spent a few weeks surveying timber licences near Upper Arrow Lake south of Arrowhead.

Swannell wrote to Ada at the beginning of June.

> Arrowhead Lumber Co. Sawmill near
> Galena Bay
> Upper Arrow Lake
> June 2, 1903
>
> Received a lot of mail, the first, today. Explain my whereabouts? Halfway Creek, one foaming mill race for 20 miles [32 km]. Ten timber berths strung along it—to be surveyed by me. Trail (only blazes at best) a yard [1 m] under water occasionally. Creek 500 feet [150 m] wide at our last camp and rising 3 feet a day. Had a confab with myself. Result: today we'll all hit out leaving 250 lbs [115 kg] of grub cached in a cedar 4 miles [6.4 km] up.
>
> You should have seen the picture we saw last night. In the morning a gravel flat with 5 channels. By night a chaotic maelstrom of yellow foam, trees floating, rather racing—root & top. Roots battering along and piling up in an abattis [rampart of felled trees bound together placed with their branches outward] across the creek—a spurt of yellow foam, a crash like artillery firing salvos—and down would go the big jam to the lake. A thunder of boulders pounding lakeward under the yellow swirl.
>
> Our tent at noon was four feet [1.2 m] above the creek bed, by nightfall only a scant foot [30 cm] above water level. We cleared out at 4 a.m. this morning. Men's lives are too valuable for squandering up Halfway, so it's goodbye until low water.

Swannell wrote one more letter from the Upper Arrow Lake area, describing the difficult conditions encountered as he surveyed there. He also commented on the additional responsibilities he now had to handle as head of the survey crew.

> Saturday, 27 June 1903
> Dog Creek—Upper Arrow Lake
>
> Miserable wet outside, impossible to work so this goes down as our Sunday. Will have to work Dominion Day as well as we've lost 2 days by rain this survey. Now, according to promise, something of our life here.
>
> 6 a.m. it's "Roll out, tumble out, all aboard for breakfast." At 7 a.m. "all set, boys!" and off we make for work. The remainder of the day I stand at the transit with the ring of the axes ahead. It's wearisome now on one's feet for 10 hours. I have to be on the qui vive all the time directing operations as well as taking notes. If I sit down a moment I'm promptly

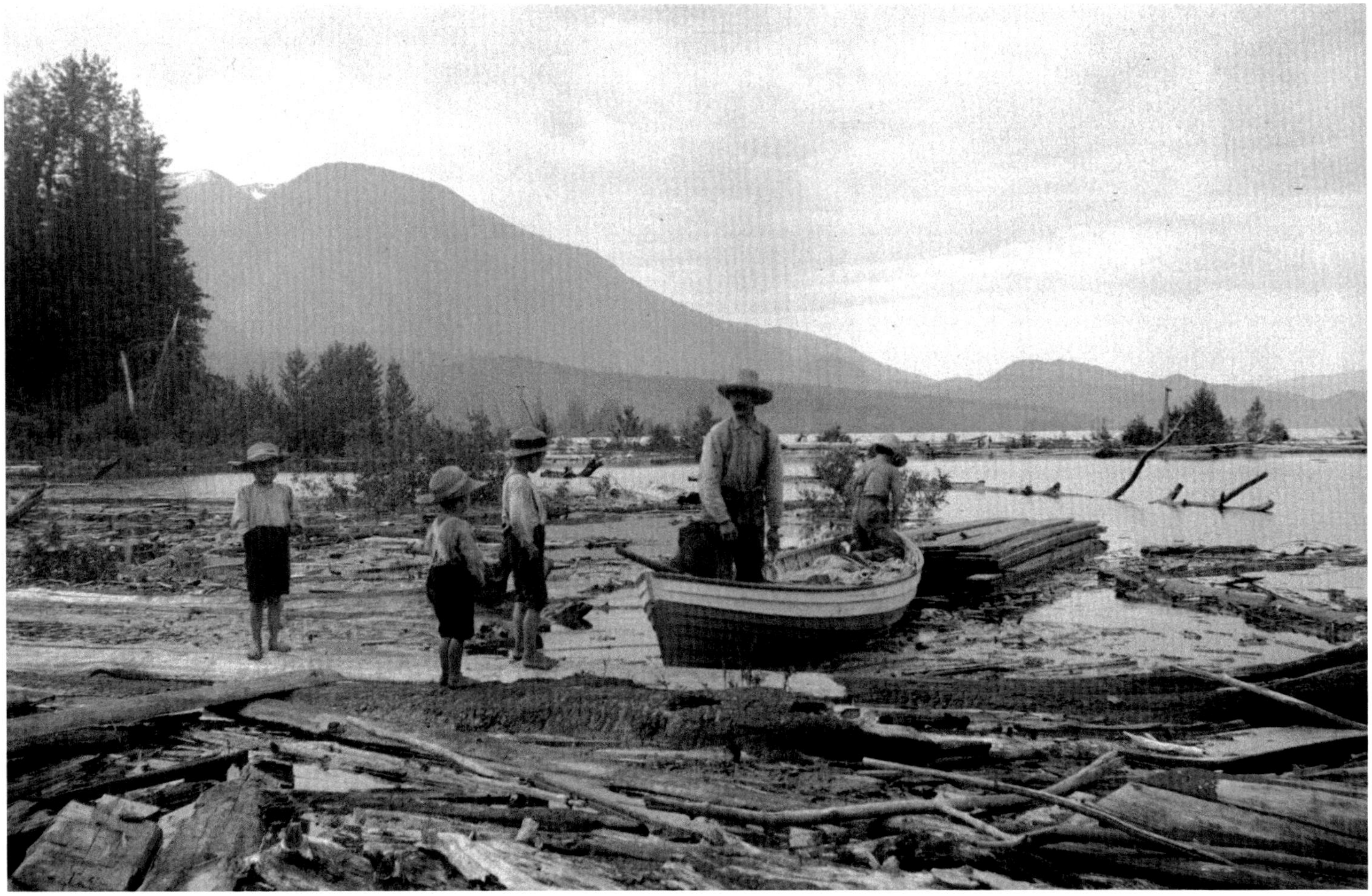

❖ Shaw and children at Hill Creek at high water. Swannell met this family during the Arrow Lakes timber survey. Image I-59869 courtesy of Royal BC Museum, BC Archives.

brought to my feet by "Line!—Point for hub." So you see, I'm regularly glued to the transit.

Mosquitoes bad, sand flies wuss, bulldogs atrocious. Regularly chew me up. I grieve to state I have been lazy o' nights and neglected to ply the sewing needle. So my sleeve is an affair of rags and tatters like the Holy Roman Empire and my bare elbow provides a splendid opening in business for my friends the skeeters. I try to get the men home by six—didn't get in till 9 the other night—whereat there was much grumbling.

The country is terribly rough and densely timbered so my daily average of line is only 2/3 mile [1 km]—of course it worries me to have the job drag along. Another cause for anxiety is the danger of making a mistake. A mistake at your work isn't of much moment, but my reading an angle a few degrees or even minutes wrong may cost $300–400. And I read a great many angles in a day, so you may imagine what care I must exercise.

I am getting along fairly well with the men. Only had one incipient mutiny which I managed to quell. Mac had lots of trouble, 7 men quit in two weeks. Said they weren't going to make slaves of themselves. We are comfortable here—have taken possession of a deserted shack. [In a letter fragment written late in 1903 Swannell commented, "Frowning must be part and parcel of the trade, for Mac always glowers at his men with looks as dour as a gathering storm."]

ARROWHEAD

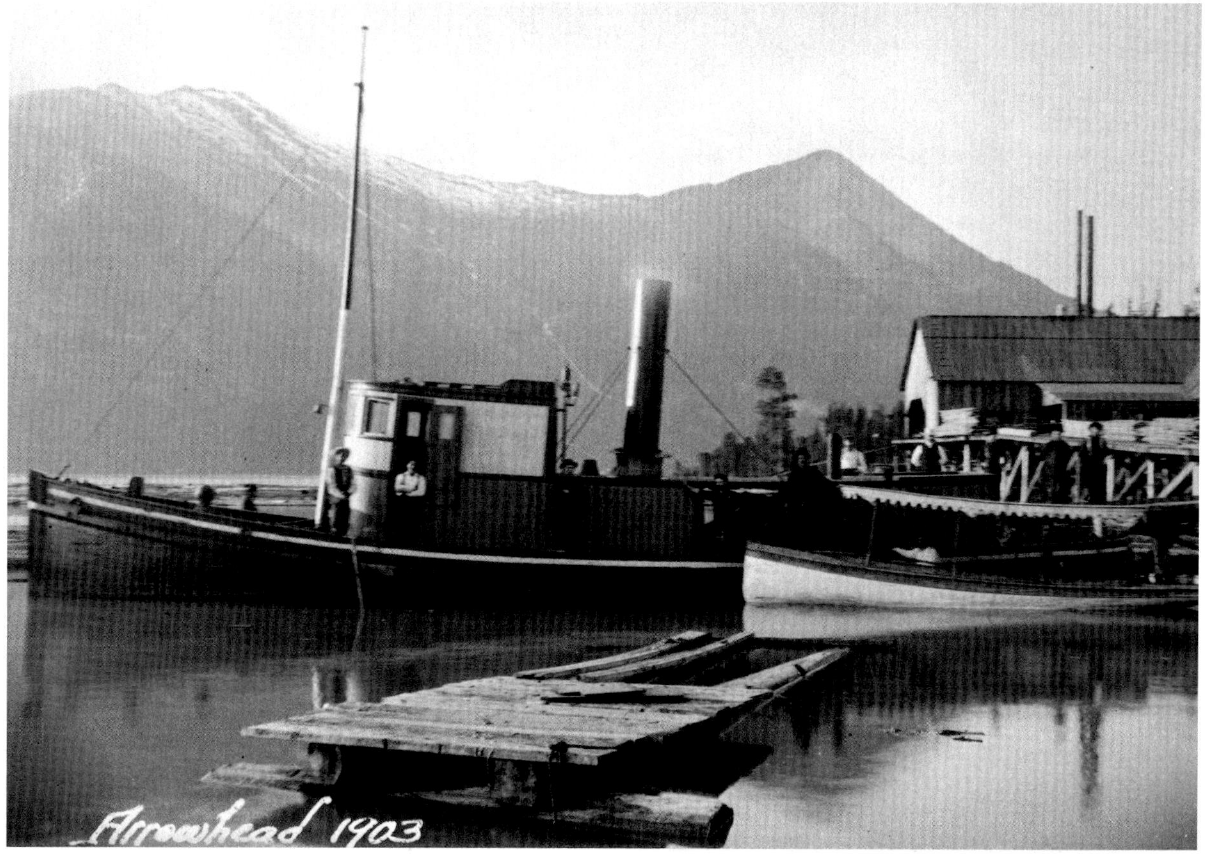

❖ Arrowhead wharf. The photographs that Swannell took of the Arrowhead waterfront are an important record of the community for they are among the earliest ones taken there. A fire in 1906 destroyed much of Arrowhead. Image G-06442 courtesy of Royal BC Museum, BC Archives.

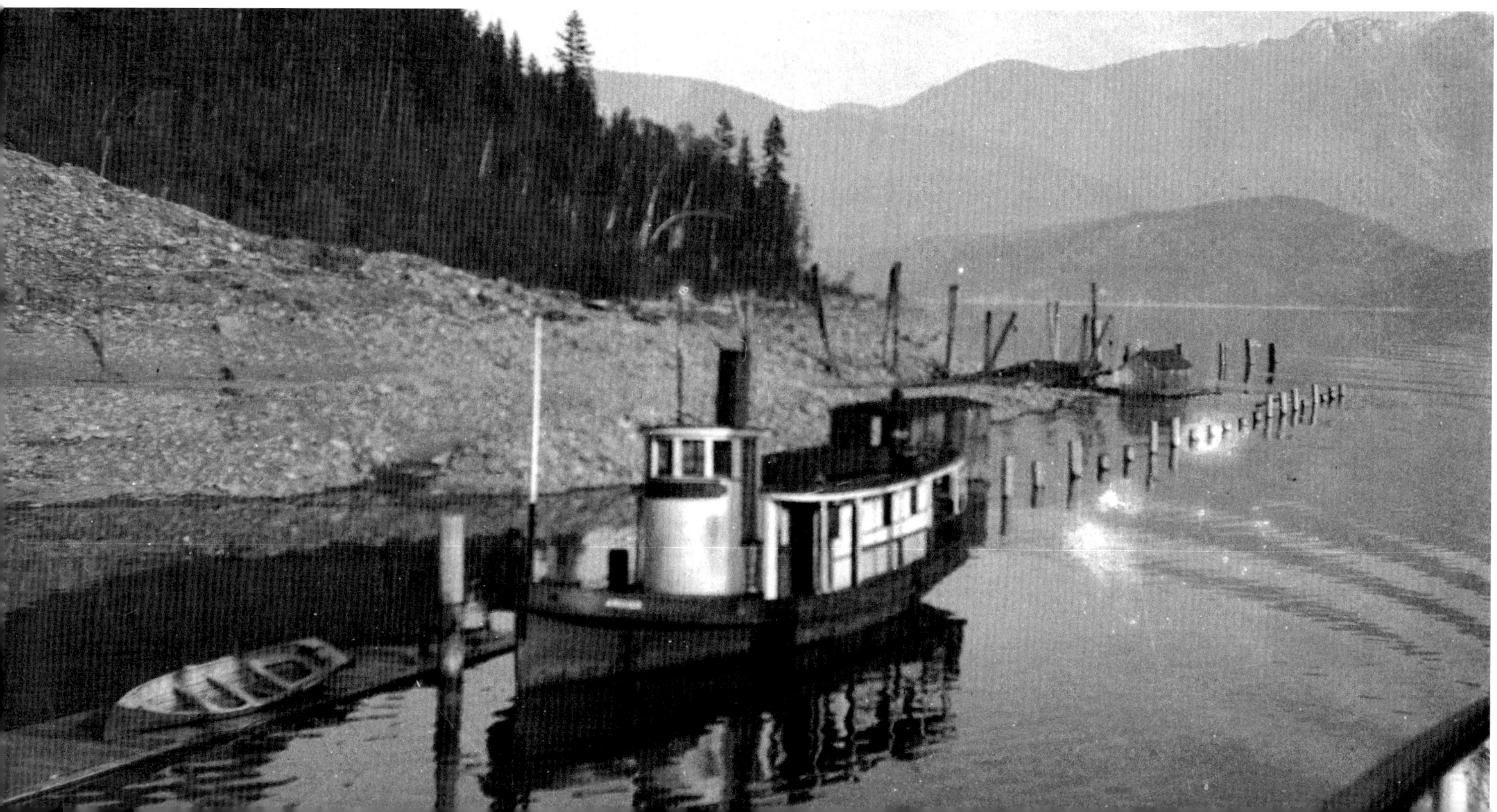

❖ Wharf at Arrowhead. Image I-58056 courtesy of Royal BC Museum, BC Archives.

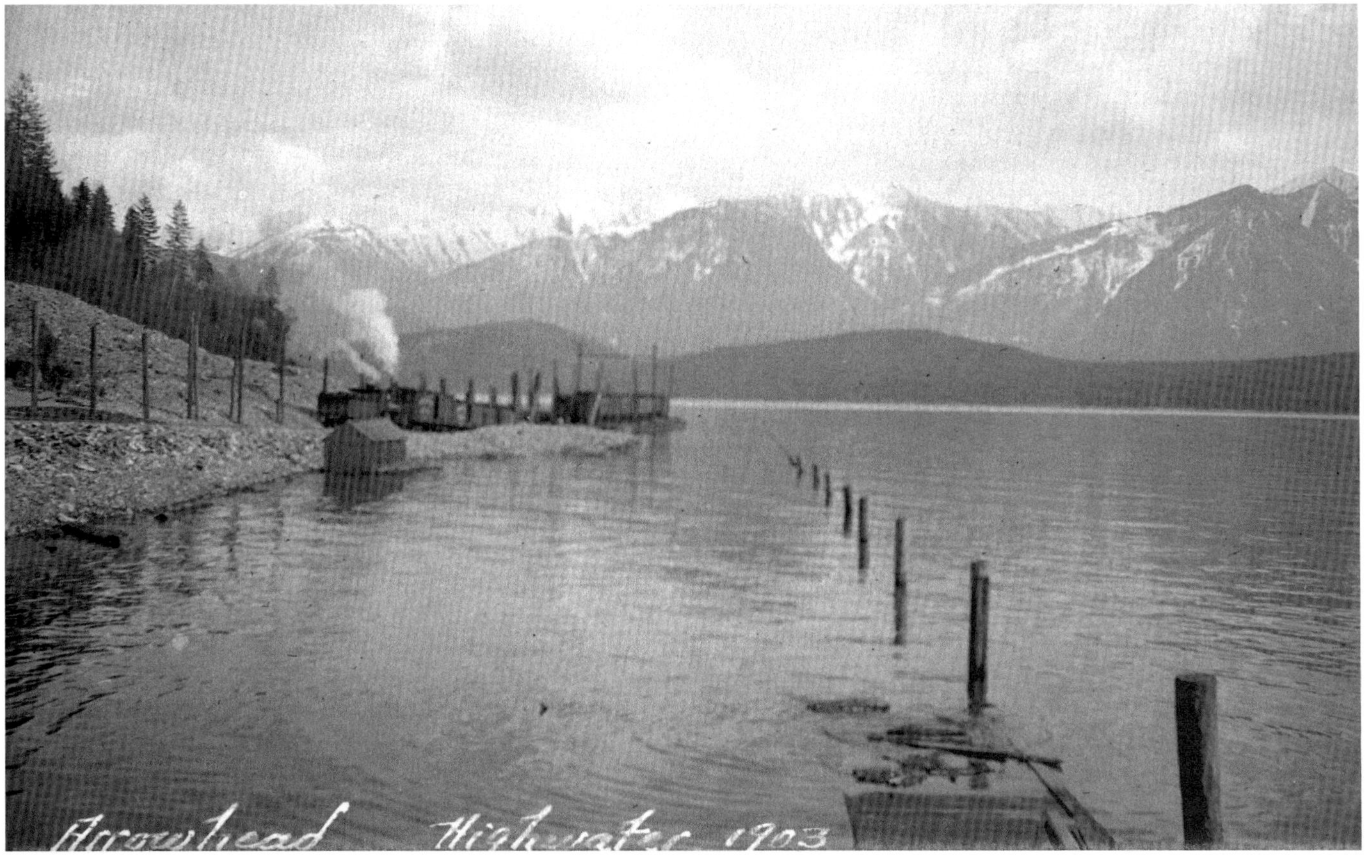

❖ Arrowhead at high water. The train cars are arriving at the car ferry, which is visible in the background. Image I-59868 courtesy of Royal BC Museum, BC Archives.

❖ Arrowhead Lumber Company sawmill. Swannell and McGregor surveyed timber licences for this company in the spring of 1903. Image I-33982 courtesy of Royal BC Museum, BC Archives.

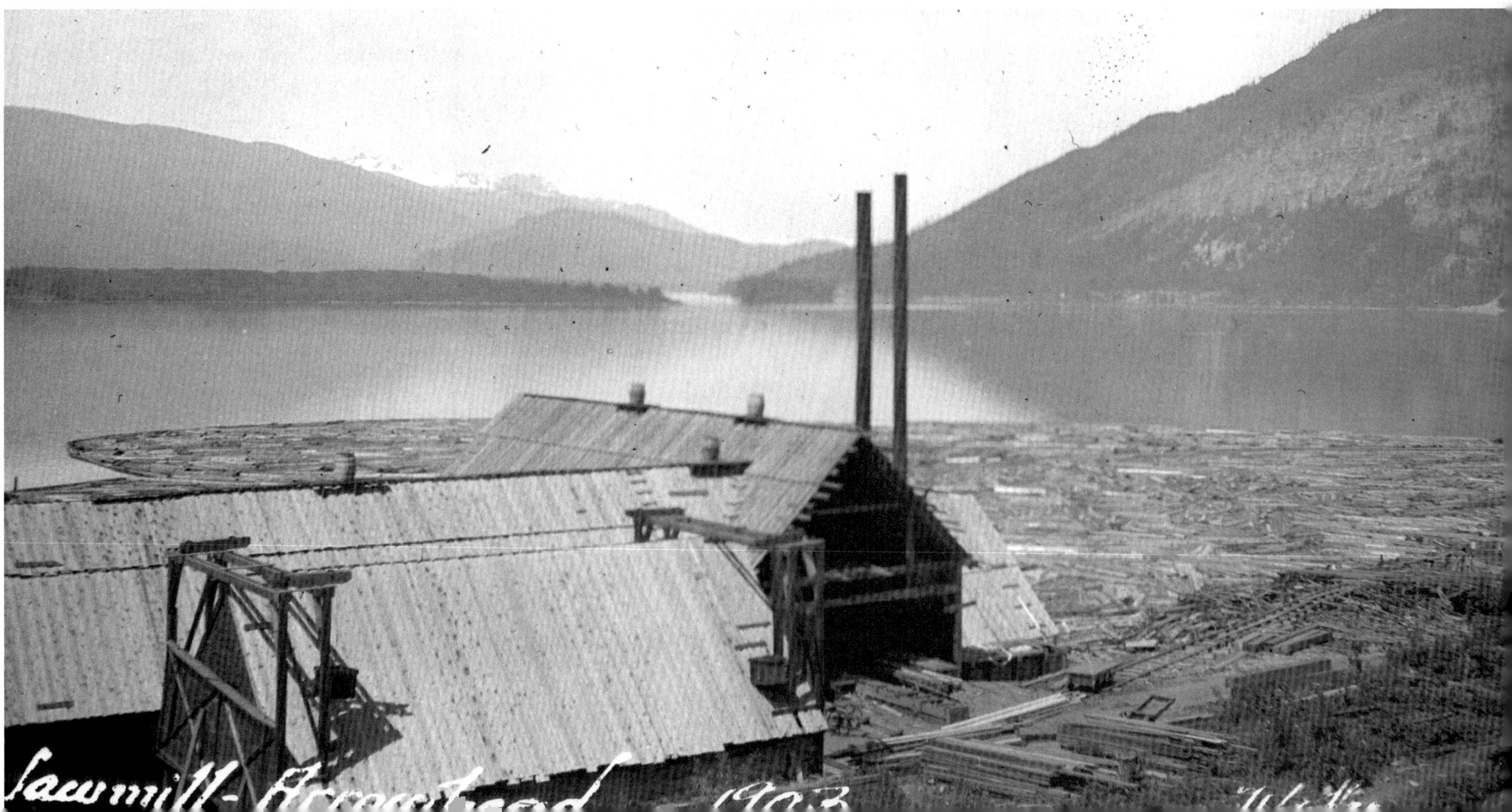

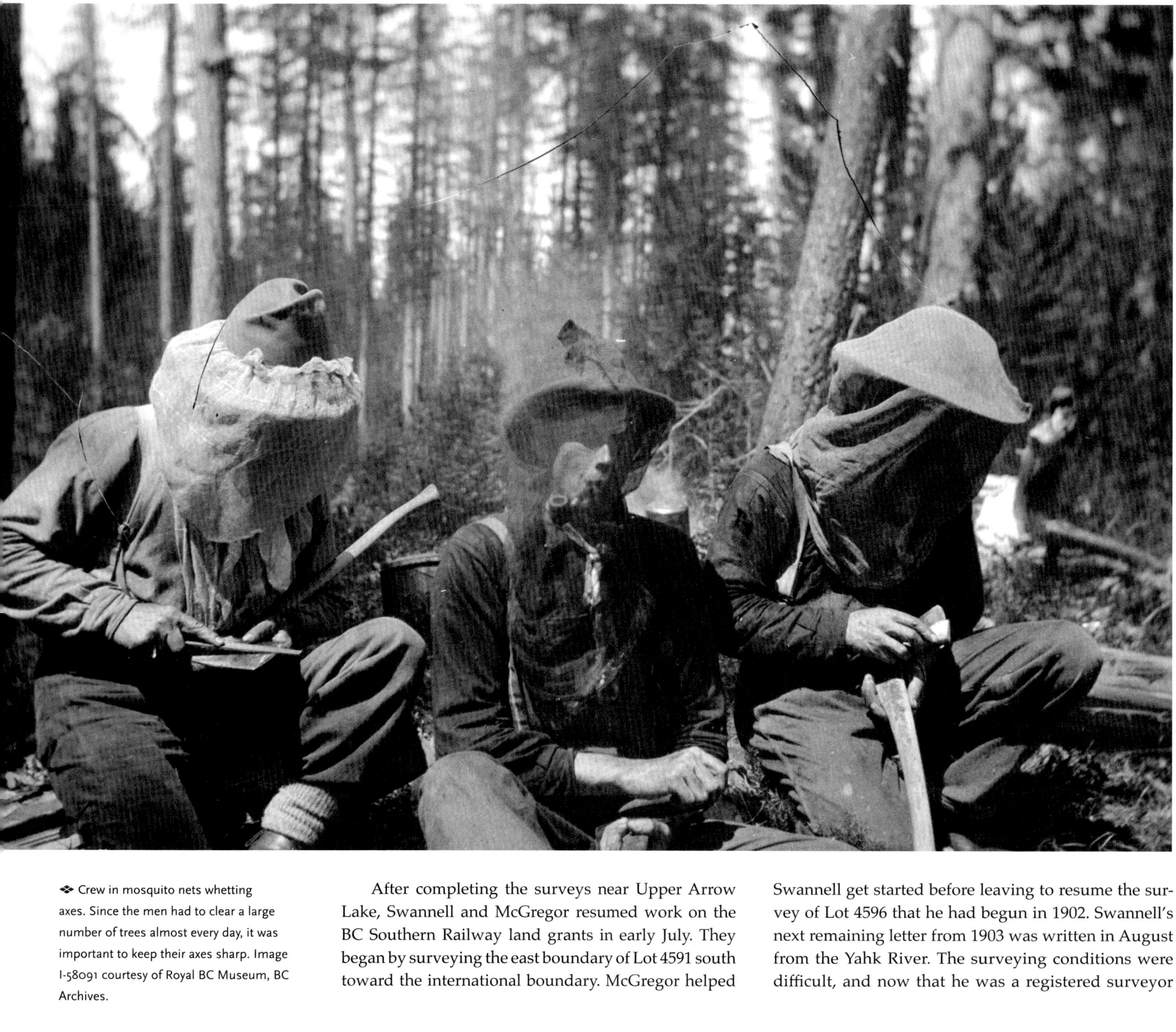

❖ Crew in mosquito nets whetting axes. Since the men had to clear a large number of trees almost every day, it was important to keep their axes sharp. Image I-58091 courtesy of Royal BC Museum, BC Archives.

After completing the surveys near Upper Arrow Lake, Swannell and McGregor resumed work on the BC Southern Railway land grants in early July. They began by surveying the east boundary of Lot 4591 south toward the international boundary. McGregor helped Swannell get started before leaving to resume the survey of Lot 4596 that he had begun in 1902. Swannell's next remaining letter from 1903 was written in August from the Yahk River. The surveying conditions were difficult, and now that he was a registered surveyor and completely in charge, he was responsible both for the surveying and for managing the people who worked on the crew.

Headwaters Yahk River
Aug. 13, 03

The unforeseen has happened again and I shall shortly have an opportunity of getting this letter out to you.

Those men of mine are simply ravenous and our spell of wet weather made them worse. The cook swears we get away with 30 lbs [14 kg] per diem. So the flour played out and I sent Benson across country to Gold Creek to intercept Ivor Thomas. He left last Monday and last night we heard the welcome tinkle of the horse bells, and by dint of much chopping of fallen timber got the cayuses into camp. Benson had shot a deer and some grouse en route, so the men simply gorged themselves to repletion.

Oh! Haven't I a gay old time with a grousing gang, bad weather and worse country last week. Sullen faces and grumbling behind my back all day. Just like children working men are—afraid of starvation, kicking at beans & bacon as if I could get them the delicacies of the season. Groused because I worked 'em in the wet, grumbled because they've only had one holiday out of three Sundays. If the grub convoy hadn't connected they'd have mutinied to a man, but not old Benson. That Worthington is the worst of all—ought to have more sense for he sets up being an English gentleman's son—now Ada, everything is sunny again for I gave them a holiday again. We've three weeks grub in camp. Ivor Thomas, Benson and the cook, old Van Eck, I exempt from all this. They've acted splendidly so far.

Mileage to date—17 miles, 20 chain [27 km] south of railway. By the way my men are discarding their tatters—no less than 8 pairs of "Big Horn" unmentionables and blue shirts flaunting around camp.

Thomas goes back out in a few days to get in some grub at Yahk Summit (the map as usual was all wrong, which is why we needed grub bad). He will also get grub cached at the forks of Bull River for next job.

I expect, indeed must, finish here in a month. We will then hit out for Wardner; and Benson will superintend crossing the Kootenay while I take a run to Cranbrook to order supplies, etc ...

Over a dozen grouse, 3 quarters of venison and lots of trout over at the cook fire, so we've struck a game country at last—and saved will be the bacon!

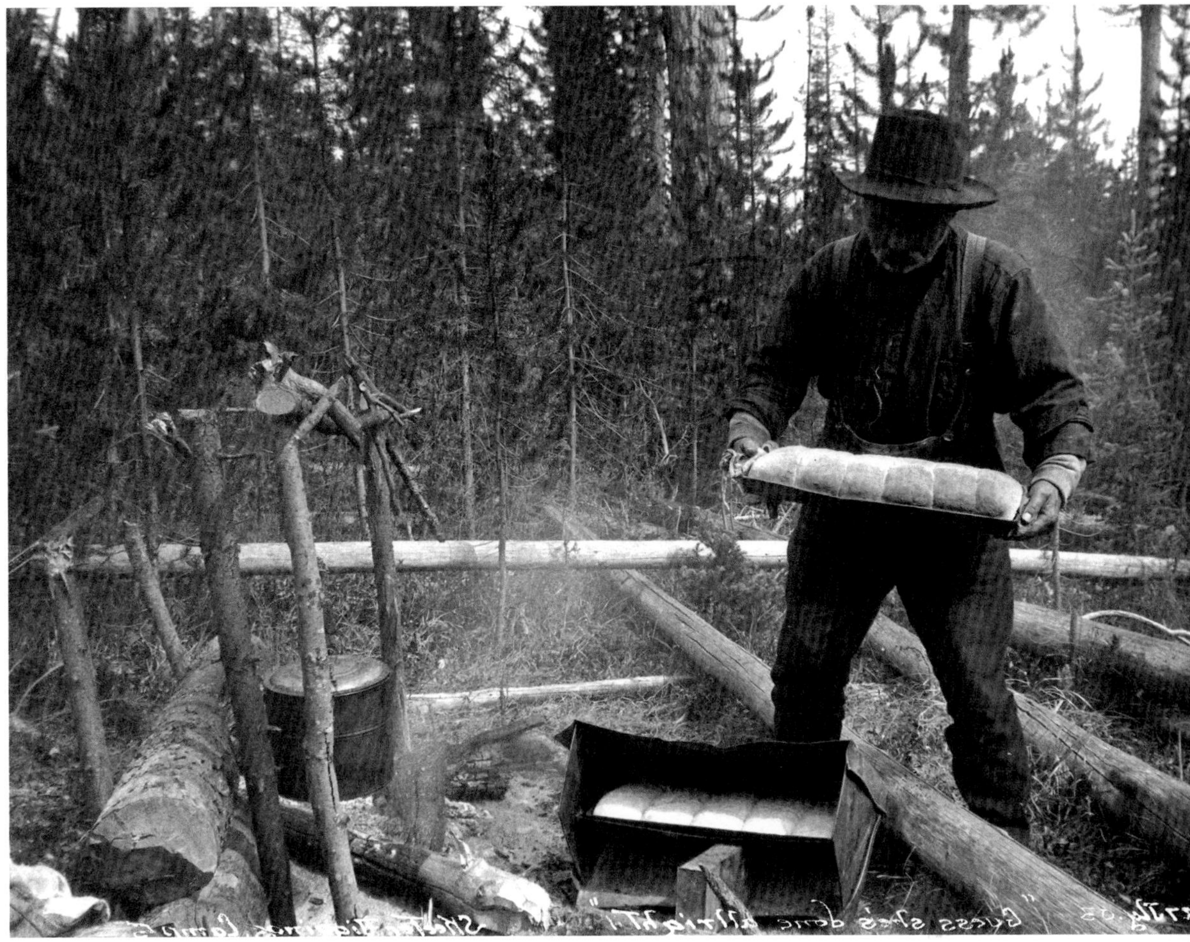

Making bread in reflector oven. This was the most common method of baking bread in a field camp. Image G-09200 courtesy of Royal BC Museum, BC Archives.

Swannell's crew surveyed Lot 4591 over 50 kilometres to the international boundary. Since the terrain was not too steep he was able to survey the actual boundary of the lot. After the crew completed their work on September 10, Swannell went to Cranbrook, where he wrote to Ada. In the fragment that remains, he describes their surveying.

> It was snowing and Benson & I on the last day were so cold and miserably wet that we couldn't have hung out much longer, when we struck the International Boundary.
>
> In 34½ miles [56 km] I got 8 minutes of azimuth off true south, but only missed the boundary between Idaho and Montana by 159.2 feet [48 m], a small error in 34½ miles of rough country, and part of time very short sight when the smoke from forest fires was dense and we couldn't check back.
>
> That was Sept. 10th. Camp was 4 [miles, 6 km] west along the Boundary Commission trail in Idaho (1/4 [mile, 400 m] from Monument).
>
> Next day Thomas and I hit out with a blanket apiece, he to get the horses, I bound for Wardner. The trail shewn on map from Bdy to Summit is a fraud, not even blazed. Rain & hail all day, wading the Yahk many places in preference to forcing our way through the dripping alder thickets & windfalls. We managed to get to our cache (at the two small lakes at head of Gold Cr) by 7:30 (20 [miles, 32 km]).

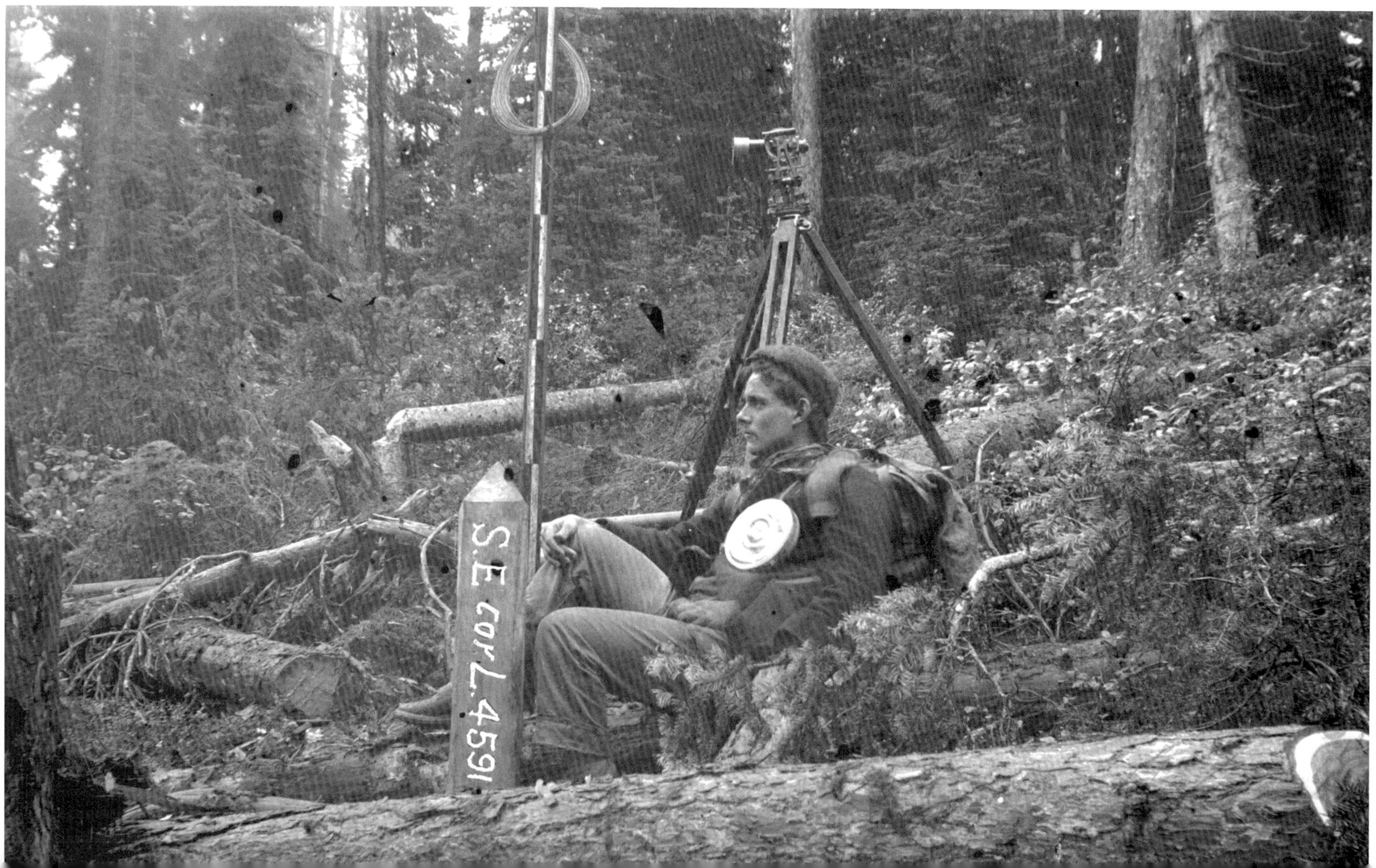

❖ Swannell at the southeast corner of Lot 4591 at the international boundary. This was the final station surveyed for the lot. Image I-33522 courtesy of Royal BC Museum, BC Archives.

Had killed 15 grouse en route, so that by 9:30 we were fed & dried out. Turned out early on Saturday as it was too cold to sleep. Thomas left me 7 miles [11 km] further on. Saturday I walked 33 miles [53 km]—was so tired in the afternoon that I could only walk half an hour at a spell. Lost trail down Plumbob Ck & so was overtaken by dark 4 [miles, 6.4 km] from Wardner. Ploughed along in the dark till 9:30 & then gave it up, as there were no stars to steer by. I shivered around a fire all night. At dawn found I was only 100 yds [91 m] from the trail. Got in at 7:30 a.m.

From the middle of September to the middle of October, Swannell and his crew surveyed a traverse

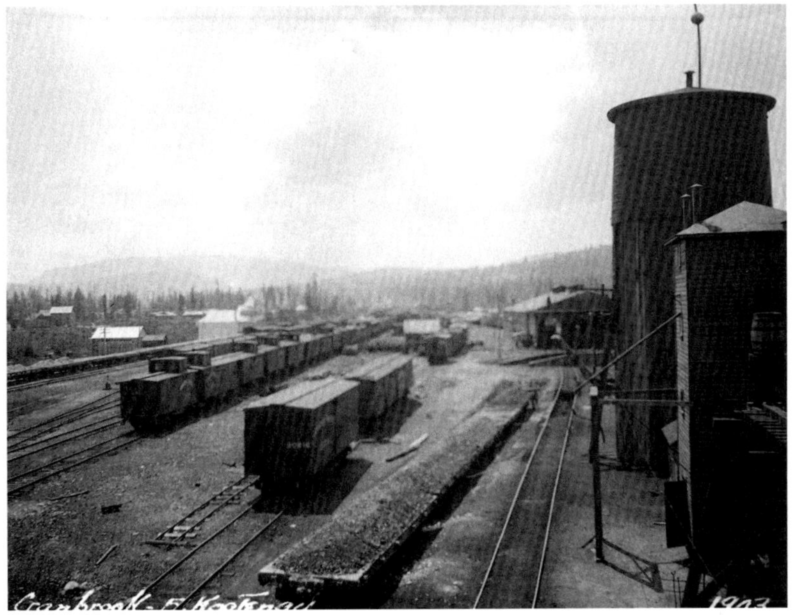

❖ Cranbrook started as a divisional point for the BC Southern Railway and supplanted Fort Steele as the main city in the East Kootenay region. The water tower in the picture still remains but has been moved to a different location. Image I-58257 courtesy of Royal BC Museum, BC Archives.

❖ Below: Cranbrook. Image I-58258 courtesy of Royal BC Museum, BC Archives. For 2104 image see page 152.

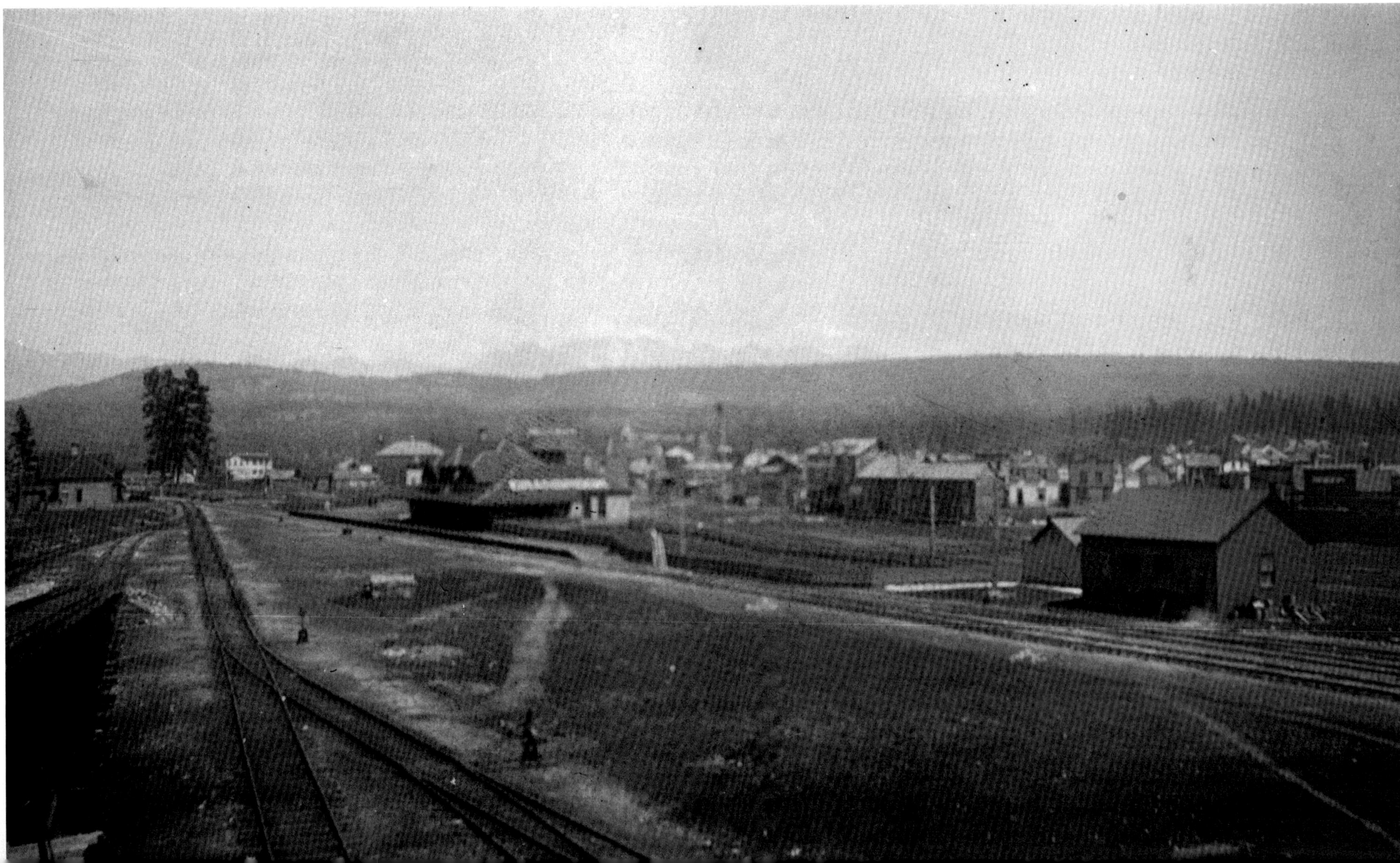

Crow's Nest Pass Lumber Company

❖ Storage boom at Wardner. Logs for the CPLC mill are being stored in a back channel of the Kootenay River. Crossing the river in the middle is the railway bridge. Constructed in 1898, it was the longest on the BC Southern line. The large rock outcrop by the bridge is a readily identifiable landmark that is still visible today. Image I-59699 courtesy of Royal BC Museum, BC Archives. For 2014 image see p.153.

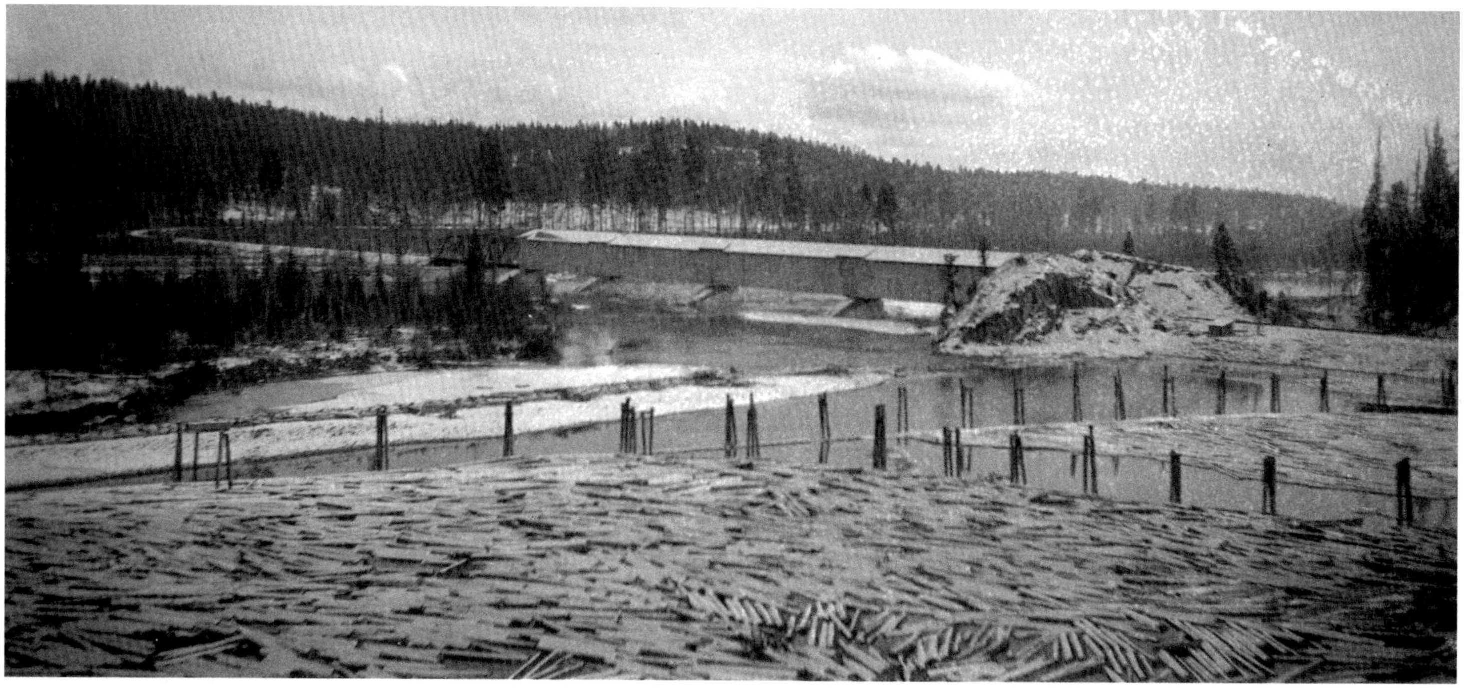

❖ Below: CPLC sawmill at Wardner. The BC Southern rail line, visible on the right, went behind the mill. Image I-33955 courtesy of Royal BC Museum, BC Archives.

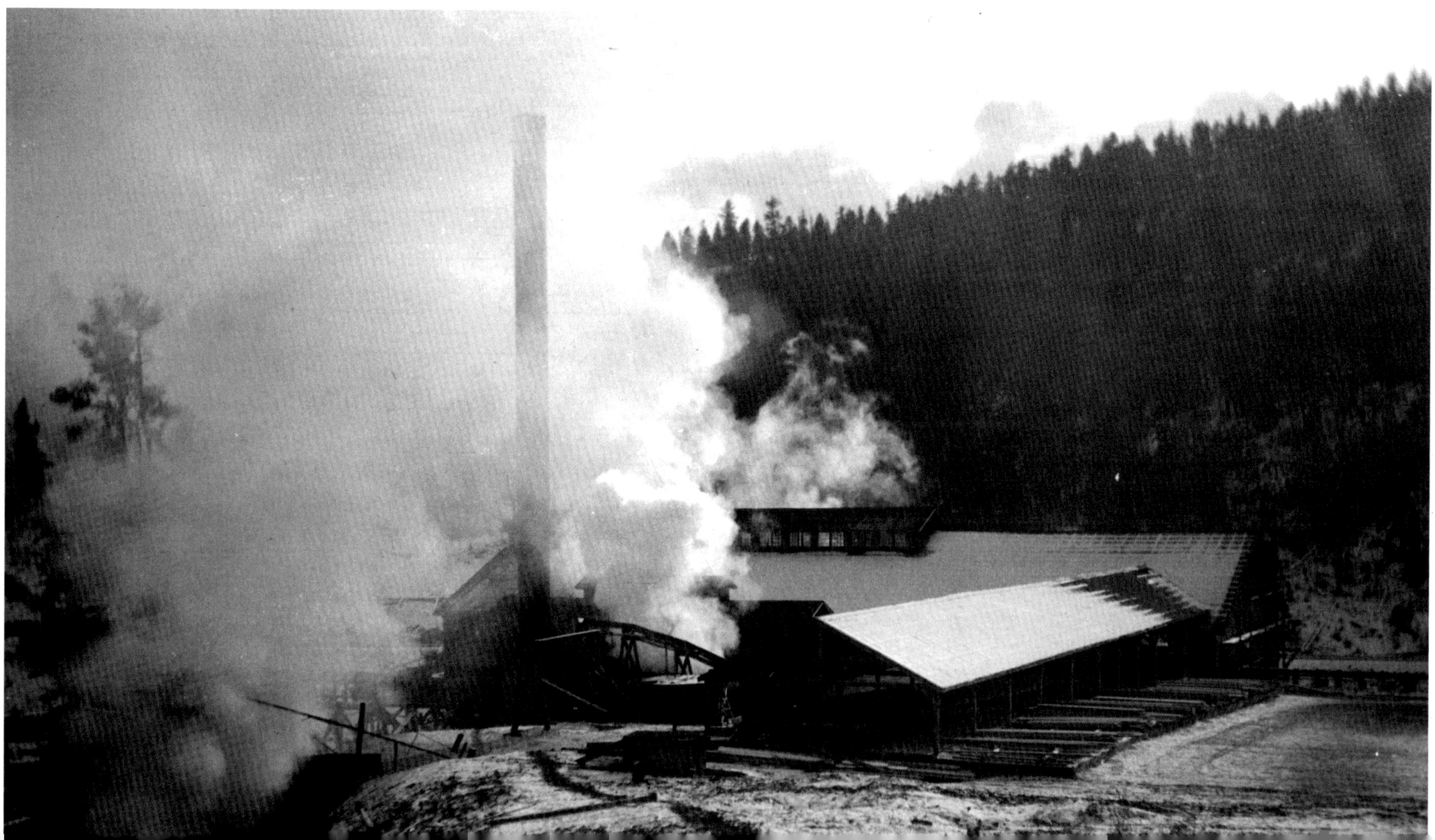

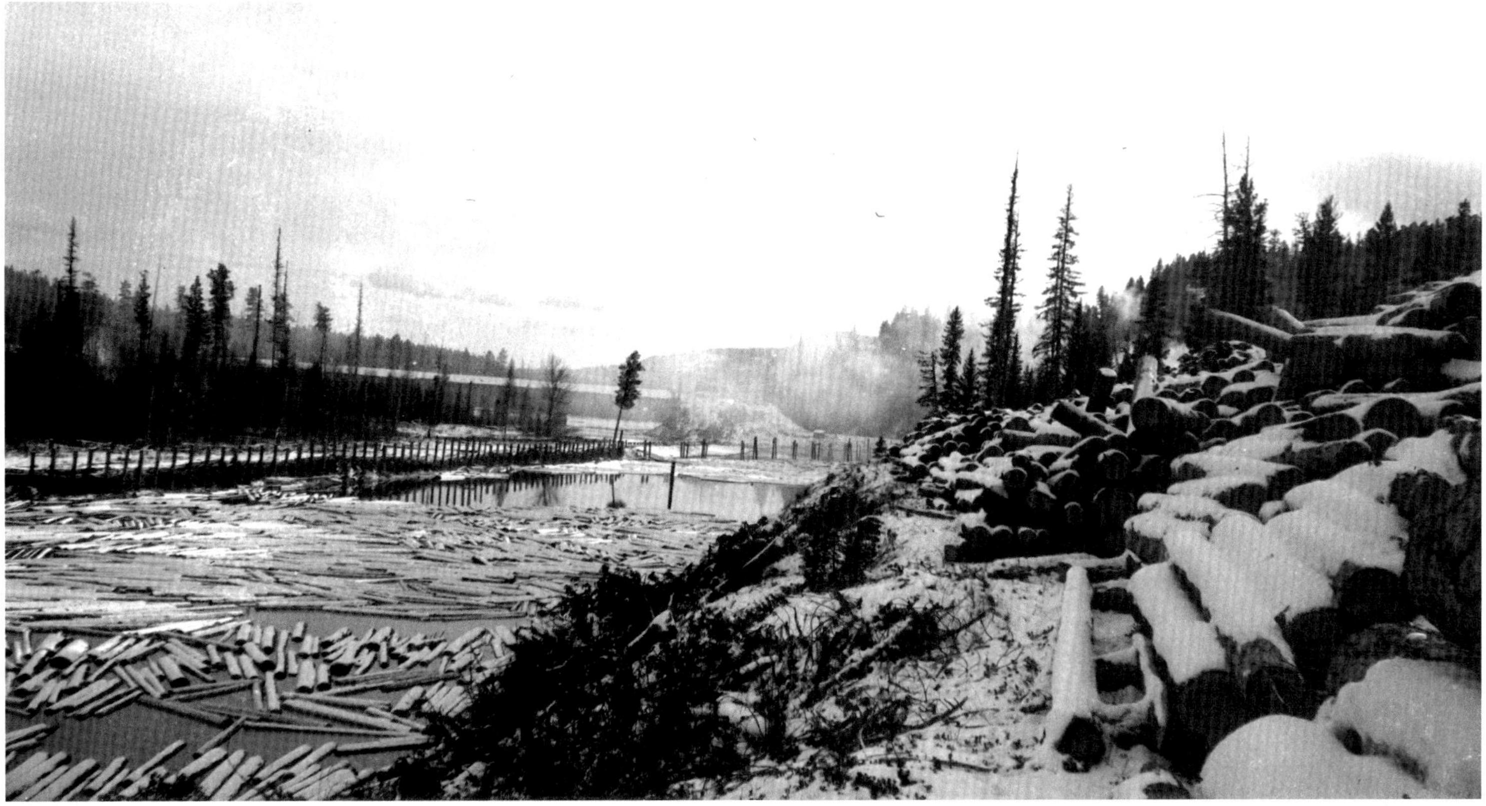

❖ Log booms at Wardner. Image I-33936 courtesy of Royal BC Museum, BC Archives.

❖ Below: Kootenay River at Wardner. This picture is taken from the east side of the Kootenay River. The railway bridge is on the left. The Crow's Nest Pass Lumber Company's sawmill is in the background. Image I-33935 courtesy of Royal BC Museum, BC Archives.

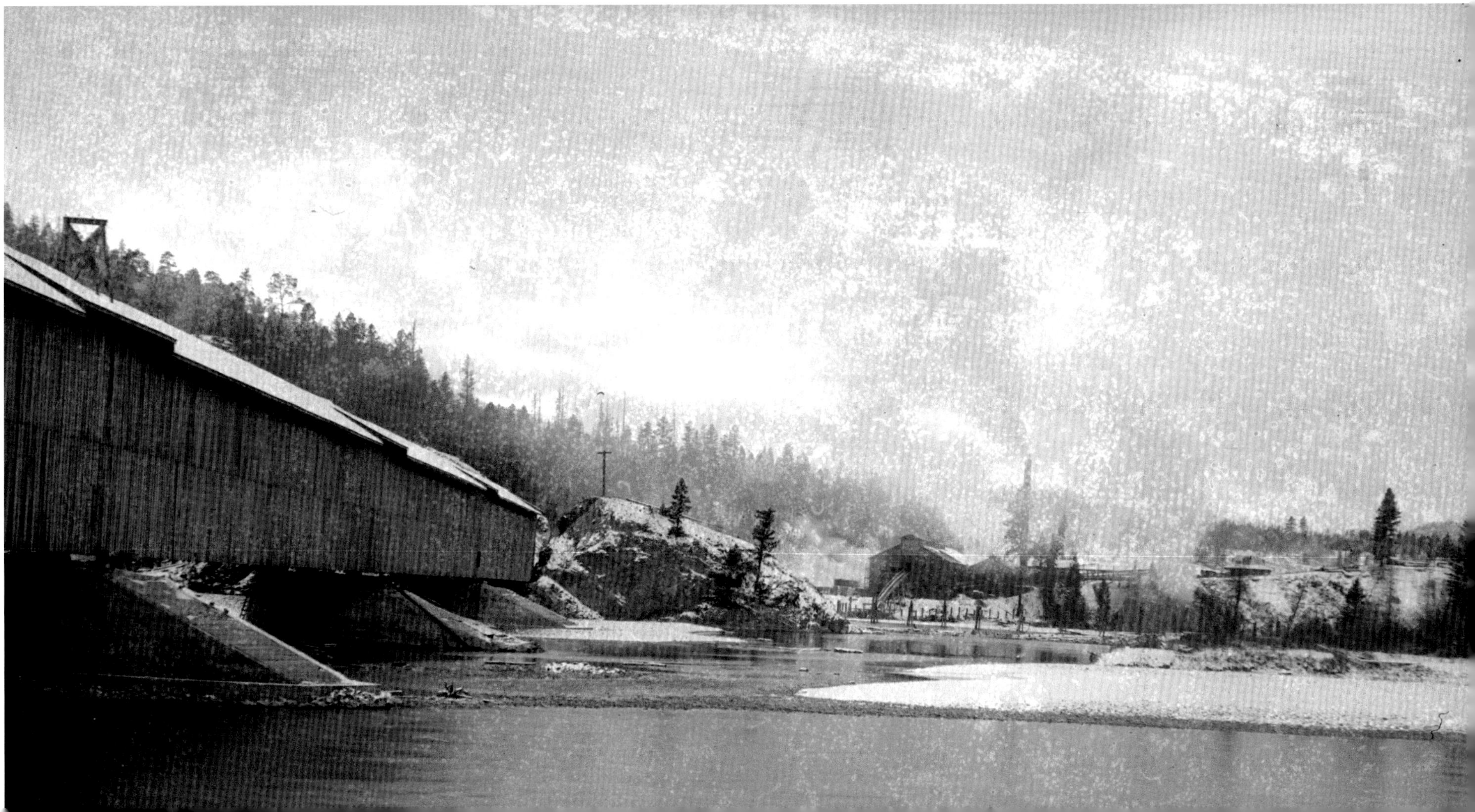

up the Bull River. In his report he wrote, "Passing Dibble Creek at 13 miles [21 km] and Quartz Creek at 20 miles [32 km] the Forks [of the Bull River] were reached at about 25 miles [40 km] from the railway, through rough, burned mountains, culminating in high peaks on either side. From the Forks we followed the east branch as far as the eastern boundary of Lot 4590, which we marked by half a mile of line on each side of the river." Swannell then returned to Wardner.

> Wardner, E. Kootenay
> 20 October, 1903
>
> My dearest Ada:
>
> ... Am afraid I've another scolding in store for me, as I've made another big walk and am pretty well played out today—37 miles [49 km] over the mountains on an execrable trail. You see we've worked Sundays and all up the Bull, so my only chance at field notes was at night—used to work till 9 o'clock every night. Am behind in spite of that, so calculated that by hitting out light ahead of the outfit I could save two or three days. So here I am, at your service.
>
> Had a pretty good trip, although it's very late in the year to be in the Rockies. We had a great time with the horses—no feed—nothing but canons & windfalls. Thomas scrambled 'em through somehow. Was nearly killed the other day up on the East Fork—horses were swept downstream, jammed among the boulders. Got them out all right luckily, but all our blankets were soaking.
>
> However, I shall spin you all sorts of yarns about this trip on my return.
>
> Have taken 50 photos—will get them developed in Cranbrook as you've enough to do. I ran into a whopper of a bear the other day on a sidehill. He stared at me & seemed undecided as to attack or flight. I got the Kodak out, took him—remembered about everything except the diaphragm. Afraid I muddled it, and hadn't time to get another shot at him, as he shambled off into the bushes.
>
> I shall meet my outfit at Elko in a couple of days and tackle that confounded line to the north. We're sure to have a bad time with snow, but I expect to finish it off in a fortnight—and then hurrah for Victoria!

Next, Swannell surveyed 5 kilometres of the actual east boundary. "Returning to the railway we established the southeast corner of Lot 4590, whence we ran three miles north along the eastern boundary and high up among bare mountains above timber line."

By late October, Swannell was down from the mountains and in the Elk River valley along the BC Southern rail line, where he wrote Ada.

> Elk Hotel, George Hoggarth, Prop
> Elko BC
> 28 October 1903
>
> I got in here very tired and wet tonight.
>
> We managed to get the line slammed through to the top of a very high ridge. The Rockies are rather a different proposition to tackle than Cedar Hill, you know. Last night the weather changed so Benson & I made a flying trip up to measure a triangulation base and get the survey gear down. We had a tough three hour climb—for the last hour right into the clouds. It was a forlorn hope and a mighty risky job for the clouds were so dense we couldn't see 100 feet [30 m]. We made it by 11 o'clock however.
>
> It was cold, icily cold amid the swirling mist with the wind sweeping down the bare ridge. We pretty nearly caved in from cold and wet. Got in the lee of a crag and tried to make a fire with some scrub timber. Benson was so chilled he couldn't hold the axe so I rustled wood while he tried to rub warmth into his hands. The matches were wet but we dried them out in our hand.

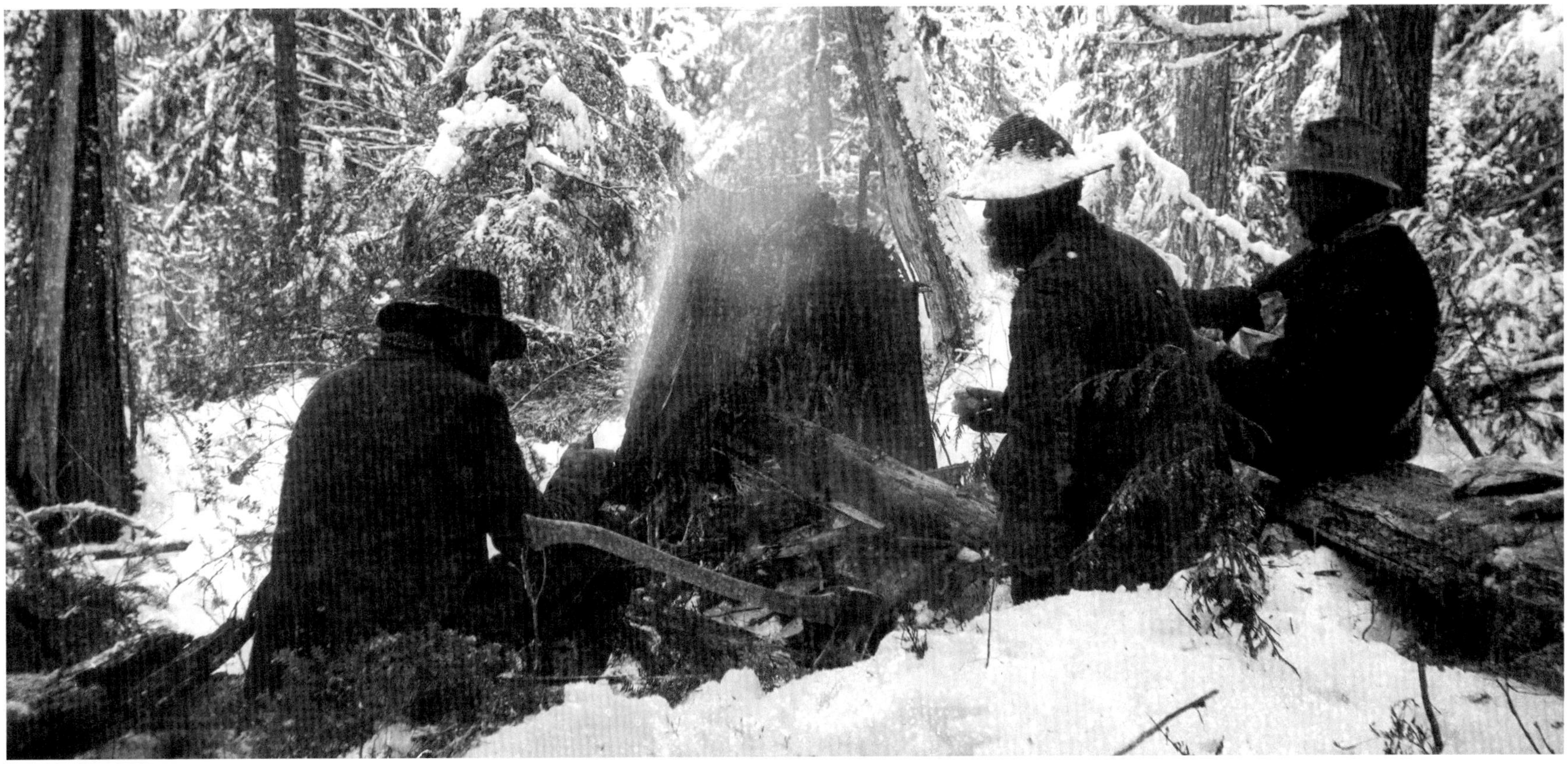

You'd have laughed to see the time we had lighting them, our fingers were so numb. Finally, by tearing leaves out of my field book we got our fire and got thawed out enough to enable us to get below the storm clouds.

Tomorrow the outfit pulls out for Sand Creek. Benson, Pete & I made [it] here tonight after a four hour walk … Will go by rail to Jaffray tomorrow and start right in at the Sand Creek traverse.

Am writing Mac advising him to abandon the contract as it is too late in the year. A frightful hailstorm and a gale of wind outside now. Lucky Benson & I aren't up on that peak, eh? Snow up there half a foot [15 cm] deep by now.

From October 31 to November 11, Swannell and his crew finished the 1903 field season by surveying another traverse, this one going for 11 kilometres up Sand Creek to the eastern boundary of Lot 4590. In a letter to CPR officials, McGregor wrote, "[Swannell] reports to me that he finished the Sand Creek traverse and ran about half a mile [800 m] of the eastern boundary of Lot 4590 in very bad weather and deep snow which was constantly increasing so as to make work impossible." On this traverse Swannell made an error in his surveying that resulted in an incorrect location of the lot boundary in the Sand Creek valley. The mistake would be discovered by surveyor John McLatchie (PLS #39) in the summer of 1904 and would lead to a controversy that would last for many years. But Swannell was not the only surveyor to make an error in his work that summer. J.H. McGregor made a mistake in one of his survey lines in Lot 4596. However, he discovered his error while doing his office work over the winter. In February, McGregor went into the field with a small crew of men on snowshoes and spent a week correcting the portion of his survey line that needed to be fixed.

Last lunch camp on Sand Creek traverse in mid-November. Image I-58102 courtesy of Royal BC Museum, BC Archives.

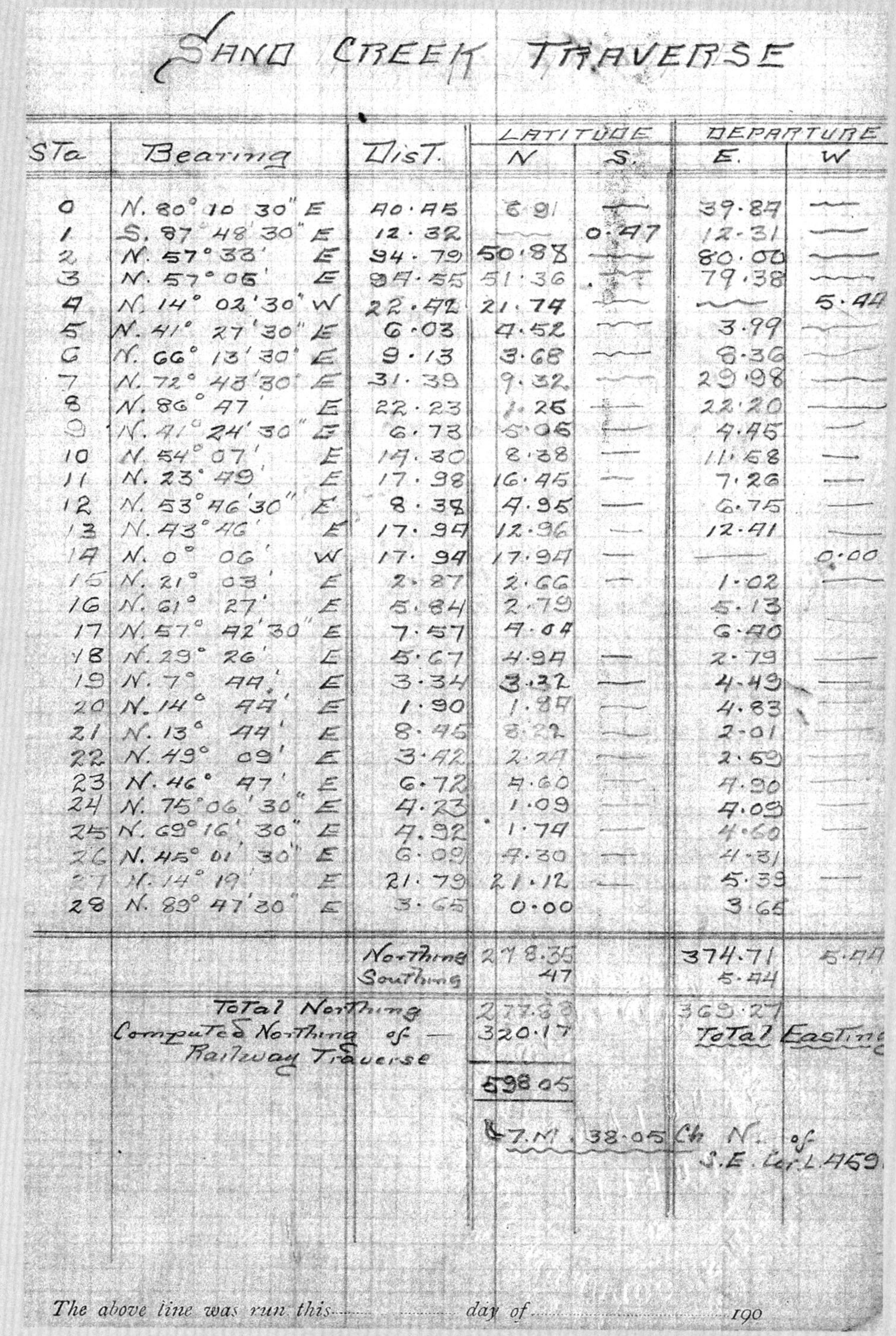

SAND CREEK TRAVERSE

STa	Bearing	DisT.	LATITUDE N	LATITUDE S.	DEPARTURE E.	DEPARTURE W
0	N. 80° 10 30" E	40·45	6·91	—	39·87	—
1	S. 87 48 30" E	12·32	—	0·47	12·31	—
2	N. 57° 33' E	94·79	50·88	—	80·00	—
3	N. 57° 05' E	94·55	51·36	—	79·38	—
4	N. 14° 02'30" W	22·42	21·74	—	—	5·44
5	N. 41° 27'30" E	6·03	4·52	—	3·99	—
6	N. 66° 13'30" E	9·13	3·68	—	8·36	—
7	N. 72° 48'30" E	31·39	9·32	—	29·98	—
8	N. 86° 47' E	22·23	1·25	—	22·20	—
9	N. 41° 24' 30" E	6·73	5·06	—	4·45	—
10	N. 54° 07' E	14·30	8·38	—	11·58	—
11	N. 23° 49' E	17·98	16·45	—	7·26	—
12	N. 53° 46'30" E	8·38	4·95	—	6·75	—
13	N. 43° 46' E	17·94	12·96	—	12·41	—
14	N. 0° 06' W	17·94	17·94	—	—	0·00
15	N. 21° 03' E	2·87	2·66	—	1·02	—
16	N. 61° 27' E	5·84	2·79		5·13	—
17	N. 57° 42'30" E	7·57	4·04		6·40	—
18	N. 29° 26' E	5·67	4·94	—	2·79	—
19	N. 7° 44' E	3·34	3·32	—	4·49	—
20	N. 14° 44' E	1·90	1·84	—	4·83	—
21	N. 13° 44' E	8·45	8·22	—	2·01	—
22	N. 49° 09' E	3·42	2·24	—	2·59	—
23	N. 46° 47' E	6·72	4·60	—	4·90	—
24	N. 75° 06' 30" E	4·23	1·09	—	4·09	—
25	N. 69° 16' 30" E	4·92	1·74	—	4·60	—
26	N. 45° 01' 30" E	6·09	4·30	—	4·31	—
27	N. 14° 19' E	21·79	21·12	—	5·39	—
28	N. 89° 47'30" E	3·65	0·00		3·65	
		Northing	278·35		374·71	5·44
		Southing	47		5·44	
	Total Northing		277·88		369·27	
	Computed Northing of Railway Traverse		320·17		Total Easting	
			598·05			

T.M. 38·05 Ch N. of S.E. Cor. L 459[illegible]

The above line was run this .... day of 190

❖ Sand Creek traverse. This is the page from Swannell's field book to which the judge referred. The judge referred to the error that Swannell made in his calculation for Station 19. The departure should be .449 instead of 4.49. The departure for station 20 is also incorrect. Courtesy Land Title Survey Authority.

Lot 4590, East Kootenay District

As a registered British Columbia land surveyor, Swannell held legal responsibility for his surveys. In his letters to Ada in 1903 he expressed concern about his new authority, particularly if he made a mistake in his surveying. In the large lots that he was surveying in the Kootenays, a mistake early in the survey could result in a large error. In his career Swannell was known to be a precise, accurate surveyor. Yet in his early years he made some mistakes in his work. The most serious one occurred in 1903 during the survey for the eastern boundary of Lot 4590.

On the Sand Creek traverse Swannell made an error in his calculation. By the time he reached what he determined to be the eastern boundary of the lot he was actually 34.23 chains (approximately 690 metres) to the west of the true description of Lot 4590, making this lot smaller. The mistake was missed not only by Swannell but also by the surveyor general's department in its final approval of the plan.

In 1904 the CPR hired John McLatchie (PLS #39), a veteran surveyor from Nelson, to finish their land grant surveys by running the eastern boundary line of Lot 4589. Before beginning this work he was also instructed to run a traverse survey up Lizard Creek to the eastern boundary of Lot 4590. This was similar to the work that Swannell had done in 1903, except that McLatchie came up to the lot boundary from the east on a drainage of the Elk River, instead of from the west on a drainage to the Kootenay River. There was a large volume of merchantable timber in the Lizard Creek drainage. Two Herchmer brothers, Henry William, and Laurence Kirby Sherwood, had obtained timber licences in the area. (Both men were lawyers from nearby Fernie and Laurence later became mayor of the town.)

McLatchie surveyed to the place that he determined to be the eastern boundary of Lot 4590 in the Lizard Creek valley. From there he ran a line south along the boundary to Sand Creek. There he found Swannell's survey post, 34.23 chains farther west. He also found that the north–south distance was in error. McLatchie completed his survey by making a loop, returning via Sand Creek to a different point on the railway line. He found that his survey checked for accuracy.

When McLatchie reported the findings of his survey, the BC government's deputy commissioner of Lands, W.S. Gore, quickly dismissed the information. In a letter to McLatchie on November 9, 1904, he stated, "This matter has been the subject of consideration by the Department, and I beg to inform you that the Hon. The Chief Commissioner of Lands and Works has decided that the boundary line as established by Mr. McGregor under authority of the Government shall be the true and unalterable boundary whether the same be found correct or not. I do not know that it is necessary to enter into any argument for the reasons leading to this decision, suffice it that they appear ample to warrant such decision."

The CPR was not about to forego the land to which it believed that it was entitled. The company contended that it should receive the amount of land that was described in the Crown grant for Lot 4590 and not be bound by the boundary line surveyed by Swannell if it was incorrect. In 1906, the company began advocating the BC government for acceptance of McLatchie's survey. On June 17, J.H. McGregor wrote to the surveyor general providing details. "Our Mr. Swannell traversed up Sand Creek from a station on the railway—the latitude and departure of which he had calculated from the railway traverse notes on file in your office. He travelled a computed distance of easting and then ran a partial boundary line north and south across the valley. He had no means of closing his survey and was not authorized to resurvey along the line of the railway traverse already made. It is therefore possible that there is an error in his work." McGregor admitted, "It appears that Mr. McLatchie was authorized to incur the expense of a closed survey and as he reports a correct closing it is probable that his work is correct." At the same time he added a note of caution. "As I pointed out verbally to you some two years ago it would be very difficult without seeing Mr. McLatchie's notes, to decide whether the probable error lies in Mr. Swannell's traverse of Sand Creek, in the older traverse of the railway, or in his computation from the notes of the older survey." In concluding his letter McGregor stated that "a comparison of the Swannell plan and notes with the McLatchie plan and note (both of Sand Creek valley) should locate the supposed error very quickly."

On October 29 the surveyor general wrote to J.S. Dennis of the CPR, stating that the Herchmer brothers were pressing for a settlement of the eastern boundary of Lot 4590. He noted that "the licenses were issued to Messrs. Herchmer and Herchmer upon the distinct understanding that no encroachment was made upon Lot 4590, now they seem to wish to adopt a boundary line in continuance of the Swannell line." Dennis replied with a nine-page letter summarizing the events that had occurred since Swannell's survey.

The resolution of the boundary continued to drag on, so on August 15, 1907, the assistant to the second vice president of the CPR wrote to the premier, Richard McBride, urging him to become involved.

> I would ask that you kindly arrange to have this matter taken up by the Chief Commissioner of Lands & Works with Mr. McKay, the Surveyor General. I have personally discussed this matter with him two or three times and feel quite satisfied that the Chief Commissioner will find that the Surveyor General rules that the survey of this line as made by Mr. McLatchie across the Lizard Creek part is correct and that there is no ground for the protest against the acceptance of this survey filed by Messrs. Herchmer Brothers of Fernie.

> We are particularly anxious to get this long outstanding matter disposed of and I will be exceedingly obliged if you will have it dealt with at the earliest possible date and advice sent to me as to the final decision.

In October 1907 the deputy minister of Lands wrote a letter stating that McLatchie's line was accepted as the correct one and that the timber licences of the Herchmer brothers overlapped the boundary of Lot 4590. The following year the surveyor general made the common sense decision to have another BC land surveyor make an independent survey of the Lot 4590 boundary in the Lizard and Sand Creek area. W.S. Drewry (PLS #14) from Nelson was chosen, but unfortunately, for unknown reasons, he never completed the survey. In 1910, Alfred Cummings (now BCLS #10) surveyed the timber licences for the Herchmer brothers and used Swannell's survey for the eastern boundary of Lot 4590.

The BC government now decided to reverse its decision by accepting Cummings's and Swannell's surveys and gazetting the Herchmer timber licences based on that boundary line. (By publishing or gazetting this decision in the province's *British Columbia Gazette*, the government would officially proclaim its decision.) On September 26, J.S. Dennis wrote a letter to William Pearce, the CPR's western consultant, in response to a memorandum Pearce wrote outlining the company's position. Dennis stated, "I will take the matter up personally with the Hon. Mr. McBride on my next visit to the coast and endeavor to get this long outstanding question finally disposed of." On October 4, Dennis wrote to surveyor general E.B. McKay protesting the government's action. He noted, "The Crown grant having issued without the survey the Crown grant would govern. If the survey differs from the Crown grant it should be made to conform therewith." The next day Dennis sent a telegram to McBride. "Am advised by Department Lands it is proposed to reject McLatchie's survey east boundary British Columbia Southern Block forty-five ninety and accept Cummings' survey and gazette timber licenses affecting our land. Have wired our strong protest against this action and reported fully. We consider action Department breach of faith. F.H. Hale who purchased land from us advises that he also had assurance Government McLatchie's survey would be approved. Please have matter delayed until I can see you personally." Frederick Harding Hale, a lumber merchant and former MP from New Brunswick, also wrote to McBride. Hale reminded the Premier that he had met with him personally and afterwards, he wrote, "you informed me that the survey had been accepted by the Government and would not be opened up, and that Mr. J.S. Dennis, Assistant to Vice-President of the CPR, statement to me in reference to the above matter was correct as to the McLatchie survey. With this assurance from you I advised Hon. James M.W. Hall of Boston [another lumber merchant] to pay over the money to the CPR, which he did."

On March 16, 1911, Dennis wrote to W.R. Ross, Minister of Lands for the British Columbia government, following a meeting with the minister the previous day. "We feel of course that Mr. McLatchie's survey should be accepted, and if your ruling is against that contention, we wish to have the court to decide the matter." The BC government officially approved the Cummings survey of the Herchmer timber licences and Ross wrote to Dennis on March 22. "In the matter of the dispute as to the establishment of the eastern boundary of Lot 4590, Kootenay, which has been before this Department for a number of years, I have to advise you that my decision in this matter is that the portion of the line run by Swannell as well as the points on the eastern boundary of said lot at Sand Creek and Bull River established by Swannell and filed by you, connected by right lines from point to point, shall be established as the eastern boundary of said lot, between said points ... The effect of the same will be to permit of the acceptance and gazetting of the surveys of Lots 9472, 9473 and 9474." Ross referred to his meeting with Dennis on March 15. "As intimated by you, on the occasion of your recent visit, I gather that this decision...will not be acceptable to your Company and that you contemplate taking the matter into court with a view to establishing the correctness of the traverse work performed by McLatchie." The Minister of Lands concluded by saying, "As you are aware the effect of this decision will increase by a considerable acreage the quantity of land within Lots 9472, 9473 and 9474, Group One, Kootenay, the gazetting of which will now be proceeded with in due course, and as the recognition of these surveys will, according to your contention, encroach upon Lot 4590, Kootenay, you have due notice of my decision and will govern yourself accordingly."

Seippel Lumber Company (a long-established firm based out of Dubuque, Illinois, which had timber licences in British Columbia) had purchased the properties from Hale in 1910. They now sued the Herchmer Brothers for trespass. Seippel lost the initial court case, but the company had sufficient resources and believed that its viewpoint was correct, so it appealed the case to the chief justice of the British Columbia court in 1913.

On February 16, 1914, C.J. Hunter, the chief justice of British Columbia, ruled in favour of the Seippel Lumber Company. He stated that the Crown grant that the CPR had received from the government had a clear description of the land's boundaries (metes and bounds) and a definite starting post for the lot. From this the true boundaries of the land grant could be ascertained and this took precedence over Swannell's incorrect survey on the ground. Therefore the company should receive the land to which it was entitled rather than the land surveyed. Hunter wrote, "It is beyond dispute that this station zero is a fixed point, as to the situation of which there is no controversy. It is, therefore, apparent that a competent surveyor could at once, having located point zero, run a line due north as required by the terms of the crown grant and in that way determine the plaintiffs' boundary." The Chief Justice criticised the handling of the case by the BC government. "It seems to me to be a very startling proposition indeed, that a man who has got a Crown grant, and whose boundary can be definitely ascertained beyond any reasonable doubt or controversy may wake up some morning to find his property swept away by the decisions of such officials, which decisions may apparently be given behind closed doors without any reason, without any notice, and without any appeal to a responsible civil tribunal."

Hunter commented on Swannell's survey.

> I find that the so-called Swannell survey was absolutely erroneous with the result that it lops off over 400 acres [161.9 ha] covered by the plaintiffs' grant. I think I can safely say that a casual inspection of Mr. Swannell's notes, even to the mind of a layman reveals the fact that they are absolutely and startlingly erroneous. Referring to station 19, the easting is given as 4.49, subtending an angle of 7 degrees and 44 minutes, the side of which is 3 chains 34 links. Now any school boy can at once see that it is impossible for the line subtending an angle of 7 degrees and 44 minutes in a right angle triangle to be 4.49 chains when one side bounding the angle is only 3.34, so that any official in the land office, if he had taken the trouble to glance at these notes, even in a casual way, could have seen that they were absolutely wrong, and, as a matter of fact, the line should have been .449 instead of 4.49.

It must have been difficult for Swannell, who was later known for his meticulous work, to realize that this error on a project at the beginning of his career created such controversy and took so long to resolve.

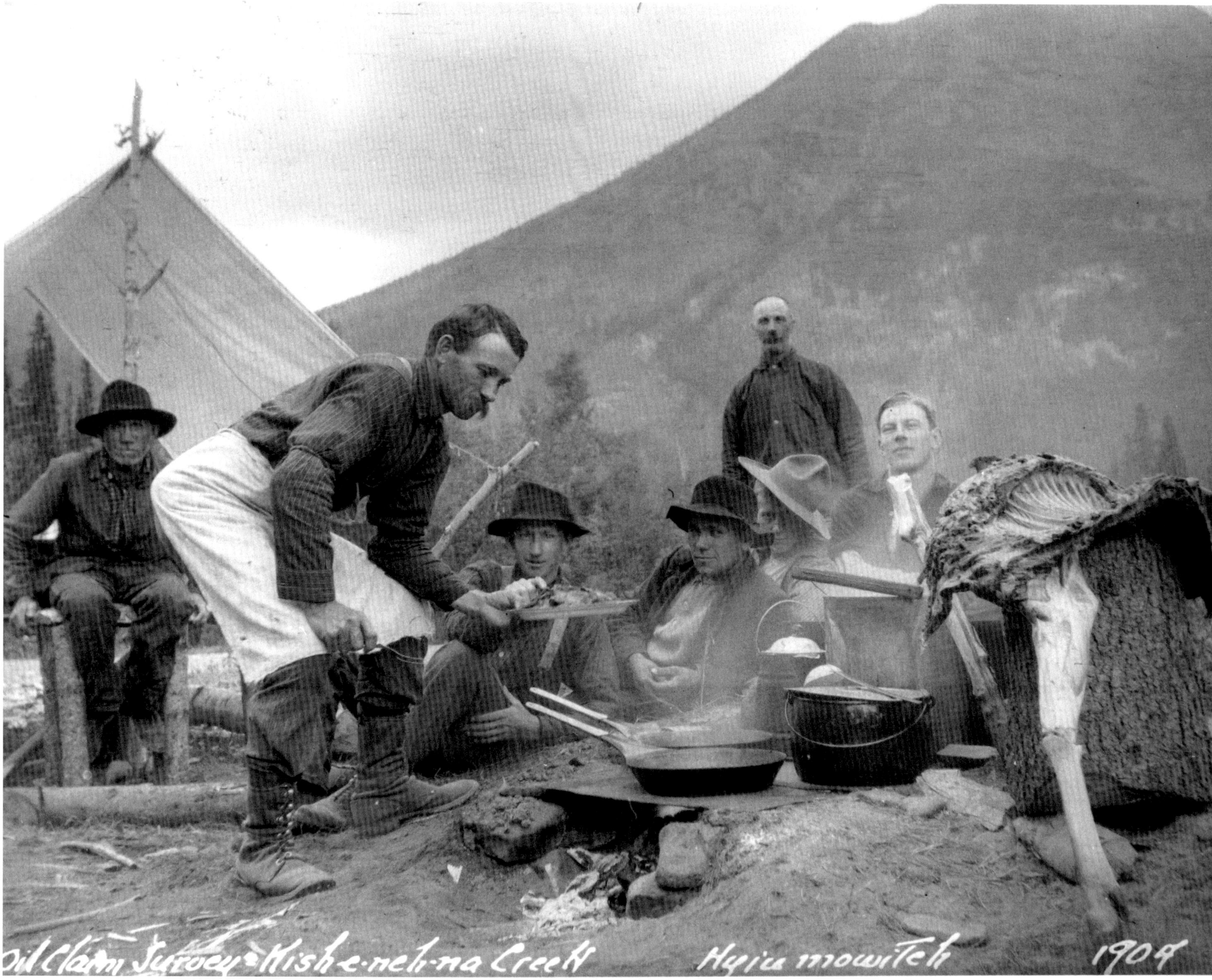

❖ "Hyiu Mowitch" (plenty of venison). Oil prospector John Watt is standing in the back. Swannnell surveyed leases in the Flathead Valley for him. Image G-03251 courtesy of Royal BC Museum, BC Archives.

1904

This would be one of the most important years in Swannell's personal life, and also a dramatic one in his surveying career. Over the winter and early spring he completed the office work for the 1903 surveys; he also obtained his Dominion land surveyor's licence, which would enable him to run surveys on land held by the federal government like the First Nations (Indian) Reserves.

In the late winter Swannell did some surveying at the Mount Sicker mining area north of Duncan.

Swannell left Victoria on April 9 to survey lots in the Cariboo region. He and his crew took the train to Ashcroft and from there travelled by stage to Clinton and on to Dog Creek.

Swannell arrived at the Gang Ranch on April 15. He and his crew spent two days resurveying some Lillooet district township lines for the provincial government and connecting them to lot surveys in the area. On the nineteenth, the crew "rode 26 miles [42 km] to Vedan's and camped." The next day they swam the horses across the Chilcotin River and camped on the north side.

For the next week Swannell and his crew surveyed a series of lots along the north side of the Chilcotin River in the Hanceville area. Then they crossed the river again and surveyed three separate lots near the Hanceville wagon road south of Big Creek. These lots were meadows that the ranch was applying to lease from the government. On the last day of April the men travelled about 48 kilometres back to the Gang Ranch. Swannell spent the first two days of May at the ranch preparing the field notes from these surveys. Then the crew surveyed two more lots west of the ranch. By May 6 he was back in Clinton, where he wrote Ada.

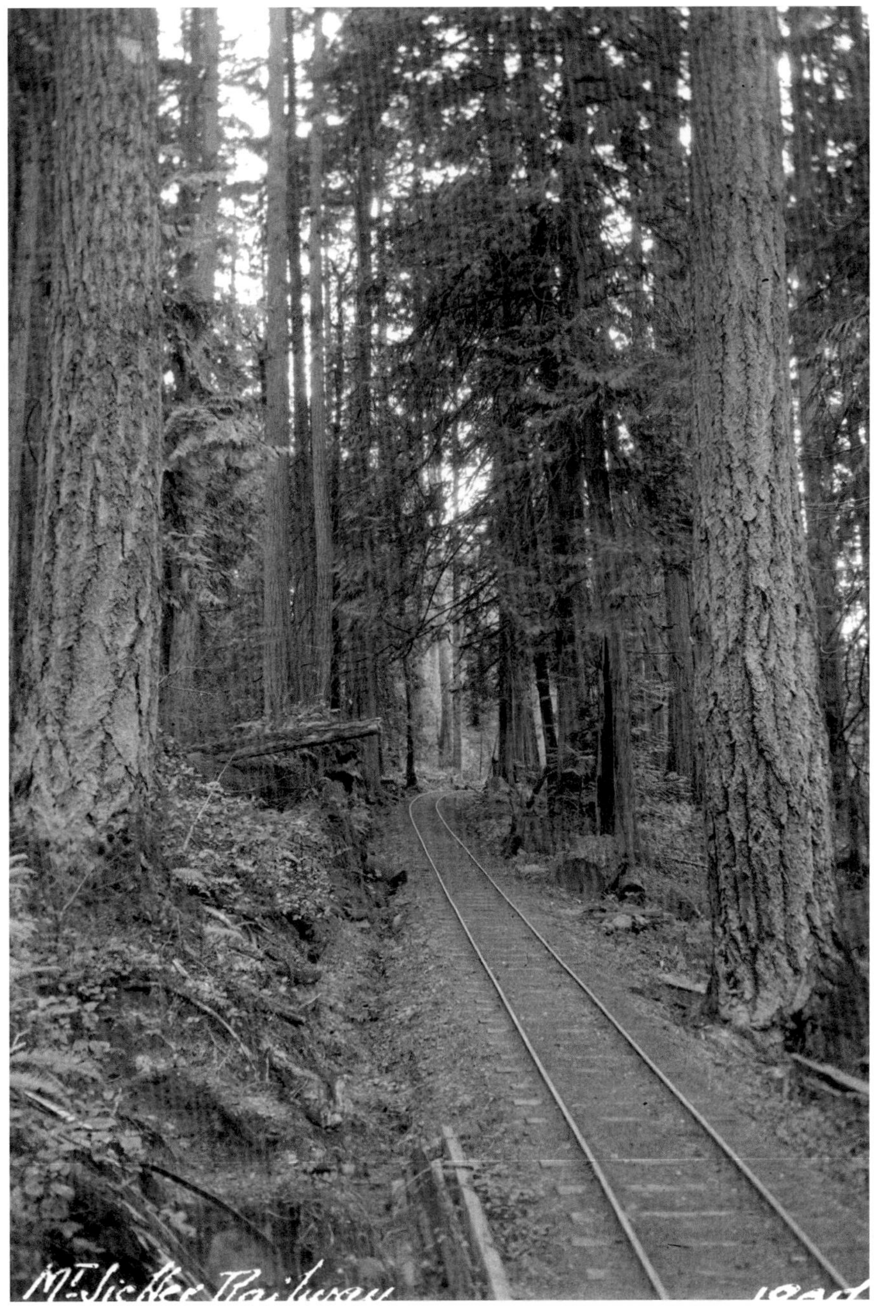

❖ Lenora and Mount Sicker Railway. This narrow gauge railway was part of the mining operations on Mount Sicker on Vancouver Island. Image I-33657 courtesy of Royal BC Museum, BC Archives.

Mount Sicker

In 1895, three American prospectors found traces of copper, gold and silver on Mount Sicker, northwest of Duncan, and staked claims. In 1897 more than sixty claims were staked around the entire mountain. The Lenora Mount Sicker Copper Mining Company began operations in 1898 and, by 1900, was the fourth-largest shipper of copper ore in British Columbia. The townsite of Lenora was established in 1900 and three mines were developed around Mount Sicker. The Lenora and Mount Sicker Railway, a steep narrow-gauge railway, was constructed from one of the mines to a smelter built at Crofton. It was the only rail line on Vancouver Island built expressly to handle ore. Since the line had grades up to 13 per cent, the railway only used Shay locomotives (heavy-duty geared locomotives used on many local logging and mining railways in British Columbia). The high grade of the ore, the close proximity to the E & N Railway line and the availability of an adequate water supply enabled the mines to be developed quickly. The mines and railway only remained in operation for about ten years, although there was some additional mining and exploration intermittently during the twentieth century.

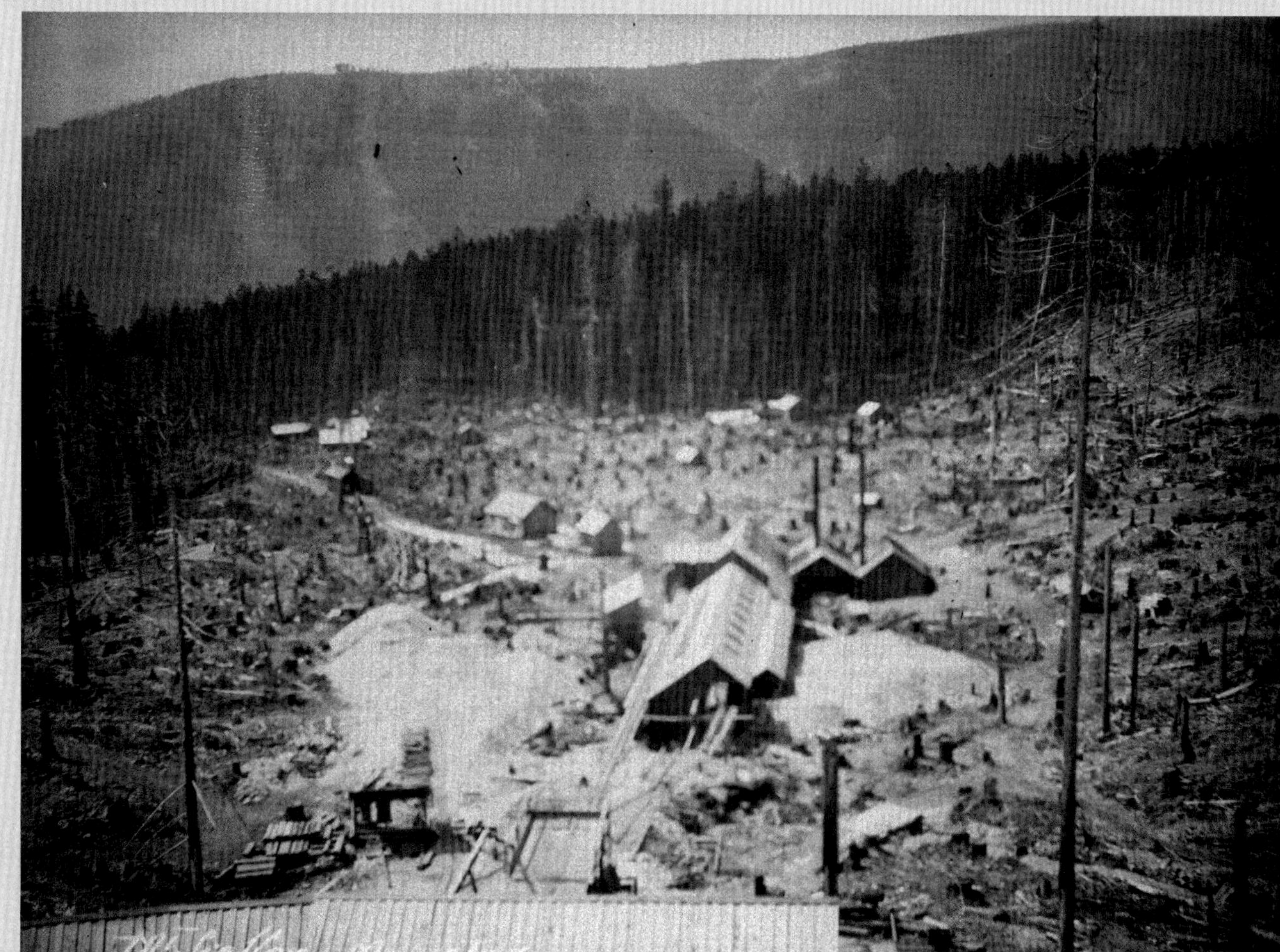

Mount Sicker. Swannell's photograph is one of the few that shows the Mount Sicker mine site in full operation. Image I-33658 courtesy of Royal BC Museum, BC Archives.

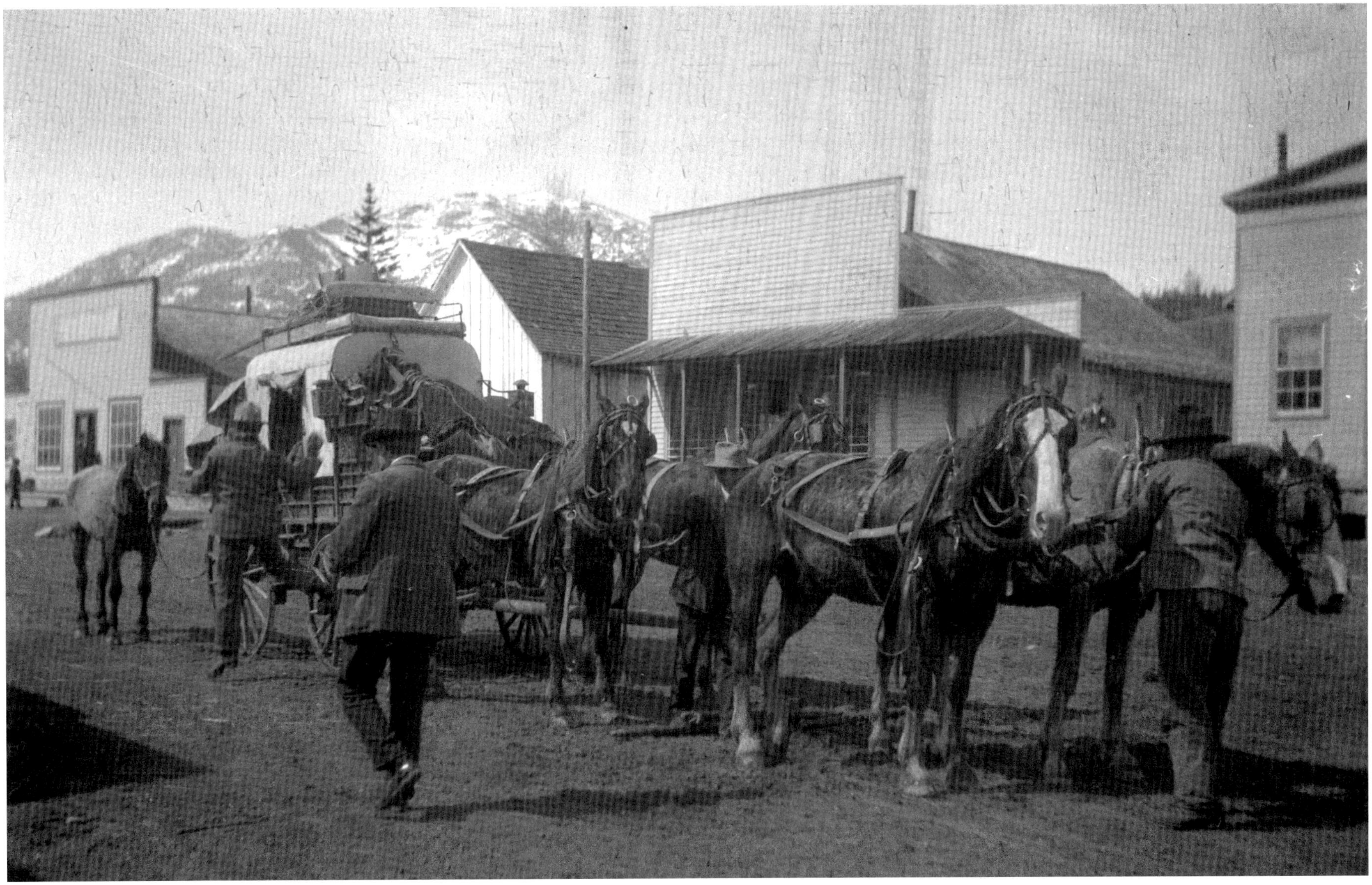

Clinton
May 6, 1904

Suffice it to say that the survey has been a No. 1, gilt-edged edition deluxe Superfine.

Of adventures there have been none except and save for the unprincipled conduct of a reprobate cayuse y-clept Huzzy under whom I suffered twice.

Item No. 1—last pitch of a steep hill, saddle slips nearly off suddenly. Huzzy, not approving of barebacked horsemanship, whirled round, pitched sideways & stood on end simultaneously. I gracefully slid off as being the only practicable course left open to me—I think I did it gracefully, but in my pardonable haste may have omitted all frills save celerity.

Next day the thong holding the coil of halter rope to the saddle horn broke, the rope mixed up with Huzzy's feet and the excitement started. This time I described a beautiful parabolic curve ... landed in the distant landscape making quite an impression on the sod with my head—which stood the test

❖ Changing horses on the Cariboo stage at Clinton. Image I-33678 courtesy of Royal BC Museum, BC Archives. For 2104 image see page 150.

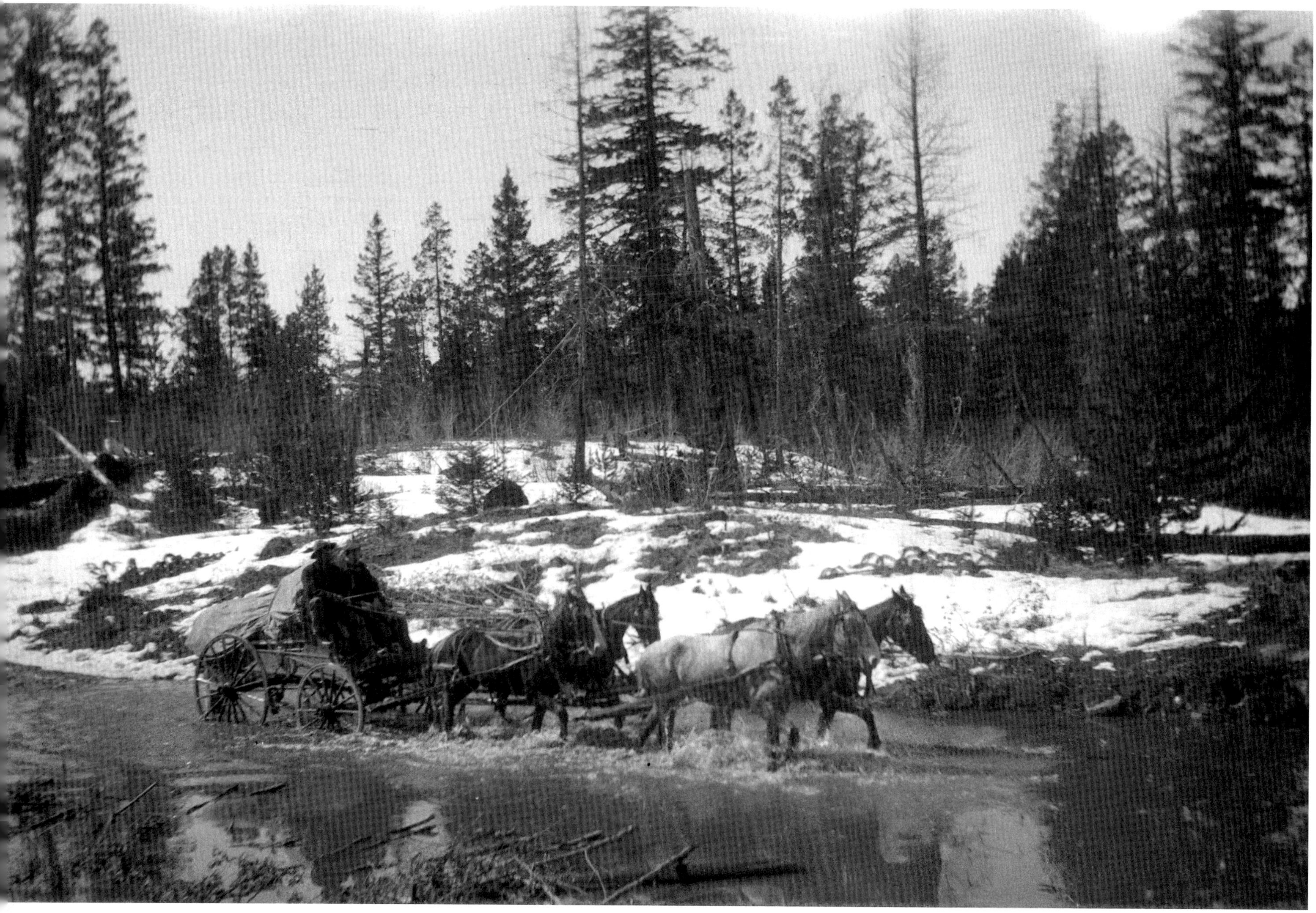

❖ The Dog Creek stage ran once a week from Clinton to Dog Creek, a small community near the Fraser River. Image I-58350 courtesy of Royal BC Museum, BC Archives.

well. Next day I had a horse guaranteed not to buck in place of half-broken Huzzy who ought to be rechristened "Fuzzy Wuzzy on a spree" (see Kipling).

Reached Clinton last night, having driven 74 miles from the Gang Ranch—was disgusted at getting a telegram ordering me up to the 150 [Mile House]. Still I expect will be back by the 24th.

Swannell spent a couple of days at Clinton. One of the men on his crew that summer was Ada's younger brother, Henry, and on May 9 he and Swannell hiked on Mt. Soues near Clinton.

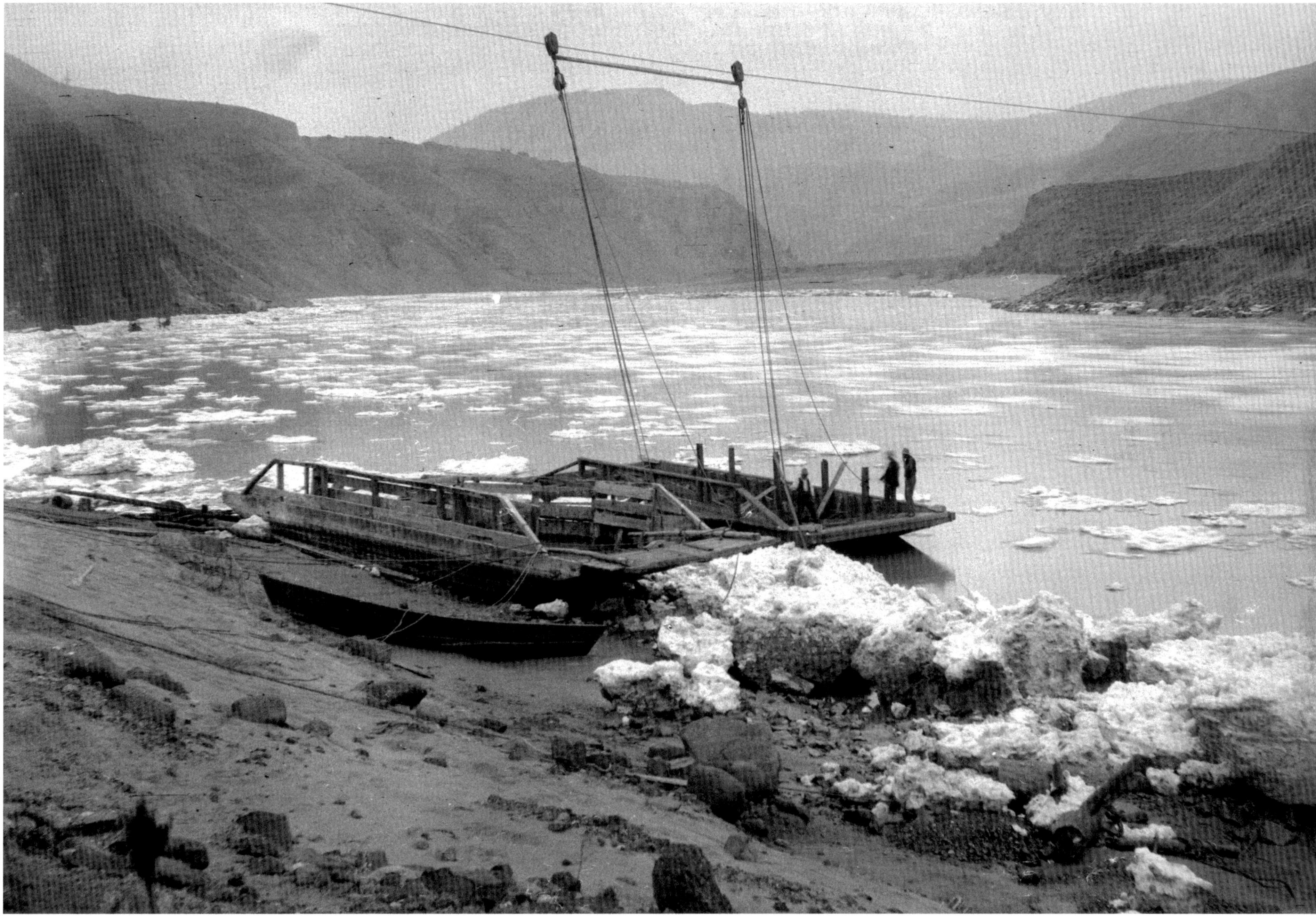

The next day, Swannell and his crew headed north on the Cariboo wagon road, stopping for the night at Boyd's roadhouse at 70 Mile House. They continued north to the Williams Lake area, where they spent about a week surveying some lots around 150 Mile House. Then they surveyed two lots near the junction of the Fraser River and the Williams Lake River. These were for a South African War grant for E.A. Carew-Gibson, a merchant with the Cariboo Trading Company. (Soldiers who served in the South African War, 1899–1902, were entitled to a free pre-emption of up to 320 acres, or they could take a cash payment. Most chose to be paid in money.) Although he had told Ada he expected to be back by May 24, Swan-

❖ Churn Creek ferry in the ice on the Fraser River. This is where Swannell and his crew crossed the river en route to the Gang Ranch. The gravel bar in the background along the river on the right is where Churn Creek flows into the Fraser. It is still visible today. Image I-33358 courtesy of Royal BC Museum, BC Archives. For 2014 image see page 148.

❖ Henry Driver on Mt. Soues. Image J-00321 courtesy of Royal BC Museum, BC Archives.

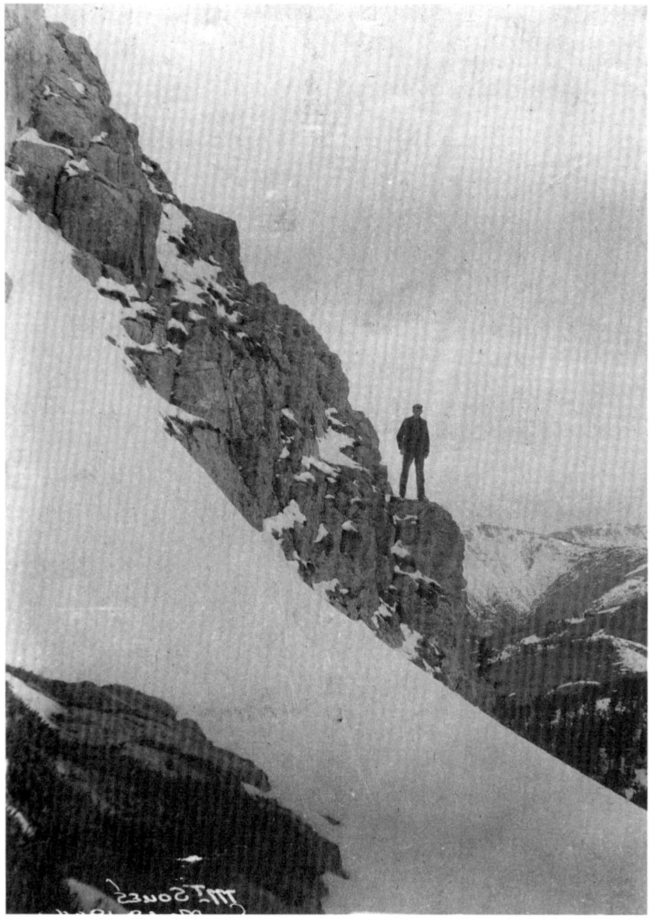

❖ Right: Salmon Canyon, Chilcotin River. Today this is known as Farwell Canyon. Image I-57591 courtesy of Royal BC Museum, BC Archives. For 2104 image see page 149.

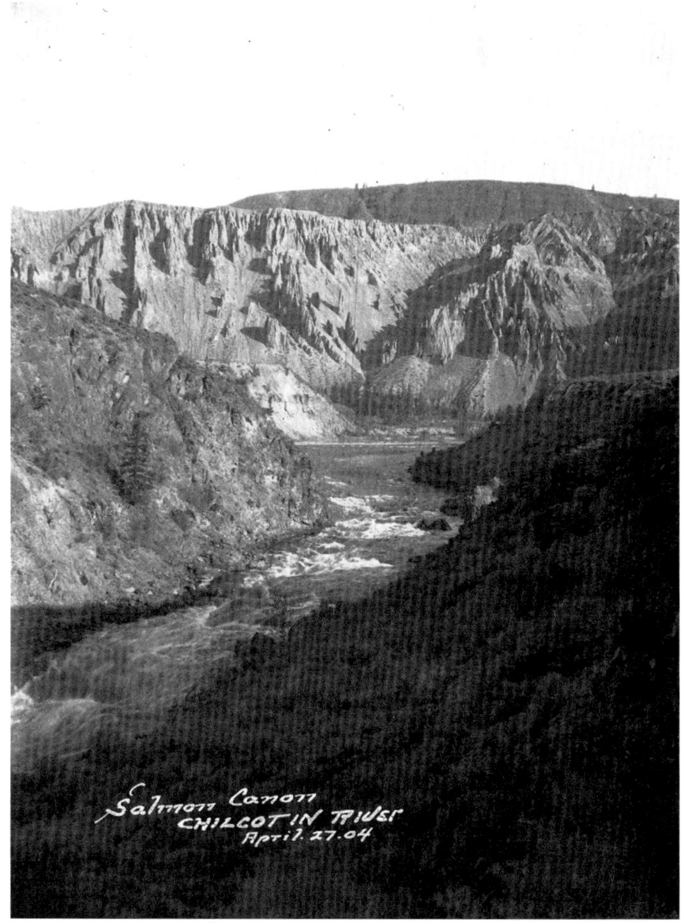

❖ Below: Pack horses swimming across the Chilcotin River. Image I-58390 courtesy of Royal BC Museum, BC Archives.

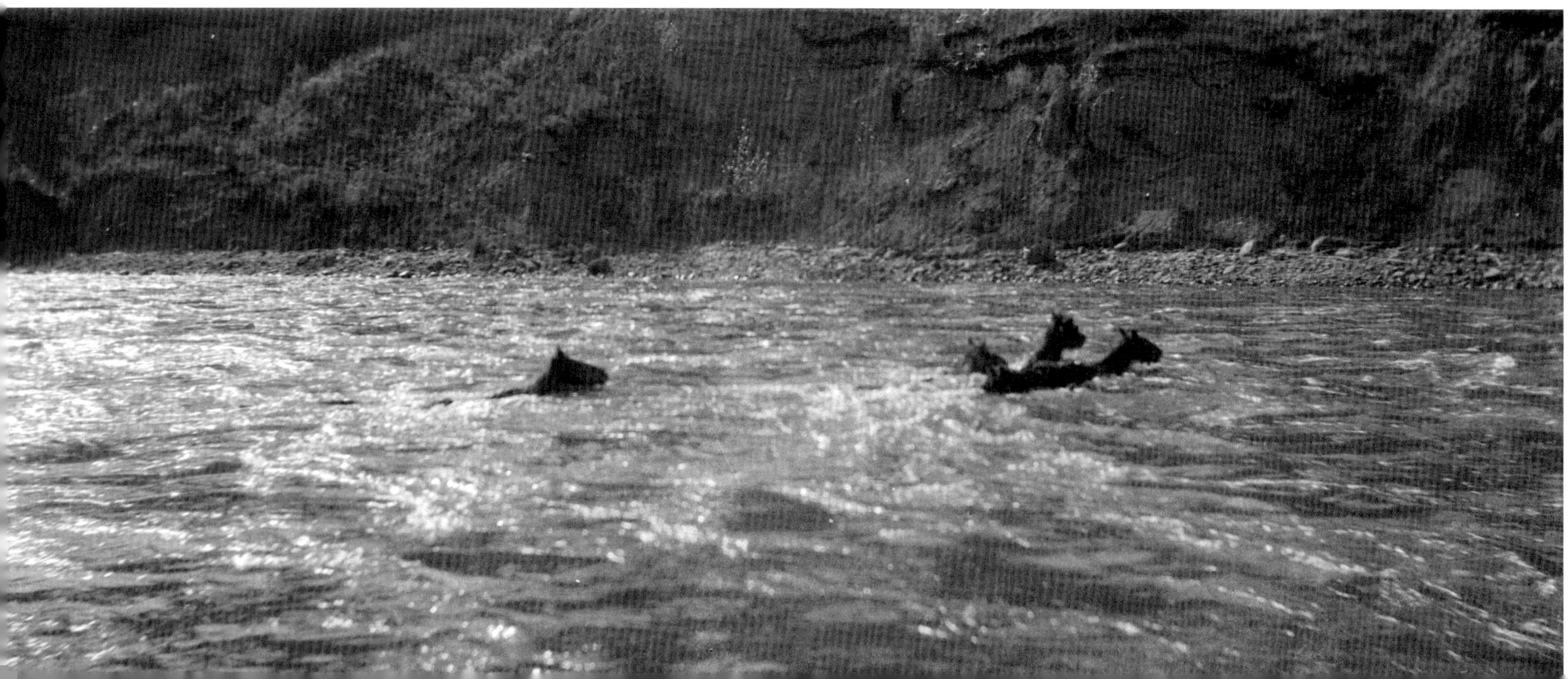

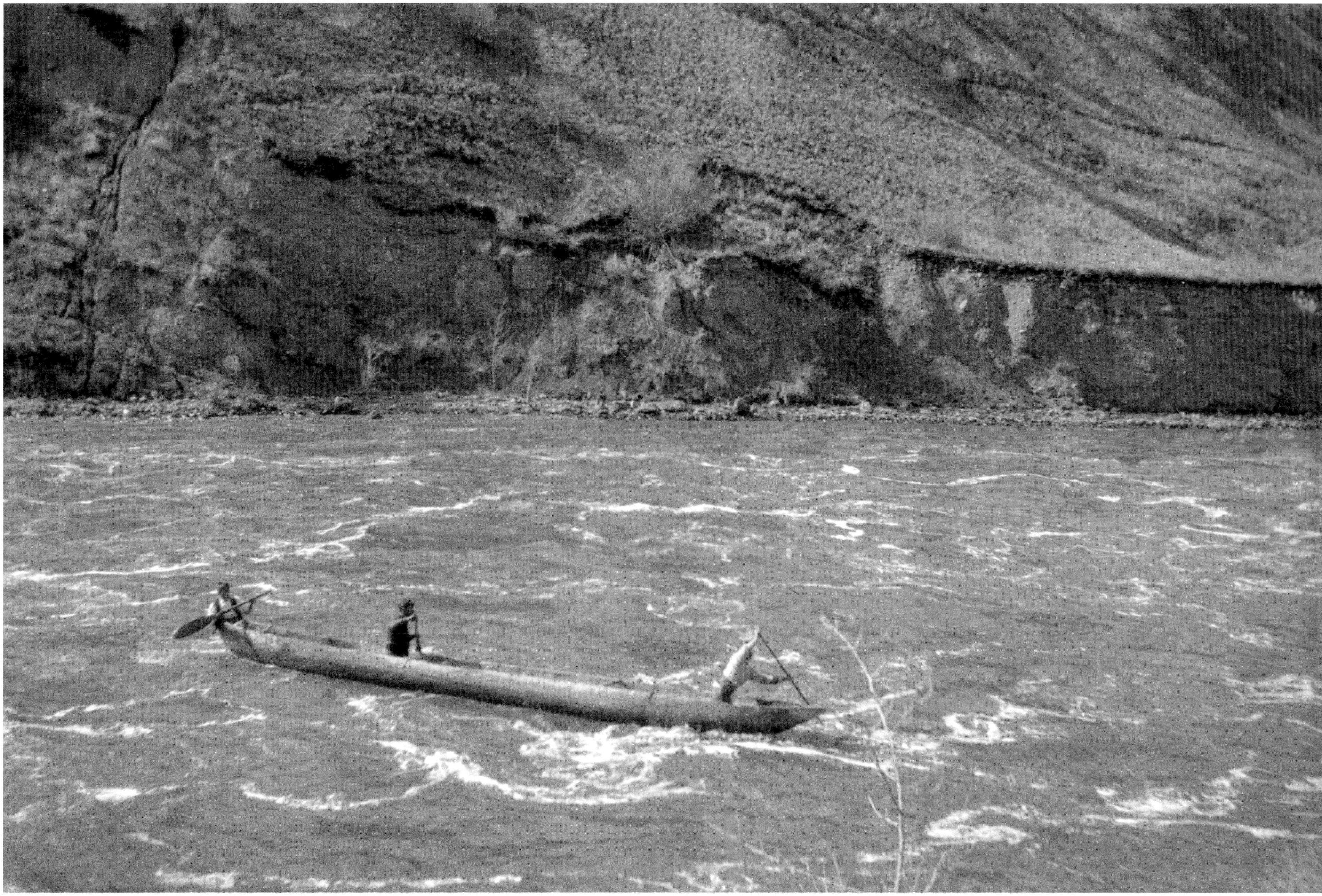

nell was still surveying some lots in the vicinity of 150 Mile House at the end of the month.

He also had to make a small survey near Cache Creek on his way back to Victoria. This was for an adjustment to the boundaries of the Bonaparte Reserve. Daniel O'Hara, a friend of former premier Charles Semlin, had a ranch adjoining the reserve and wanted a portion of the land.

The delay likely caused anxiety for Swannell and Ada, for they planned to be married on June 16. (Swannell's and Ada's friends George Copley and Nora Mabel Liddell married in April.) But Swannell arrived in time, and after his wedding he remained in Victoria for about a month.

Swannell then returned to the Kootenays for a third year of surveying. By 1904 there was only one

❖ Running the Chilcotin. Image I-59715 courtesy of Royal BC Museum, BC Archives.

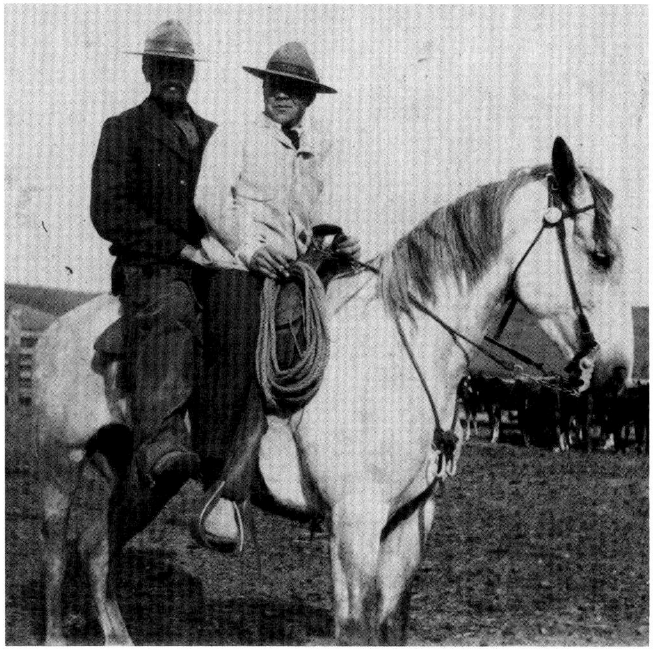

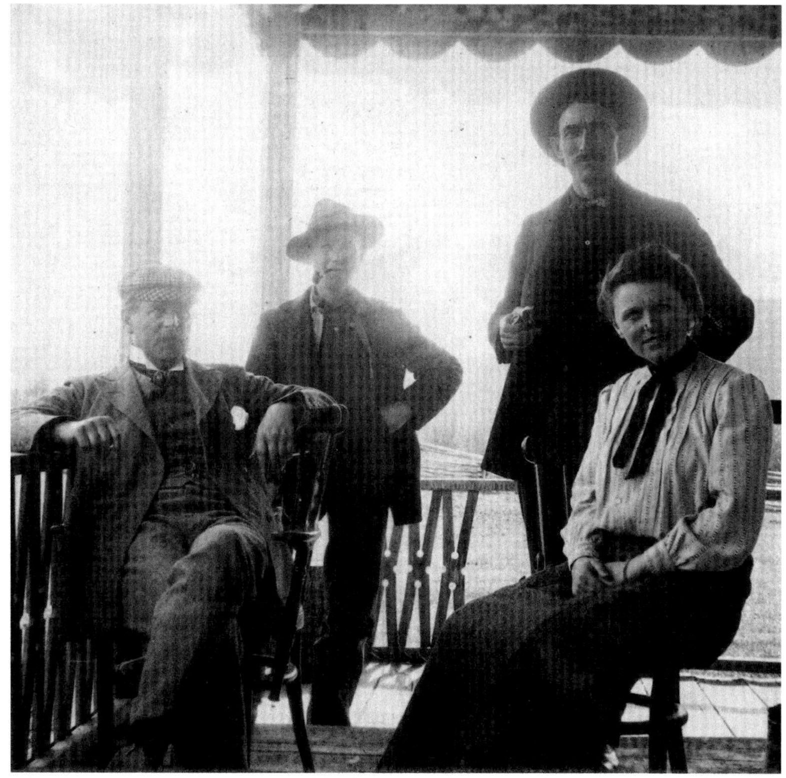

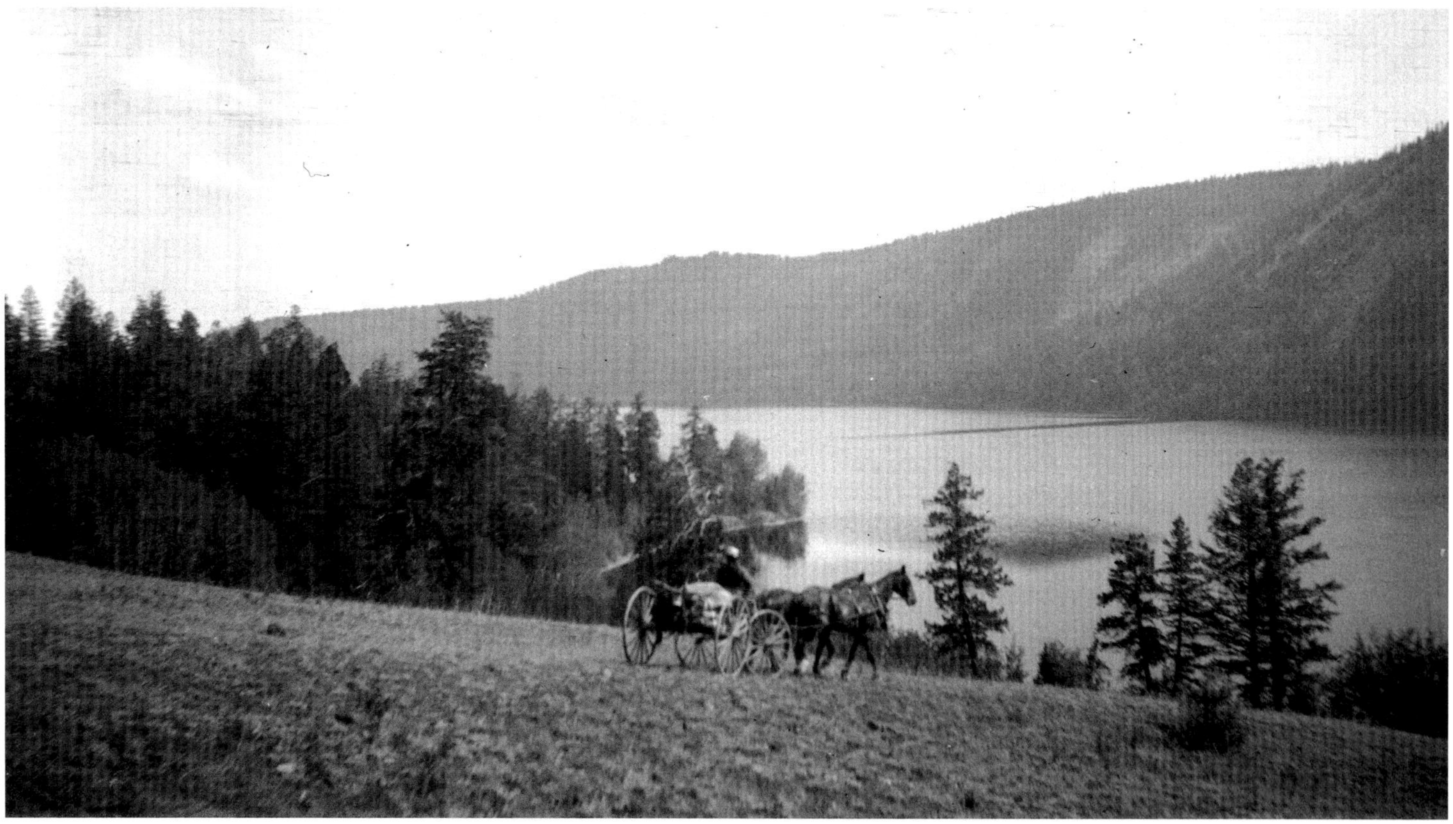

❖ Right: J.D.Prentice (front left) and Archie Knight (back right). J.D. Prentice was the manager of the Gang Ranch in 1904. He was the MLA for the Lillooet East district in 1894 and from 1898 to 1903, and a member in the cabinet of Premier James Dunsmuir. Archie Knight was a cowboy at the ranch. In 1898 he was a member of one of the four groups that drove cattle from the Chilcotin to the Yukon. Swannell took this photograph when he was surveying some lots for the Gang Ranch. Image J-00322 courtesy of Royal BC Museum, BC Archives.

❖ Left: Tsilhquot'in men on horse. This photograph was taken at the Gang Ranch. Image J-00319 courtesy of Royal BC Museum, BC Archives.

❖ This page bottom: Williams Lake. Much of the foreground later became a gravel pit. Image I-33348 courtesy of Royal BC Museum, BC Archives. For 2014 image see page 150.

Opposite page

❖ Top left: Prentice's team crossing Deep Creek. Swannell met J.D. Prentice again in May. Image I-33667 courtesy of Royal BC Museum, BC Archives.

❖ Top right: 83 Mile House. Image J-00318 courtesy of Royal BC Museum, BC Archives.

❖ Bottom: 150 Mile House. There was a roadhouse and store at this community. Image I-33356 courtesy of Royal BC Museum, BC Archives.

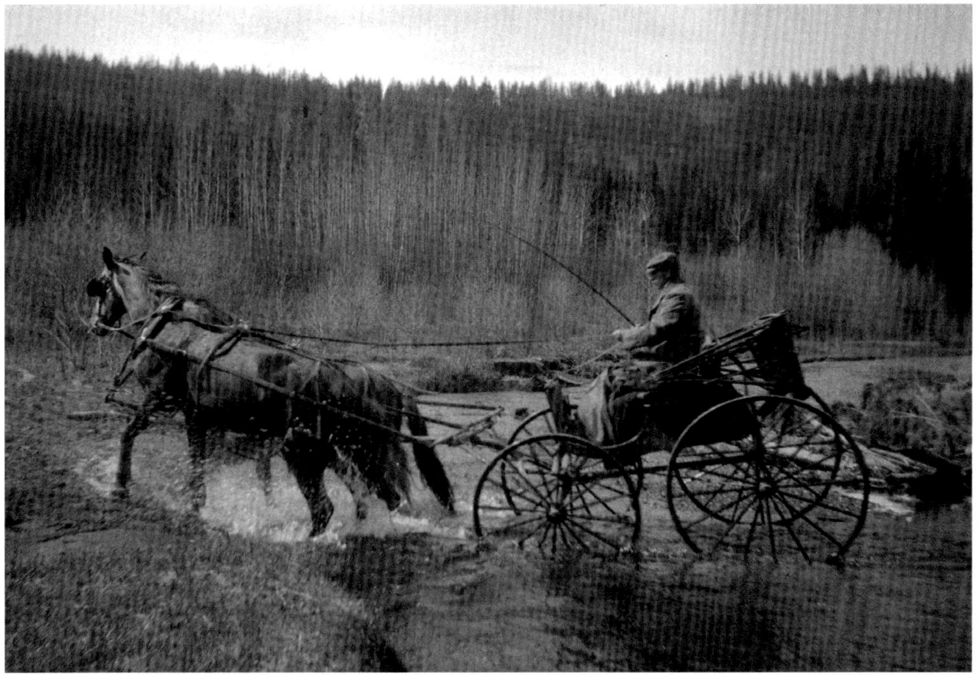

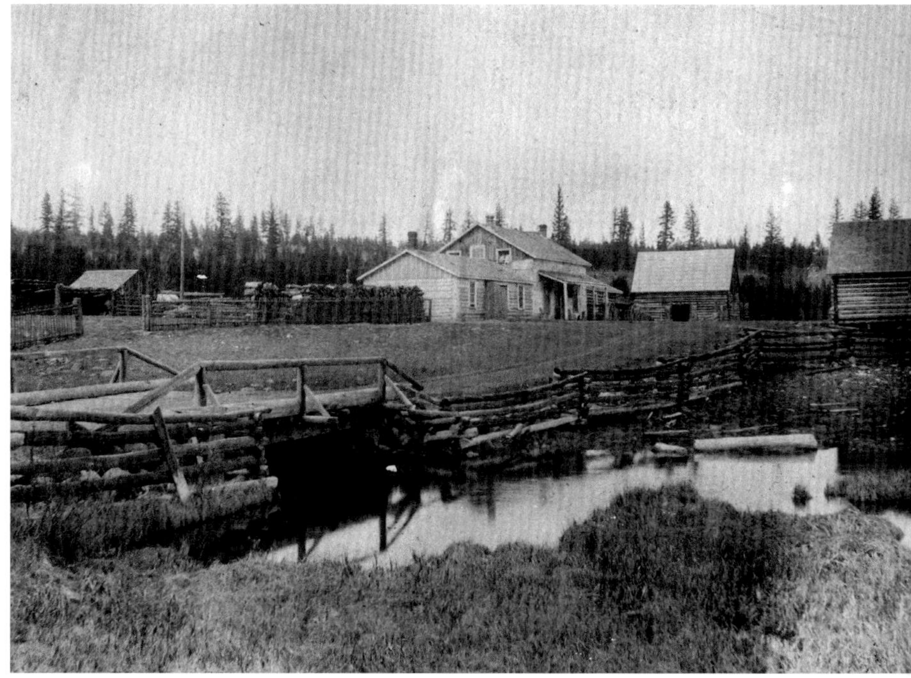

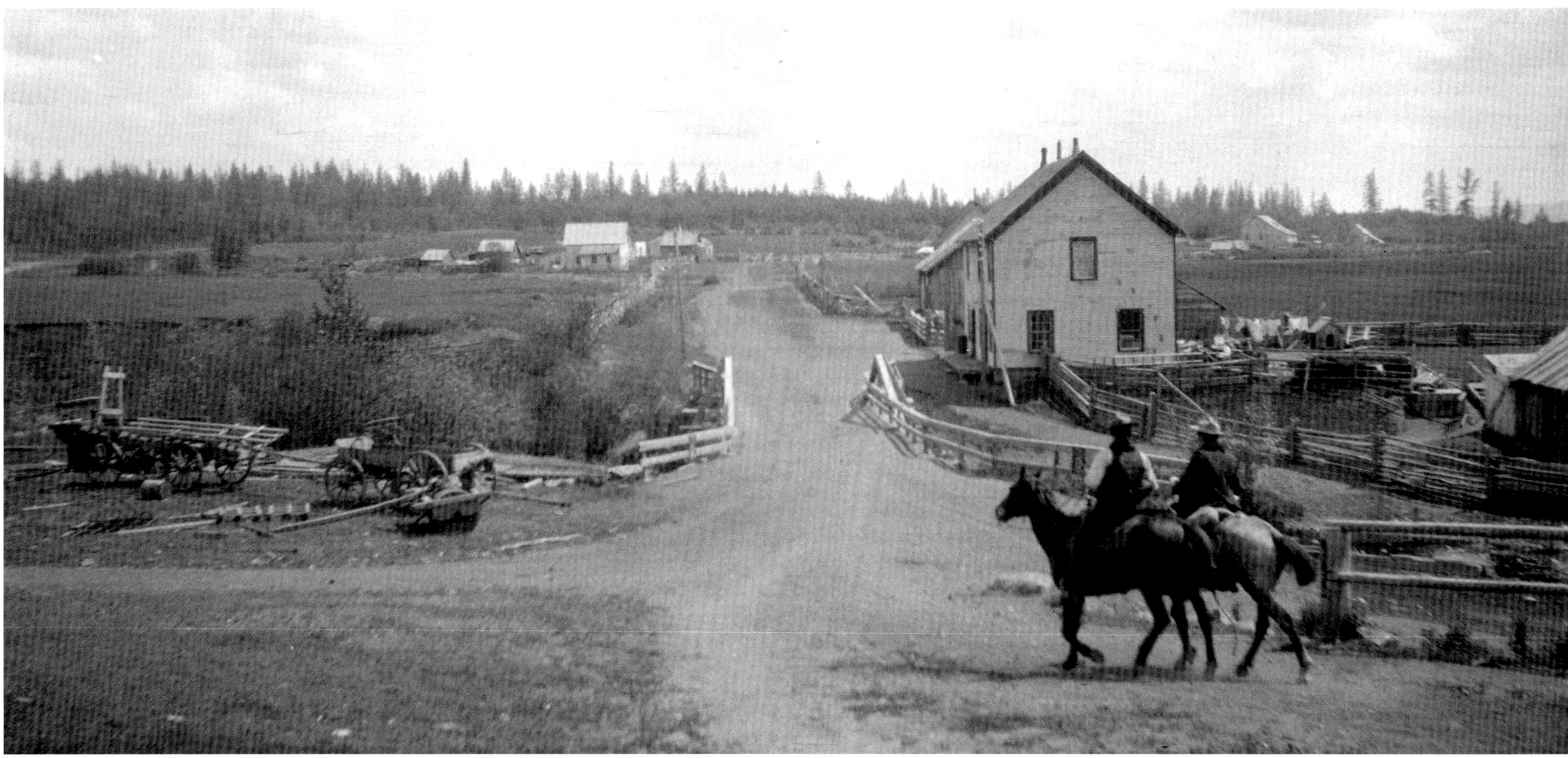

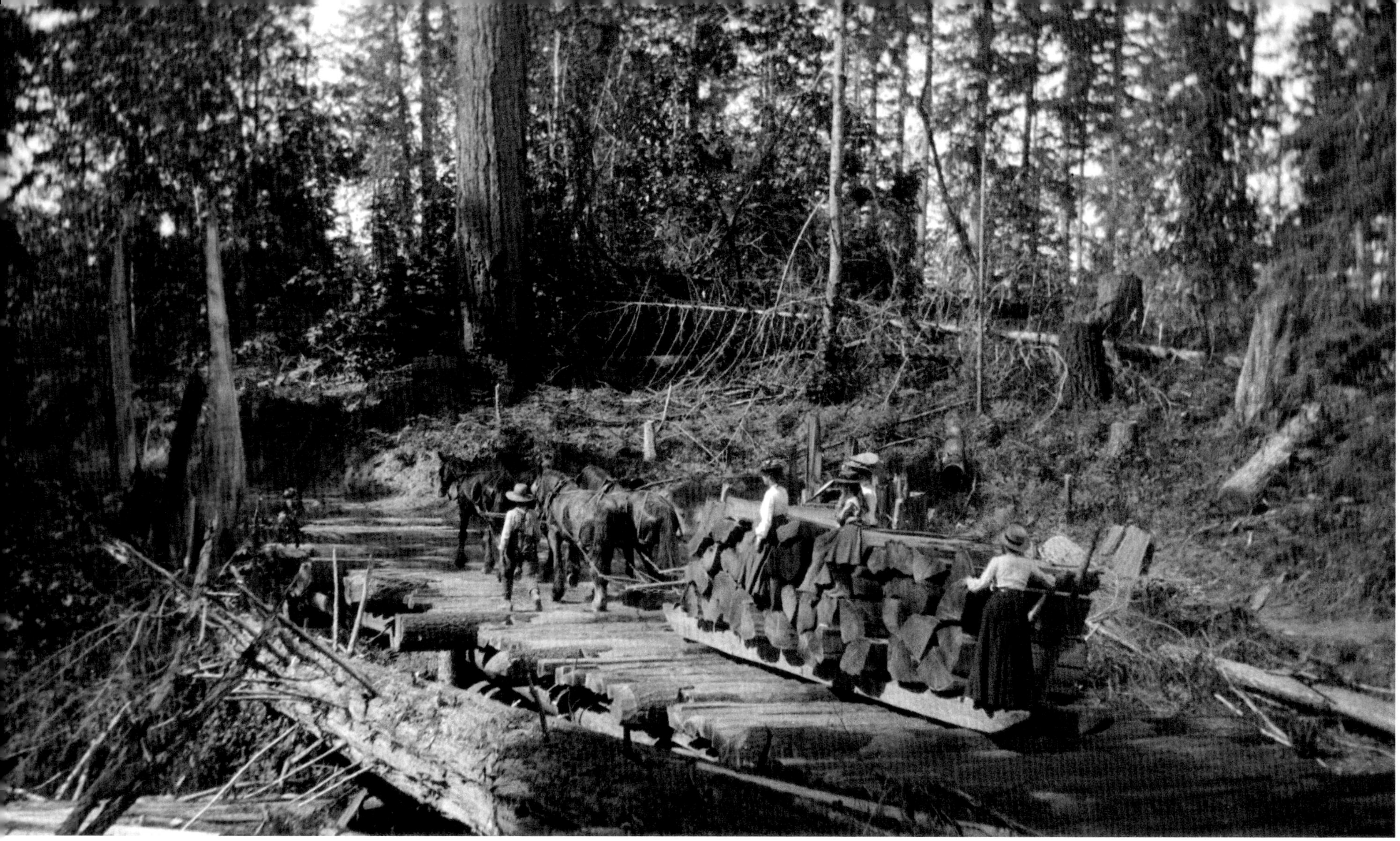

❖ Hauling shingle bolts on a sled, Mill Bay. This picture was taken shortly after Swannell's wedding. Image I-58319 courtesy of Royal BC Museum, BC Archives.

major line that still needed to be surveyed for the BC Southern Railway land grant: the east boundary of Lot 4589, which mainly went through the Flathead drainage. The CPR had realized that this line needed to be established as early as the summer of 1903. In a letter from the CPR's BC land commissioner to the assistant of the president of the company on July 20, J.L. Doupe wrote, "During my last visit to Victoria Mr. W.S. Gore, Deputy Commissioner of Lands and Works, stated that the Government are particularly anxious to have the survey of the east and south boundaries of British Columbia Southern Block 4589 completed at as early a date as possible." Doupe believed that "the desire of the Government to have the lines in question surveyed is in connection with the claims which have been filed in deficiency Block 4593." Block 4593 was one of the two Columbia and Western Railway land grants in the Crowsnest Pass region that the provincial government had disallowed. This area in the southeast corner of the province adjoined the eastern boundary of Lot 4589. There was considerable interest in the potential of the oil and coal in the former Block 4593, but before the government opened the area to mineral claims it wanted to know the location of the CPR's land in Block 4589. Doupe noted that the survey of this lot line had not been included in Gore & McGregor's contract for 1903. One of the reasons for the delay in surveying Lot 4589 was that the northeast corner was defined as beginning on the BC Southern Railway line on the boundary between British Columbia and the District of Alberta, and that point had not yet been established by a boundary commission.

Since the survey of the east boundary of Lot 4589 was not a big project, the CPR gave it to John McLatchie, who had previously surveyed in the Flathead valley and who also had experience in oil claims. The surveyor general instructed McLatchie to determine where he thought the BC Southern railway line crossed the boundary between the province and the territory. The southern boundary of Lot 4589 was the border with the United States and in 1904 the International Boundary Commission was resurveying the border, so the CPR did not need to confirm that line. The CPR asked McLatchie to include in his report any signs of oil that he found near his survey line. But before he began surveying Lot 4589, the CPR requested that McLatchie make a small survey of the east boundary of Lot 4590. This resulted in the controversy described in the 1903 chapter.

In 1904 Swannell went to the Kootenay region to survey oil leases for John Watt in the North Fork of the Flathead River drainage in the former Block 4593, in the southeast corner of the province.

It was easier to reach this remote location from the United States, so the crew took a train to Seattle and then rode on the Great Northern rail line to Belton, Montana (today this is the west entrance to Glacier National Park, but in 1904 this area was a forest preserve). From Belton Swannell and his crew followed a wagon road along the North Fork of the Flathead River for about 65 kilometres to the border.

Swannell and his crew began by surveying a set of claims in the Kishinena Creek drainage near the United States border. Starting from the Flathead River valley, the surveying took Swannell and his crew high into the Rocky Mountains until they were about 15 kilometres from the intersection of the border of the United States, Alberta and British Columbia.

Swannell and his crew then moved to Sage Creek, another tributary on the east side of the North Fork of the Flathead River. There they encountered a large forest fire that drove the men out of the area. In the only letter after his marriage that remains, Swannell described the incident.

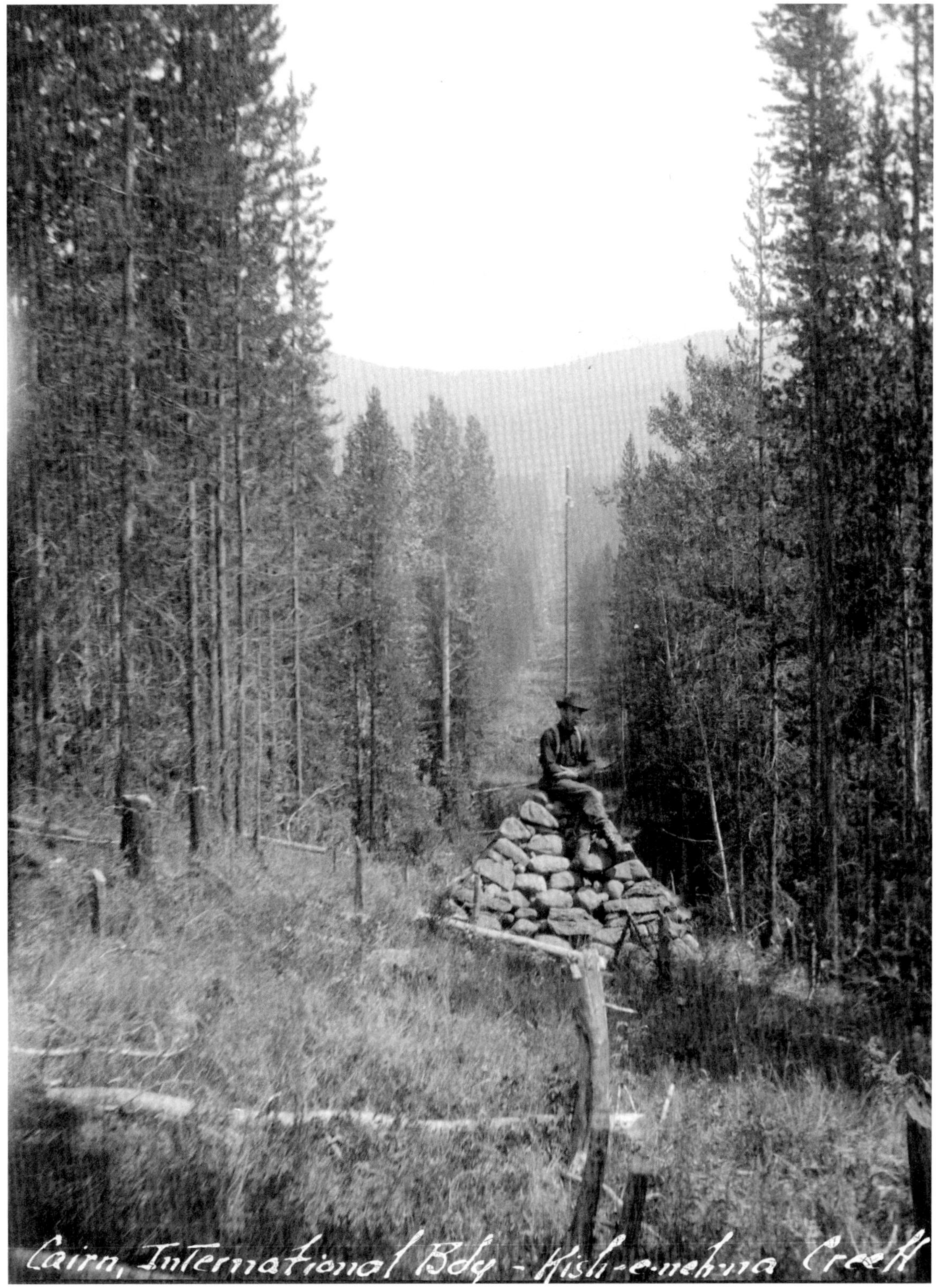

❖ Cairn on the international boundary, Kishinena Creek. Image I-58260 courtesy of Royal BC Museum, BC Archives.

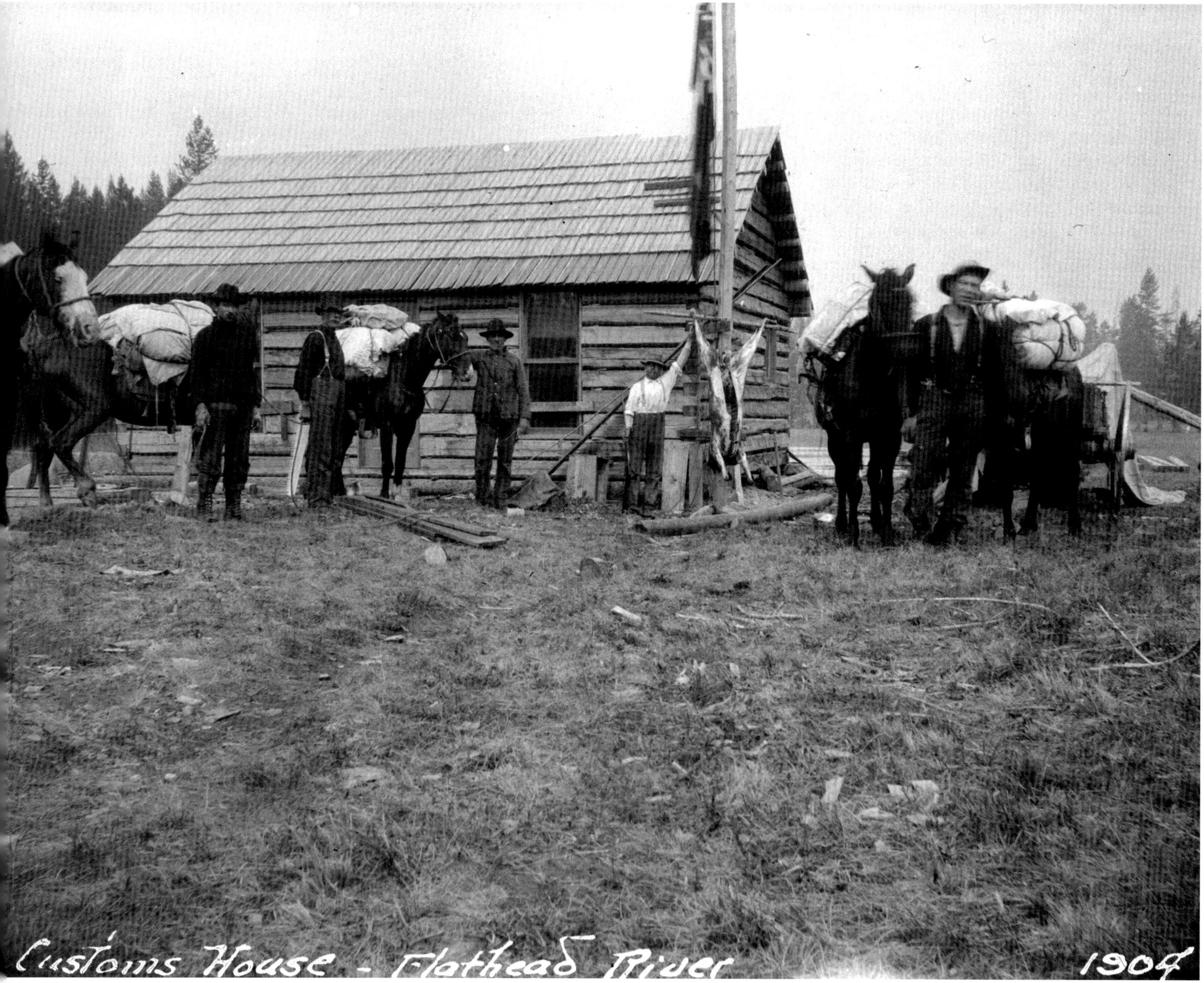

❖ Customs house, Flathead River. Image I-58059 courtesy of Royal BC Museum, BC Archives.

Kalispell, Montana
August 17, 1904
My dear wife,

Back again in Montana for a time. We had just packed in to Sage Creek when we were burnt right out by a forest fire. Lost tents, blankets, grub, pack saddles. Escaped with what we happened to be wearing—not one coat or blanket among 9 of us. I managed to fish out the transit, but the wood bark was all burnt. The field books are also safe, though charred and hardly to be deciphered. Cook badly burned—I got off scot free although I rescued the field books amid a salvo of exploding cartridges.

We froze that night around a fire in a wet spruce bottom. Next day Horace & I pushed on ahead of the others and tramped the 70 miles [110 km] into Belton in 23 hours. The whole forest for 30 miles [48 km] is ablaze—over an area of about 3000 sq. miles [7770 km²]—the roar of the fire across the Flathead and the crash of falling trees. To our west—the sun a clearcut disc of blood red, set in the murky smoke clouds by day, the sky a sullen red glow at night pierced in the west by the stabbing shoots of flame.

We have been getting a new outfit here today. I wrote Frank Bowes and wired him to buy a new chain in Seattle—no answer. So I shall have to go to Spokane myself tonight—very annoying as we are in a hurry to get in again.

Swannell and his crew returned to the Flathead and completed their survey without further incident.

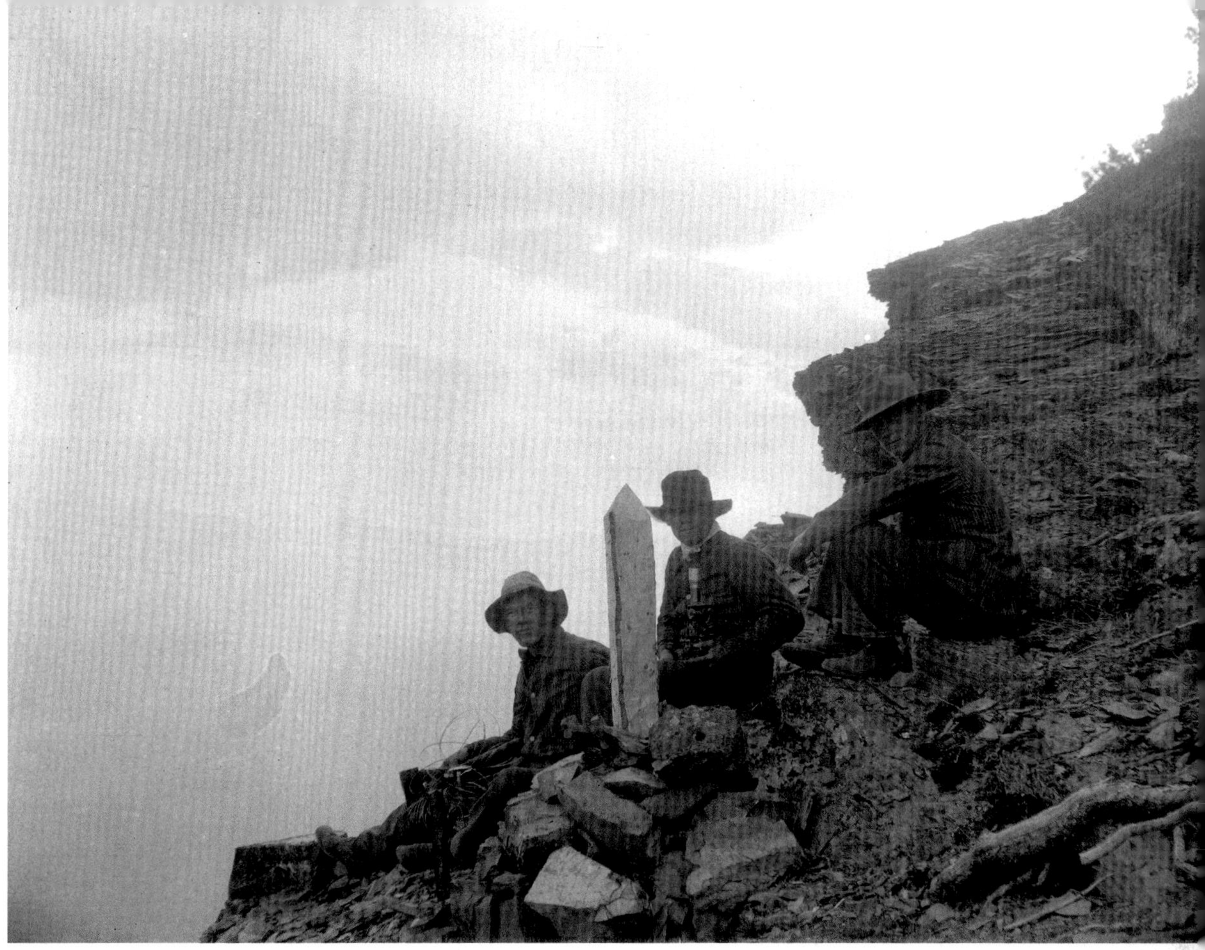

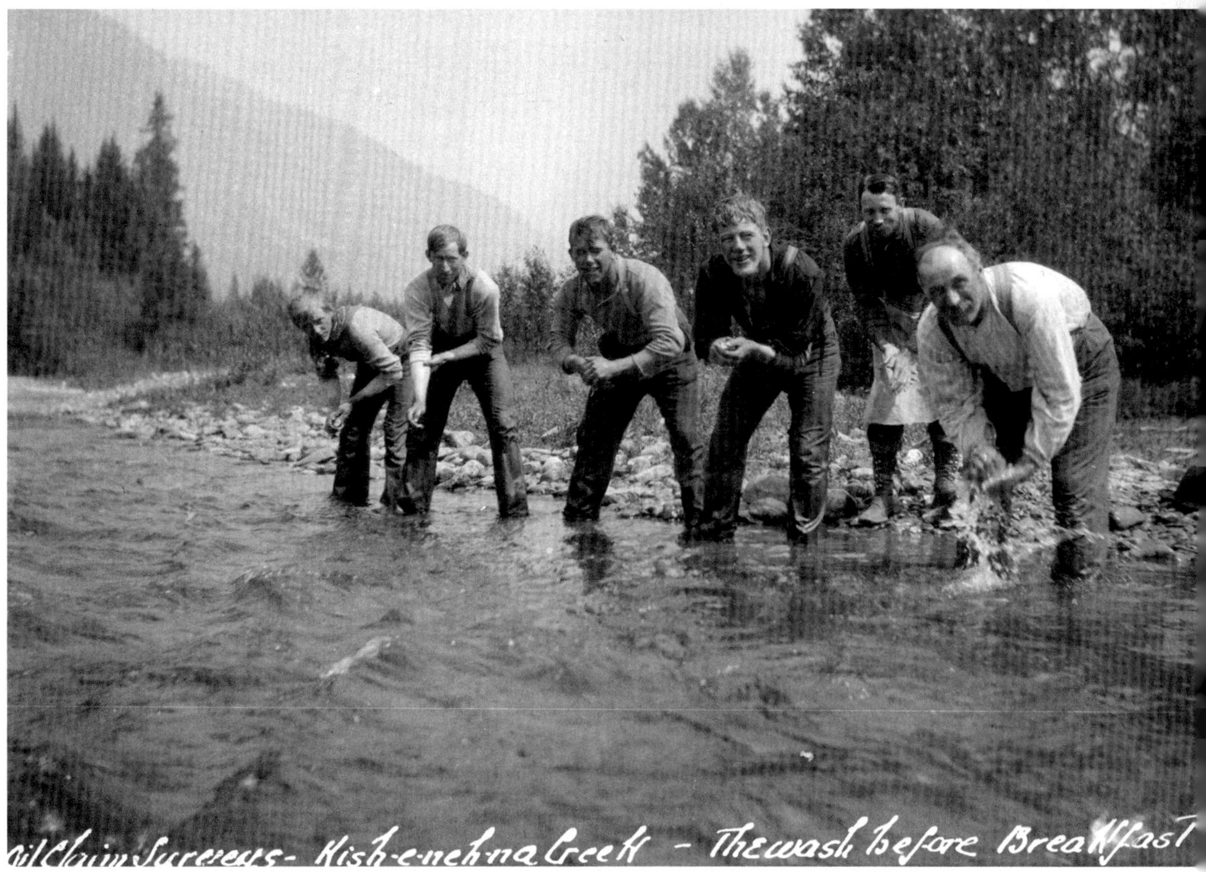

Top: A surveying station above the Sage Creek valley. Image I-57064 courtesy of Royal BC Museum, BC Archives.

Right: Kishinena Creek: the wash before breakfast. Image I-58263 courtesy of Royal BC Museum, BC Archives.

Forest Fire at Sage Creek

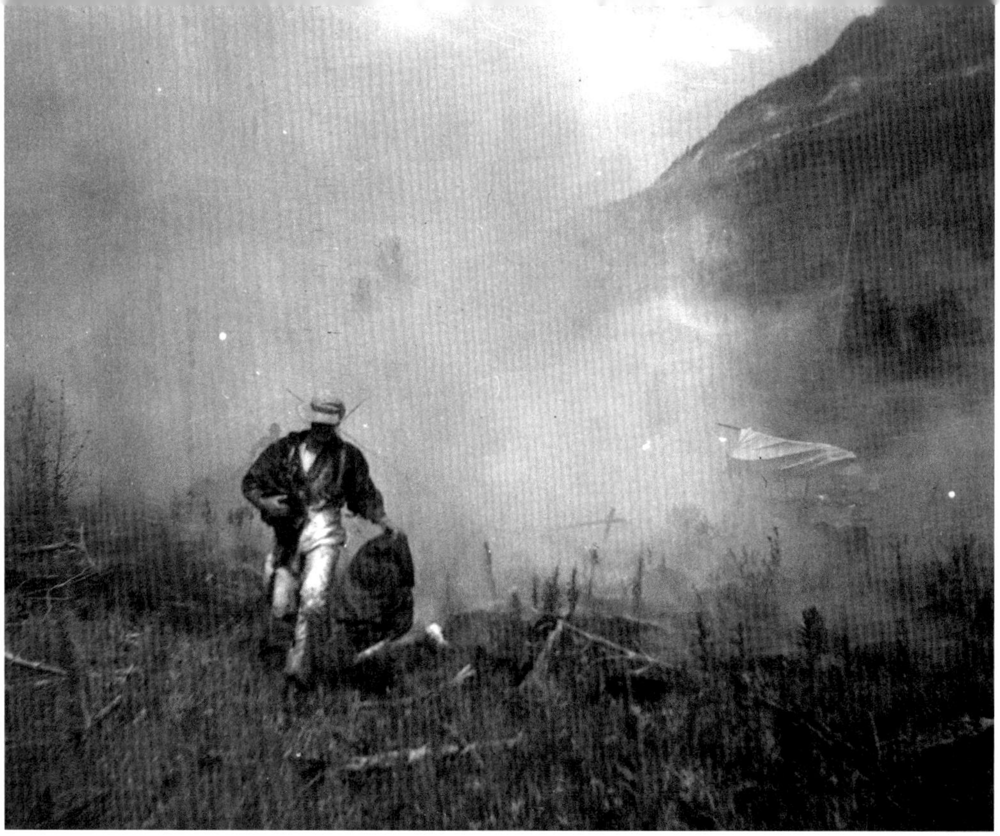

❖ Sage Creek survey camp during a forest fire. Image I-33606 courtesy of Royal BC Museum, BC Archives.

Opposite page

❖ Top: The aftermath of a fire. Image I-57056 courtesy of Royal BC Museum, BC Archives.

❖ Bottom: Big Spring burning. Image I-57050 courtesy of Royal BC Museum, BC Archives.

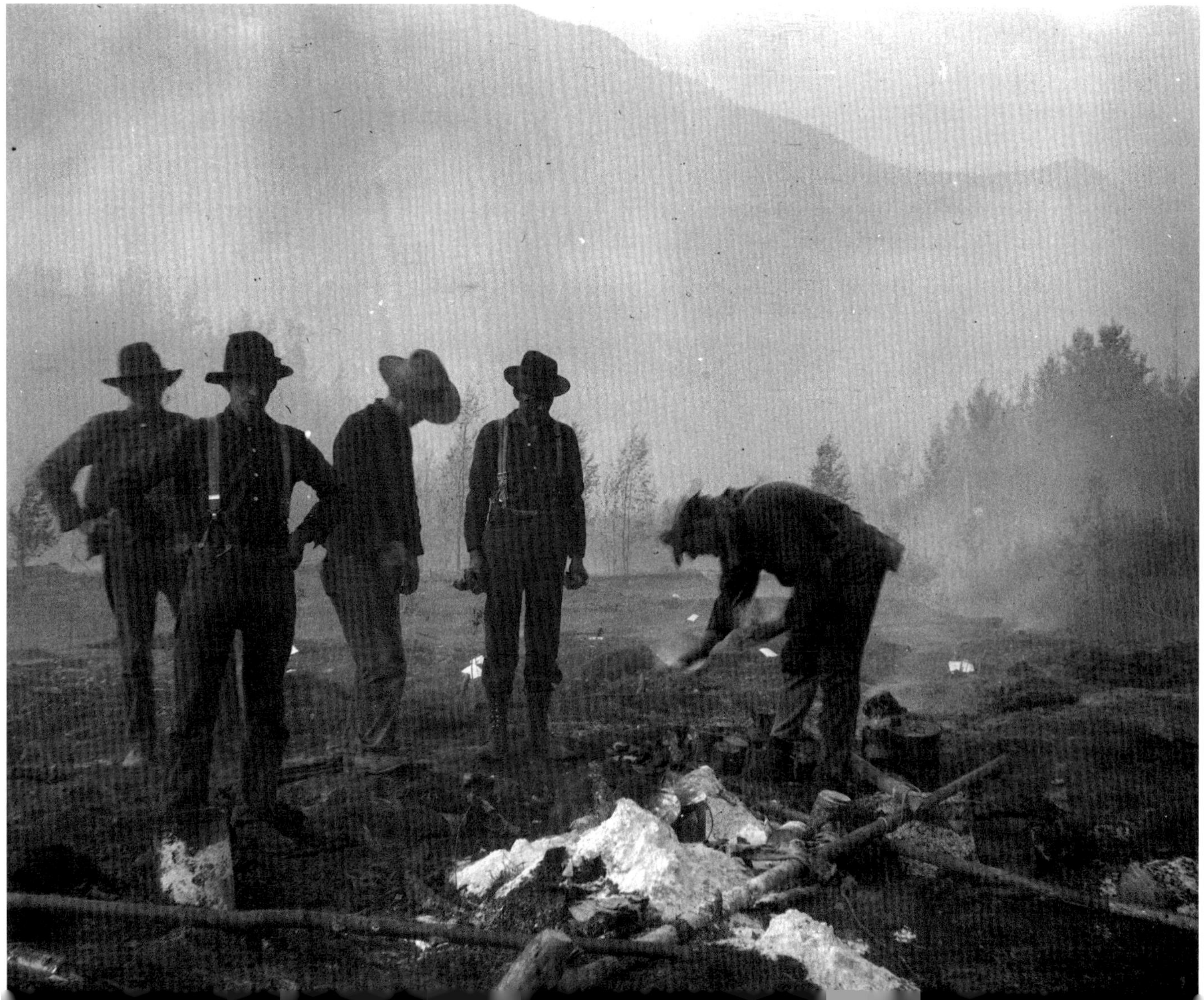

❖ Sage Creek survey camp after a forest fire. Image I-33607 courtesy of Royal BC Museum, BC Archives.

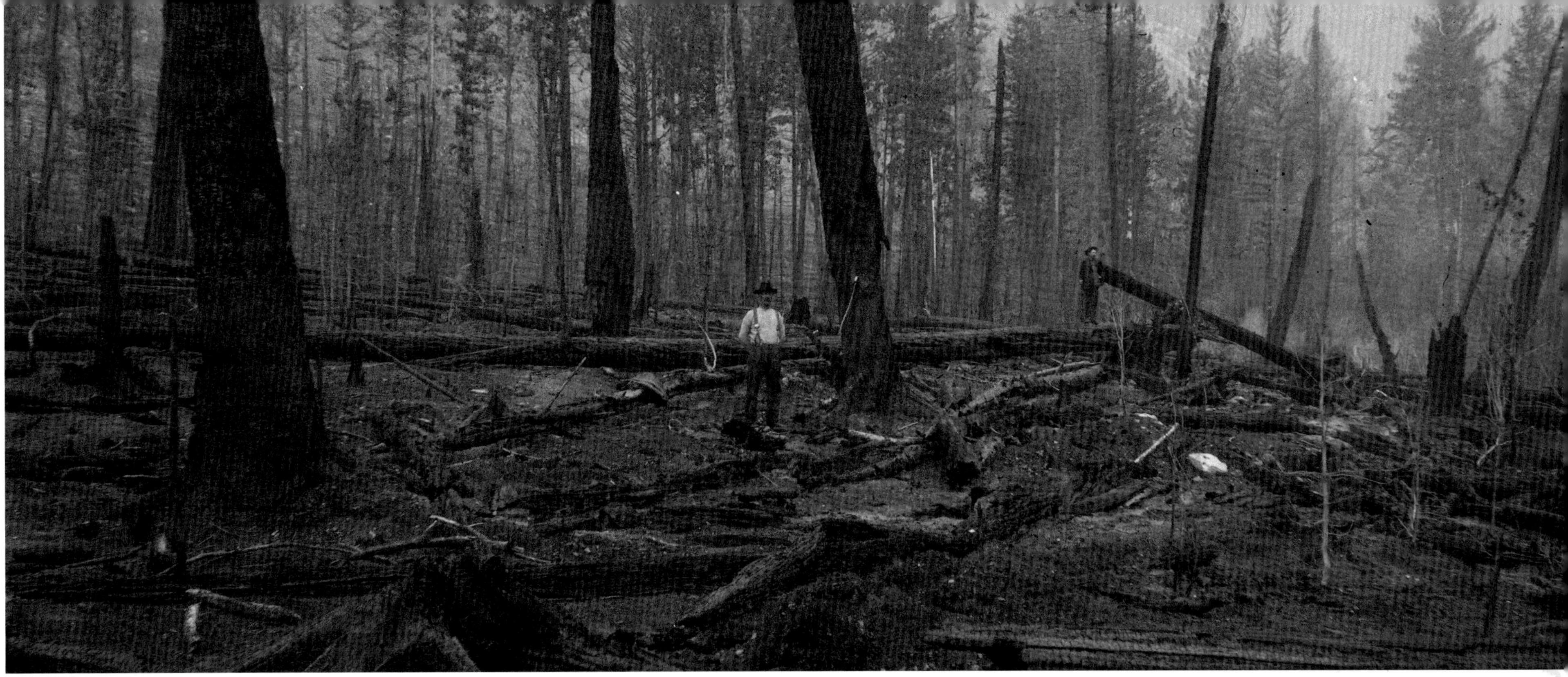

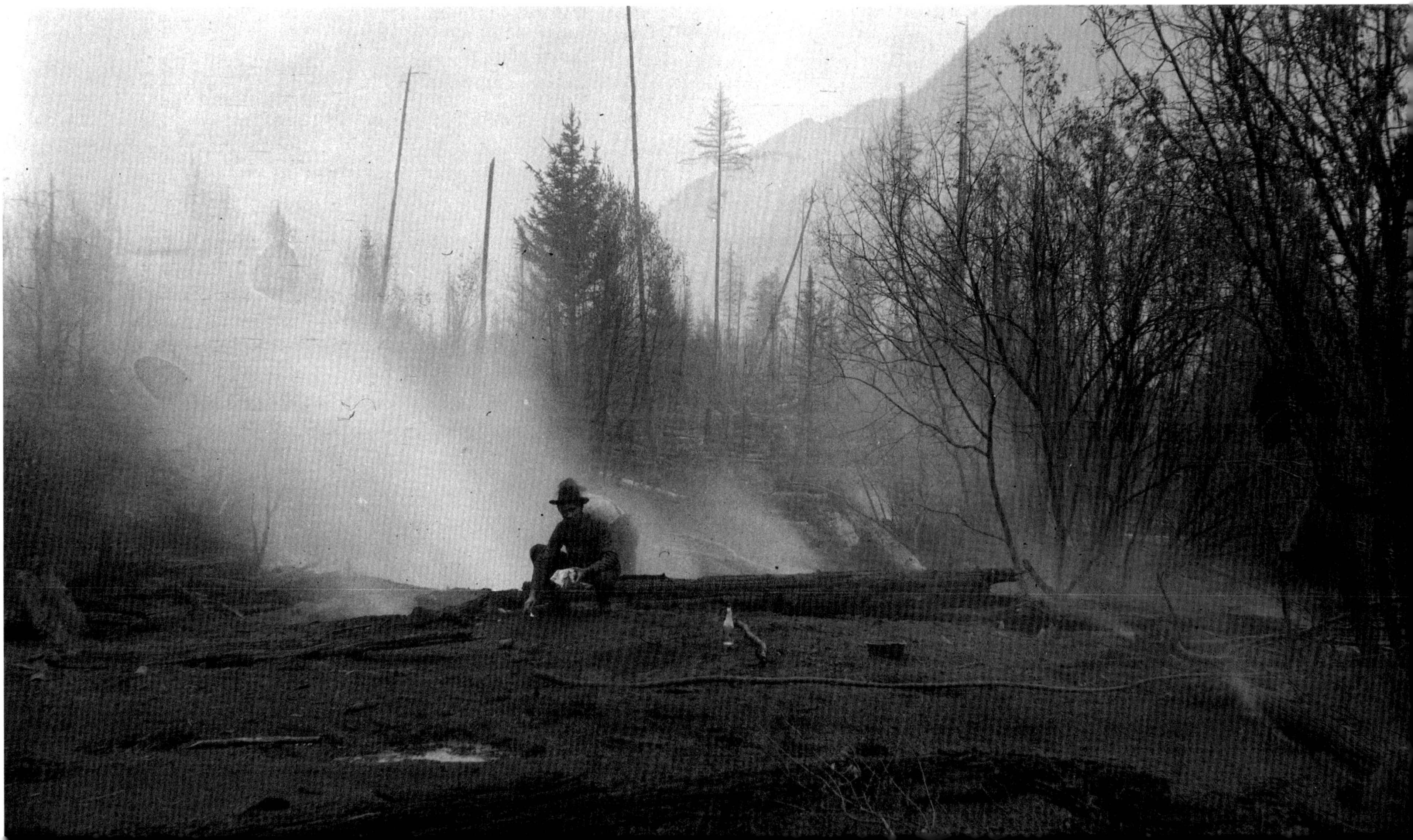

Oil in the Flathead

At the beginning of the twentieth century, the automobile began to revolutionize transportation. And with the rapidly increasing number of cars being produced came the hunt for the fossil fuel that propelled automobiles—oil.

Oil had been discovered in several places in the United States and eastern Canada, but the only well producing oil in western Canada in 1904 was at Oil City, Alberta, in present-day Waterton National Park. About 20 kilometres west, in British Columbia, there were seepages of oil at a couple of locations in the Kishinena and Sage Creek valleys of the Flathead River area. And there were also oil seepages near Kintla Lake, just a few kilometres south of the Kishinena area, in Montana. The oil seepages in British Columbia had been known and used by the First Nations people, and were examined by Alfred Selwyn, the director of the Geological Survey of Canada, in 1891. In 1903 British Columbia's provincial mineralogist visited the Kishinena and Sage Creek (originally known as Oil Creek) area and wrote a detailed report on his observations, which appeared in the 1904 government sessional papers. William Fleet Robertson noted the oil seepages and the spring near Sage Creek, but he was cautious about the potential for oil production because of the geological formations in the area. Robertson wrote, "In conclusion I would report: the actual seeping of oil in three or more places; that the existence of a body of oil underground is quite possible, though exceedingly problematic; that if such a body of oil does exist it can only be demonstrated by boring in all probability to a considerable depth and at a considerable cost."

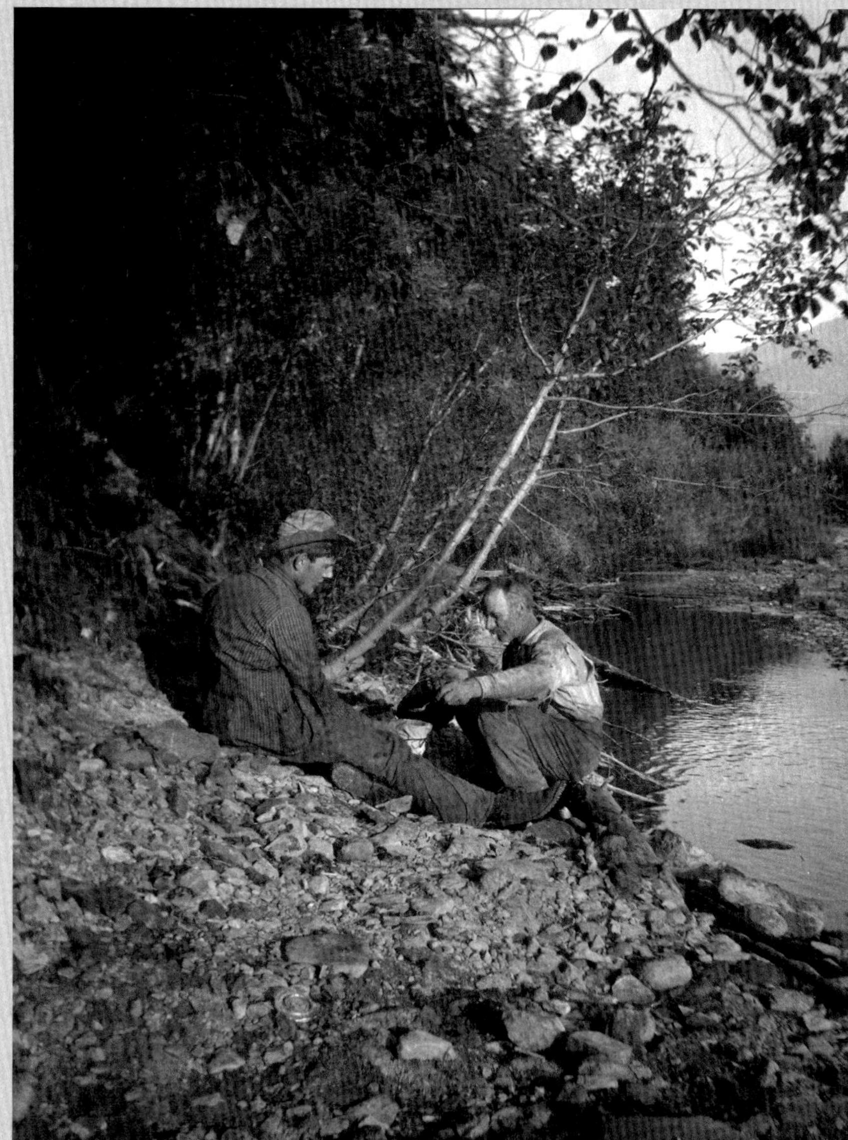

❖Skimming oil at one of the seepages on Sage Creek. John Watt is on the right. His son, Tom, who worked as a chainman on this survey, is on the left. Image I-57048 courtesy of Royal BC Museum, BC Archives.

This area was part of Lot 4593, one of the controversial lots granted to the Columbia and Western Railway. After the BC government's decision not to give this land to the railway as part of its subsidy, the chief commissioner of Lands and Works published a notice in the *BC Gazette* on June 16, 1904, that licences to prospect for coal and petroleum in this area would be accepted. The licence contained the following caution: "Owing to the number of applicants for licenses to prospect for coal and petroleum, and the peculiar circumstances surrounding the application for and issuance of these licenses, and the well known fact that the issuance has been unavoidably suspended for so many months, the Government of British Columbia finds it impossible to determine the equitable rights of the numerous applicants."

It also recognized that the mining claims in this area might result in a court case: "For the purpose of enabling all persons to go before the proper tribunal for the determination of their respective rights and priorities, this license is issued and accepted subject to such prior rights of other persons as may exist by law, and the date of this license is not to be taken or held as in any sense determining such priority."

In issuing the licence, the government stated that it "shall not be held responsible for, or in connection with any conflict which may arise with other claimants of the same ground, and that under no circumstances will licence fees be refunded."

The government added a final disclaimer. "The holder hereby waives any claim or demand against the Government, and expressly agrees not to take any steps or proceedings, or present any petitions to enforce any alleged claim or demand against the Government of the Province of British Columbia arising out of the issuance of this license."

One of the people interested in staking claims for petroleum in the Flathead was John Watt, a lumberman and prospector from Sarnia, Ontario. Watt first visited the Flathead in the fall of 1898 with a party of six that included two of his brothers. Following this trip he wrote a letter to the Department of Lands stating that each of the men had staked a section of land "about three or four miles north of the International Boundary line on a tributary of the Flathead River." They wanted to purchase this land and Watt inquired about the process of doing so.

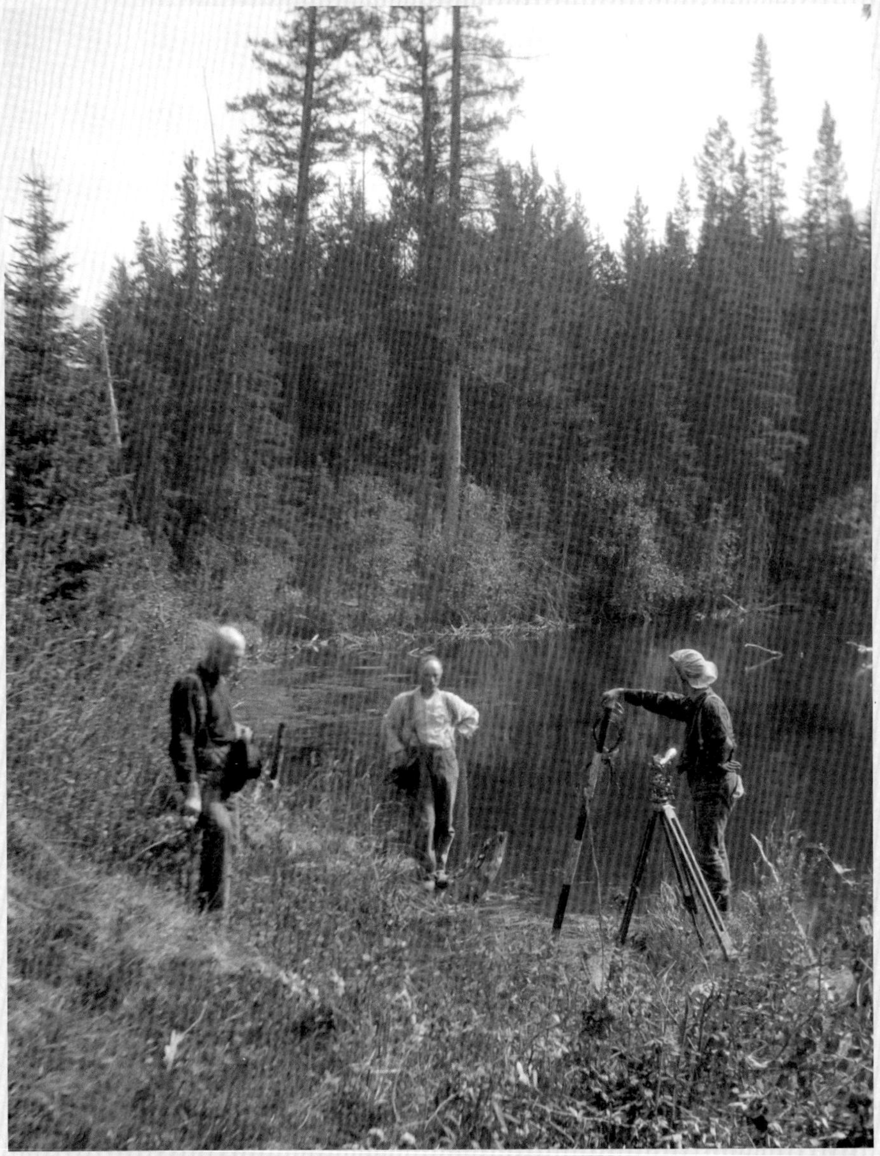

John Watt, left, is standing at one of the oil springs in the Kishinena valley. Image I-33605 courtesy of Royal BC Museum, BC Archives.

On November 1, the deputy commissioner of Lands, William S. Gore, replied that the land was "under reservation and not open to sale or settlement." Watt was not easily deterred. In late November he wrote to Gore that their party had found oil in three places, gathered some samples and had it analyzed by a professor at the University of Toronto, and that he would be sending a copy as soon as he received the report. Watt told Gore, "Now, we have found those indications of coal oil and located on the property in good faith; the land is worthless unless oil can be produced in paying quantities; what we ask is that we be allowed the privilege of further prospecting and developing the same."

In a letter dated December 6, Gore reiterated the government's position. "I beg to inform you that in pursuance of the reservation over lands in the southeast portion of the Province, the Government have, in all cases (and there have been several) declined to grant licenses to prospect for and work coal or petroleum. I regret to say that, as long as the reservation exists, applications from yourself or other persons cannot be favourably entertained."

In the fall of 1899 Watt wrote once more to Gore, enquiring whether the reservation was still in place. When the deputy minister replied that there had been no change to the status of the land in the Flathead, Watt wrote to James Smart, the federal deputy minister of the Interior. "It is rumoured that this land was reserved for the British Columbia and Southern R.R. This land is not surveyed, and as I have squatted on this land and want to perfect my title, I would like very much if you could give me the necessary information how to proceed, whether I will have to apply to the Do-

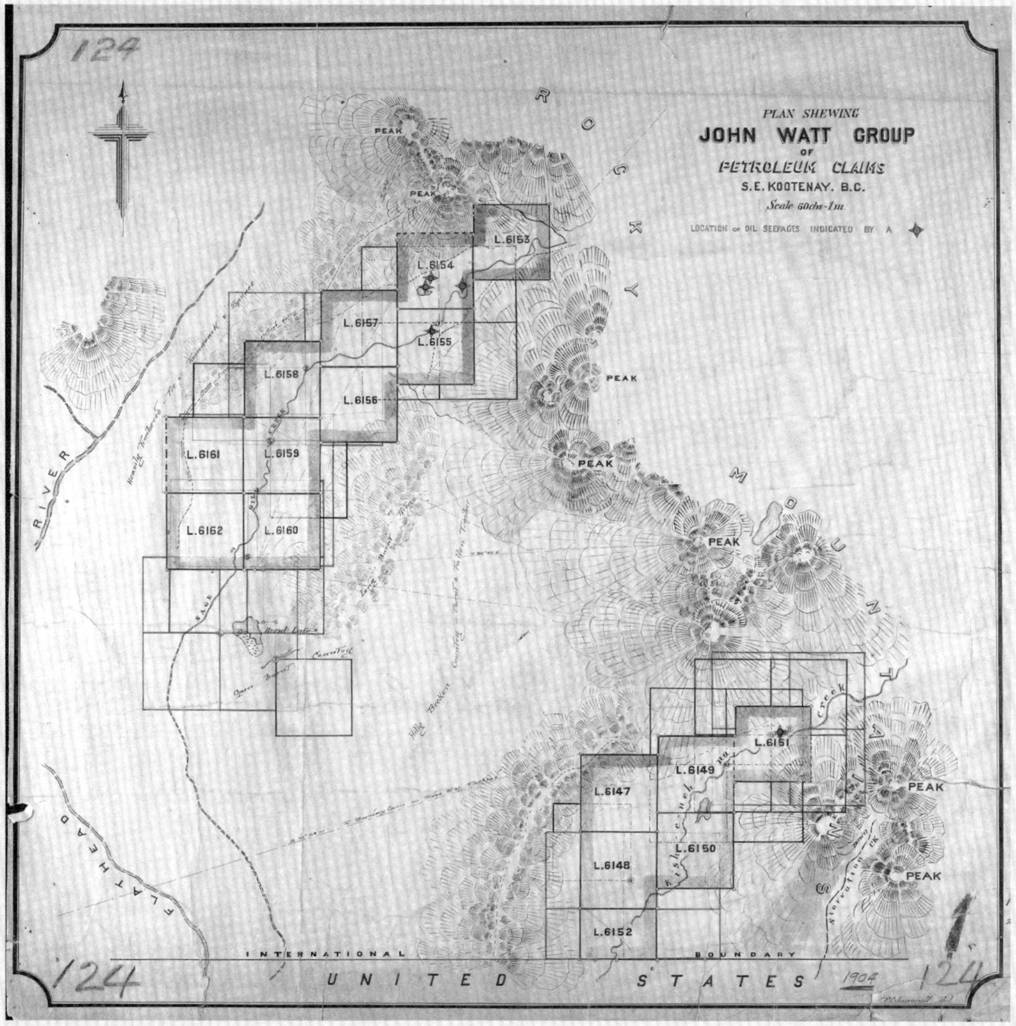

❖ Swannell made this map for John Watt to show the oil claims that he surveyed in the Kishinena and Sage Creek valleys. Many of these claims were made by friends and family of John Watt. Map CM/B/14118 courtesy of Royal BC Museum, BC Archives.

minion or British Columbia offices. I might just say that I have found strong indications of petroleum on this land and want to secure the privilege of further prospecting and developing it, if it can be procured in paying quantities." The Department of the Interior reminded Watt that this was provincial land and sent his enquiry to Gore, who once again told Watt that the lands were under reservation. Despite this Watt staked a series of claims in the Flathead in July 1900.

Although they lost the land subsidy for Lot 4593, the CPR was still interested in mineral claims in the Flathead, so the company made arrangements with John Watt to finance him. Two days after the BC government lifted the reservation on mineral claims in the Flathead, Watt applied for licences to prospect for petroleum. Since Gore & McGregor had previously surveyed for the CPR they also received the contract to survey John Watt's mineral claims. The connection between Thomas Gore and his brother William was probably important also. The company sent Swannell to the Flathead to survey Watt's mineral claims.

The lifting of the reservation in the Flathead produced a flurry of activity in this remote valley in 1904. There were several oil and coal claims staked; John McLatchie surveyed the east boundary of Lot 4589; a wagon road was constructed from Fernie through the Flathead down to the international boundary, with the CPR helping to pay for it. In late April the *Fort Steele Prospector* speculated that "unless they [the CPR] feel reasonably certain that at least part of their claim to the coveted coal and oil lands will be eventually allowed by the McBride government they would hardly undertake to join in road making."

As the government anticipated, there were overlapping and conflicting claims in the mineral licences in the Flathead valley and the issue went to the BC Supreme Court for settlement. Five groups had claims for mineral licences. One of these included Charles Hibbert Tupper, a lawyer and the son of former prime minister Charles Tupper. He was also the lawyer for this group. Other members of this group included Robert Gilmour Leckie, a well-known mining engineer and military man, and James Baker, a former MLA for the East Kootenay district, who was a partner in two different groups of claimants, even though he had retired to England.

Watt and his group won the initial ruling regarding his claim, but in 1906 they lost in an appeal made by Tupper. Although Watt was the original locator who had a valid claim, the judge believed that the evidence showed that Watt's group did not stake their initial claims properly. Watt was not ready to see his dreams of striking oil in commercial quantity in the Flathead die, so he appealed the 1906 judgement. In a Supreme Court case in 1907, in which Swannell testified, Watt again lost his bid to have his mineral claims accepted.

Still, he was not finished. Watt now took his cause to the McBride government, and in 1911 he applied to Cabinet seeking compensation for his time and expenses. When Watt became frustrated and afraid that he would not receive any remuneration, he hinted at death threats against the premier. This prompted McBride to hire some Pinkerton detective agents, who came from Seattle and spent about a month investigating Watt. In his report on February 14, an agent described the man: "Watts, being an old man of about 70 years of age, is very wroth over losing these lands and it would seem that the matter has preyed on his mind until it has affected it somewhat, and he has made threats to shoot or kill Hon. Richard McBride, whom it is presumed, he holds responsible for the loss of the lands." The agent spent a few weeks following Watt and spoke with him a few times. He reported that Watt still had a couple of bottles of oil from the Flathead claims. Watt met Swannell twice on February 28 before a meeting with the provincial Cabinet the following day. Since this was the last day of the session of the legislature, the Cabinet did not take any action. Swannell met with Watt two more times in the following week and attended the Cabinet meeting on March 8. But the executive branch decided not to make a decision regarding Watt's claims at that time. By now the agent believed that Watt's threats were not serious and that he was more interested in vindication of his cause, and McBride terminated the work of the Pinkerton detective.

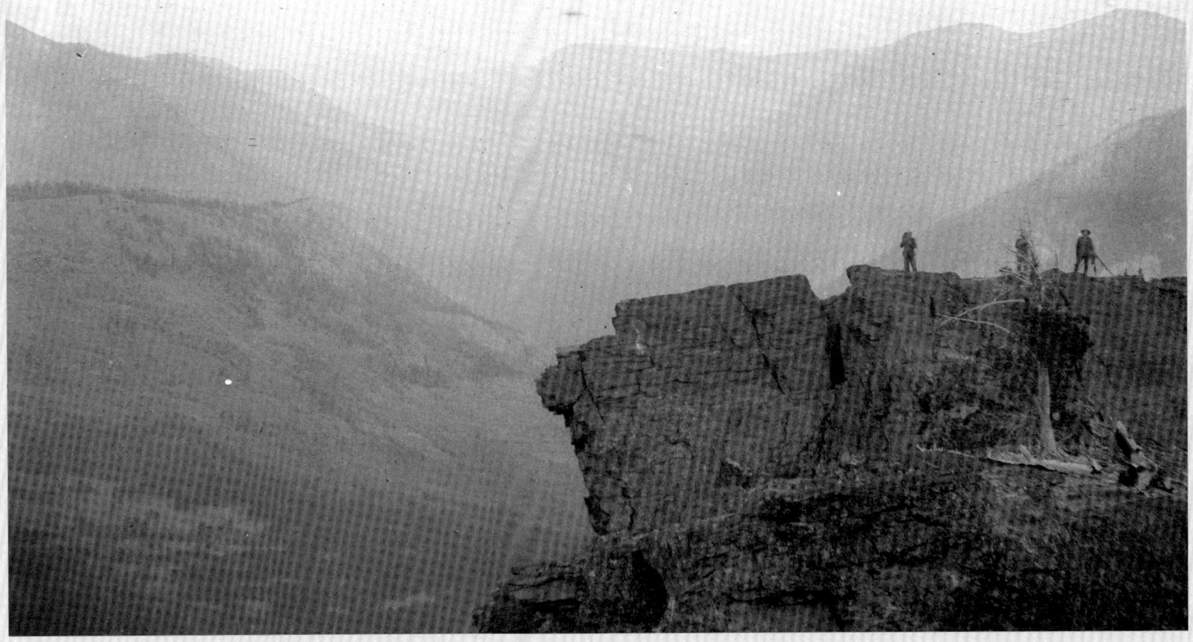

❖ At the northwest corner of Lot 6151, 900 metres above the Kishinena valley. Image I-57061 courtesy of Royal BC Museum, BC Archives.

Although Watt never regained his claims, he eventually received compensation. On February 26, 1913, the government passed Resolution 254A: "Resolved, that a sum not exceeding $30,000 be granted to His Majesty to defray the expenses of miscellaneous allowance in settlement of coal and petroleum licenses issued to John Watt in S.E. Kootenay."

McBride told the legislative assembly, "John Watt was a pioneer locator of coal and petroleum claims in South East Kootenay and these were subsequently taken by a company." He explained the reason for the government's decision to pay Watt. "Ultimately all the facts were reviewed before the Executive of the Government at great length by Watt and those appearing on his behalf. There was no possible way of having the case taken to a higher tribunal, and as Watt had shown, to our satisfaction at any rate, that he was the original locator and was entitled to the property, we, after a most careful review of all the facts, concluded that as an act of grace the Government ought to compensate him to the extent of $30,000. This was in the general public interest, and it was an action we felt we could readily justify."

A couple of wells were drilled at Sage Creek and in the Akamina valley and they found oil, but like those in nearby Oil City, these wells never produced enough to be commercially viable. In the summer of 1930, Swannell's son, Lorne, worked on a BC Forest Service inventory survey in the Flathead, and he visited and photographed a derrick that was operating at Sage Creek at that time.

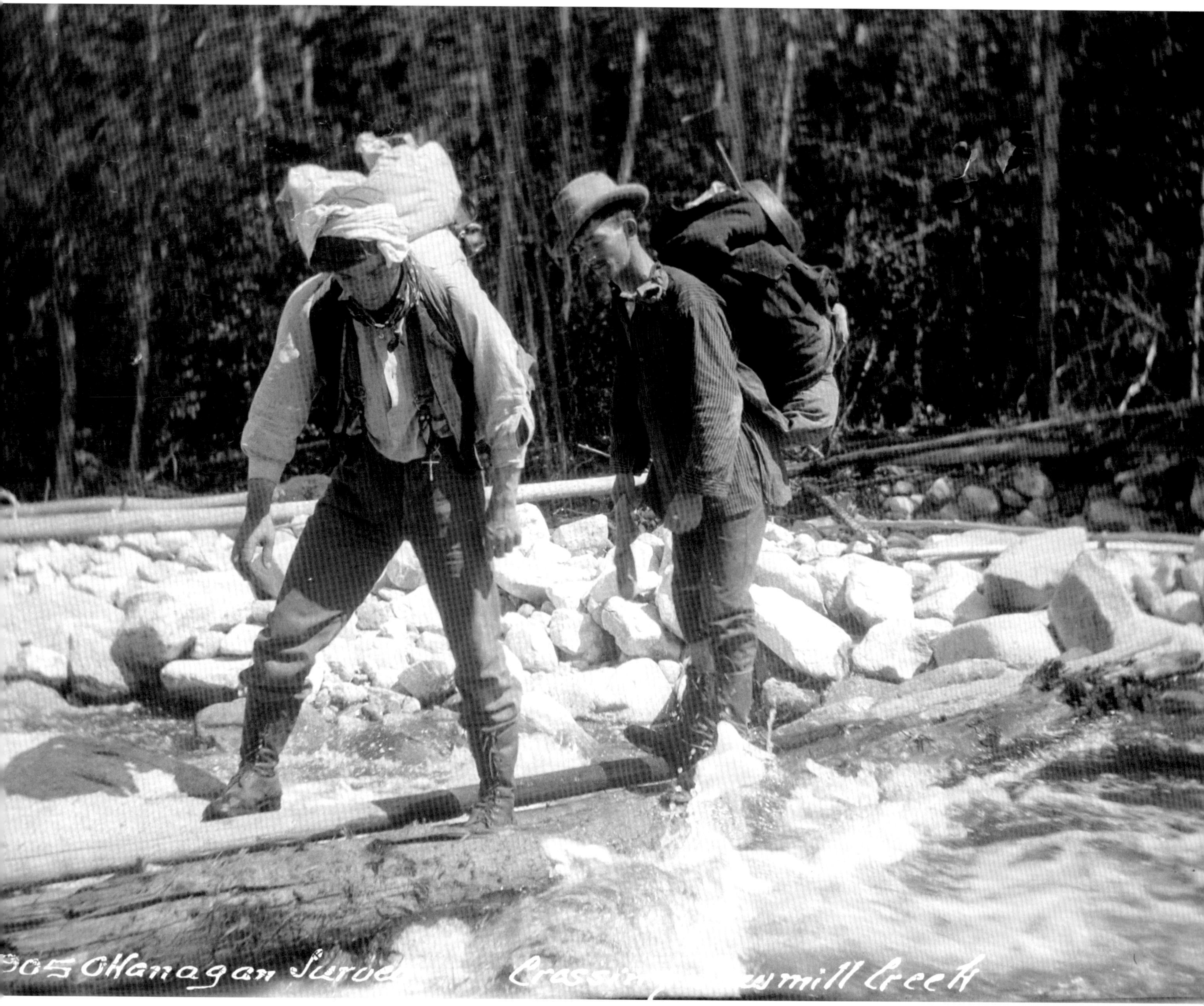

❖Crossing Sawmill Creek. Frank Swannell is on the left and George Copley is on the right. Image I-58397 courtesy of Royal BC Museum, BC Archives.

1905

In April Swannell made a small but significant survey in Victoria for Gore & McGregor.

After the cancellation of the land grants of Lots 4593 and 4594 in Crowsnest Pass, the CPR had to select land in the boundary area for the remaining portion of the Columbia and Western Railway land grant. Thirteen lots were chosen in 1904 for the "Columbia and Western land grant deficiency." Unlike the previous land grants to the CPR in southern BC, these lots would not be conveyed to the railway until they were actually surveyed, so the CPR needed this work done in 1905. The lot pattern was a checkerboard like the BC Southern land in the Kootenays, but these lots were smaller and very few of them directly adjoined. This meant that there was a considerably larger amount of surveying in proportion to the amount of land being received. In addition to having the surveys completed quickly, the CPR wanted them done as cheaply as possible. The company thought that this land would not be as difficult to survey as the Kootenay lots, so it decided to send out bid tenders to several surveyors and surveying companies.

On March 8, the CPR's BC land commissioner wrote to Gore & McGregor. "We desire to put in hand at once the survey of Columbia and Western land grant blocks. Will you kindly let me have statement as to contract price at which you would be prepared to undertake this work. The survey will be made on practically the same basis as that carried on by you in locating boundaries in British Columbia Southern blocks, and I would like your statement as to price based on price per mile for traverse and per mile on block boundaries run." McGregor replied to J.S. Dennis that they would do the surveys at the same rate as they had done in 1903. McGregor was surprised and disappointed at the response of William Pearce, the CPR's chief surveyor, on April 28. "I regret that your tender is so much higher in some cases than my estimate that no other course appears open to me but to wait until other tenders are received before recommending the allotment to any." J.H. McGregor met Dennis in Calgary and T.S. Gore wrote to William Pearce, but with limited success.

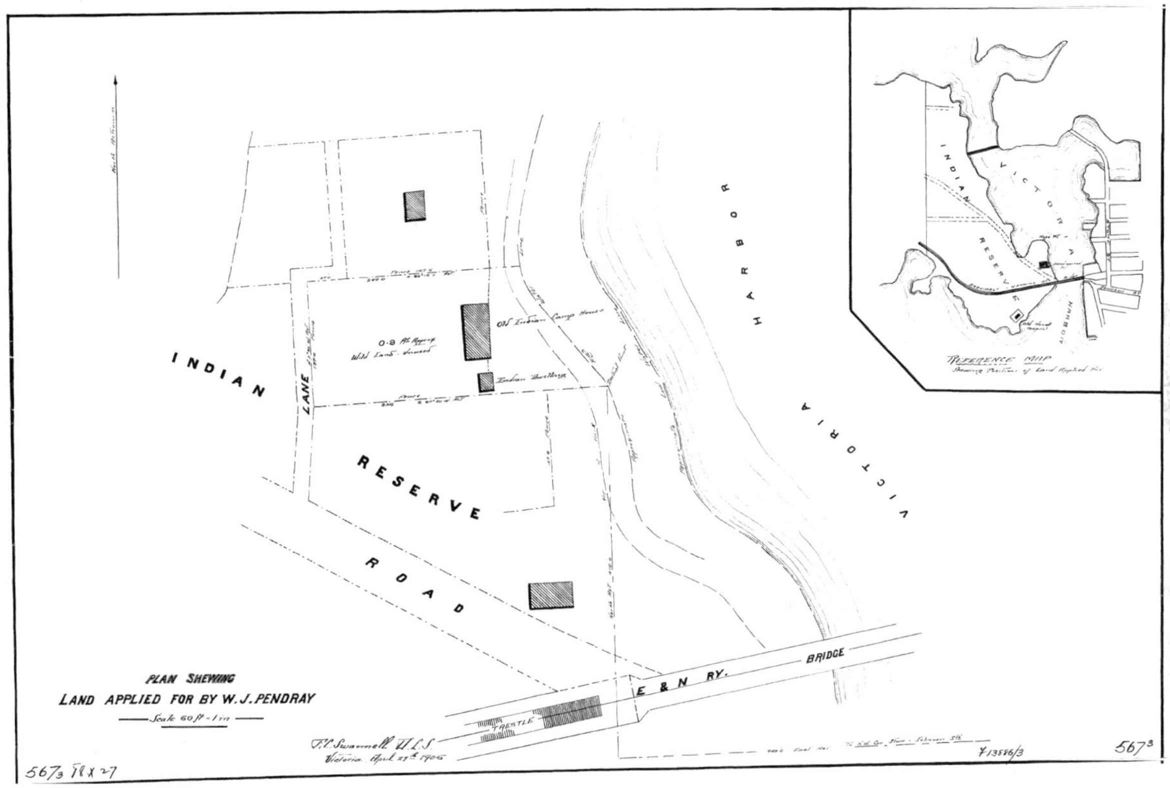

Swannell's map shows the First Nations dwellings that existed on this portion of the Songhees Reserve in 1905. Map T567 courtesy National Resources Canada.

William J. Pendray and the Songhees Reserve

William J. Pendray, founder of BC Soap Works and the British American Paint Company, which had been in Victoria for many years, needed to move his businesses since the Canadian Pacific Railway had purchased and planned to occupy the property that he was renting.

In March 1905, Pendray wrote to the Victoria city council requesting assistance. He stated, "For our new premises we will require at least one and a half acres of land, near the railways and shipping points. At present a site on the Indian reserve appears the most suitable property for our purposes in the city, and we now submit for your consideration the proposition of securing for us free, a site for our factories." In addition to council's assistance with the land, Pendray stated that "it will be necessary for us to be exempt from taxation for twenty-five years, and to be supplied with free water for the same period of time."

The Songhees reserve was located on their traditional territory in the Inner Harbour area of downtown Victoria. The provincial government had been attempting to move the Songhees to a new reserve away from the centre of Victoria for several years. In 1901 Premier James Dunsmuir proposed to the federal government (who had responsibility for Indian reserves) that they surrender two parcels of the Songhees land to the provincial government, sell the remainder and, after moving the First Nations people to a new reserve, give any remaining money to the province. The federal government agreed, but there was difficulty finding a location suitable for the Songhees, since their first choice, traditional territory on Hudson's Bay Company property at Cadboro Bay, was not available.

In March 1905, Premier Richard McBride declared, "It was clear that the Indians must retain possession of the Reserve until they were made party to an arrangement satisfactory to them. The Indians had been taught to believe that it would be foolish for them to yield their tenure without a considerable indemnity and generous terms. They could not be removed by force. The first essential was to get the Indians to acquiesce in an agreement." Victoria city council rejected Pendray's request for assistance. Despite this, Pendray continued with his request for a lease of a portion of the Songhees land, and Swannell made the survey in April.

The *British Colonist* reported about a meeting held on May 19 regarding Pendray's application.

> A large deputation of leading citizens, headed by His Worship, Mayor Barnard, Mr. S.J. Pitts, President of the Board of Trade, and Mr. R.B. McMicking, President of the Liberal Association, yesterday waited upon Mr. A.W. Vowell, Superintendent of Indian Affairs [for British Columbia], with the purpose of laying before him the matter of the application of Messrs. Pendray & Sons for a lease of a portion of the Songhees Indian Reserve on the area lying above the E & N Railway bridge, as a site for the reconstructed paint and soap works of the concern mentioned, which will shortly be required to remove from Humboldt Street.

The newspaper noted that Vowell had a copy of Swannell's survey plan for Pendray's request. In his response Vowell noted some difficulties with the application.

> The Indians, he said, were in a suspicious mood in respect to any attempts to settle the reserve difficulty. They were watching the newspapers carefully and were timorous about consenting to any arrangement which might be construed as implying a loss of their interests; and this feeling was

> accentuated by the meddlesome tactics of outsiders, who on more than one occasion, had needlessly alarmed the Indians.
>
> Then again, they viewed with disfavour an arrangement which did not imply to them any pecuniary sense. The granting of a lease for which a rental would be paid would under the present system of administering the reserve by the Indian Department result in the money being paid into the Treasury Department in trust for the Indians who would receive no direct payment. Under this condition of the affairs the Indians now living on the reserve approached the matter at present under discussion in a somewhat indifferent spirit. They feel that they should receive some direct pecuniary benefit.

The *British Colonist* reported on another meeting at the end of May.

> The Indians on the Songhees reserve have turned their faces against the proposition to lease a portion of their holdings to Messrs. Pendray & Son for factory purposes.
>
> This was made clear at a conference had with them by Indian Agent Robertson of Duncan and Mr. Pendray.
>
> They refuse to discuss the question of a grant of land in any shape or form, claiming that they have fully discussed the question amongst themselves and unanimously deciding on taking this attitude.

After this meeting Pendray found a location for his two companies at Laurel Point, not far from his residence. In 1910 the Songhees signed agreements in principle to move to the Esquimalt area, which they did the following year. J.H. McGregor surveyed the new Esquimalt reserve in 1911.

On May 8 Pearce wrote to Gore.

> I would say that all the contracts have been awarded except [Lots] 3635, 3636 and 2705. Though there was and is every disposition on the part of Mr. Dennis and self to award your firm the contracts if the same nearly approached, what it was estimated the surveys should cost, or what others would perform them for, your figures were so far above both, that there was no alternative but to follow the course that has been. I may add that the contract has been awarded on terms, which in proportion to your offer, are as $1.00 to $1.98…
>
> When I met Mr. McGregor, he advised me that he had been on the Okanagan Lake and had been up the north fork of the Kettle River some miles above Grand Forks. I therefore concluded that he should have a good knowledge of the country to be surveyed. I also advised him what I thought the country was like. When I received his tender, I was astonished at the prices tendered.

Gore & McGregor later received part of two additional lots, 2709 and 2710, to survey. The company chose Swannell to do the Columbia and Western survey while McGregor went to northern BC. The instructions sent from the CPR on May 26 stated, "It is desired unless it will interfere with your plans that you survey the westerly limit of 2709 before you do the easterly one of 2710. The object sought is to enable the contractor for 2711 to ascertain where the north limit of 2711 is, it being a fixed distance from the International Boundary."

Another surveyor working on the land grants was John McLatchie. While Swannell was surveying parts of Lots 2709 and 2710, McLatchie was instructed

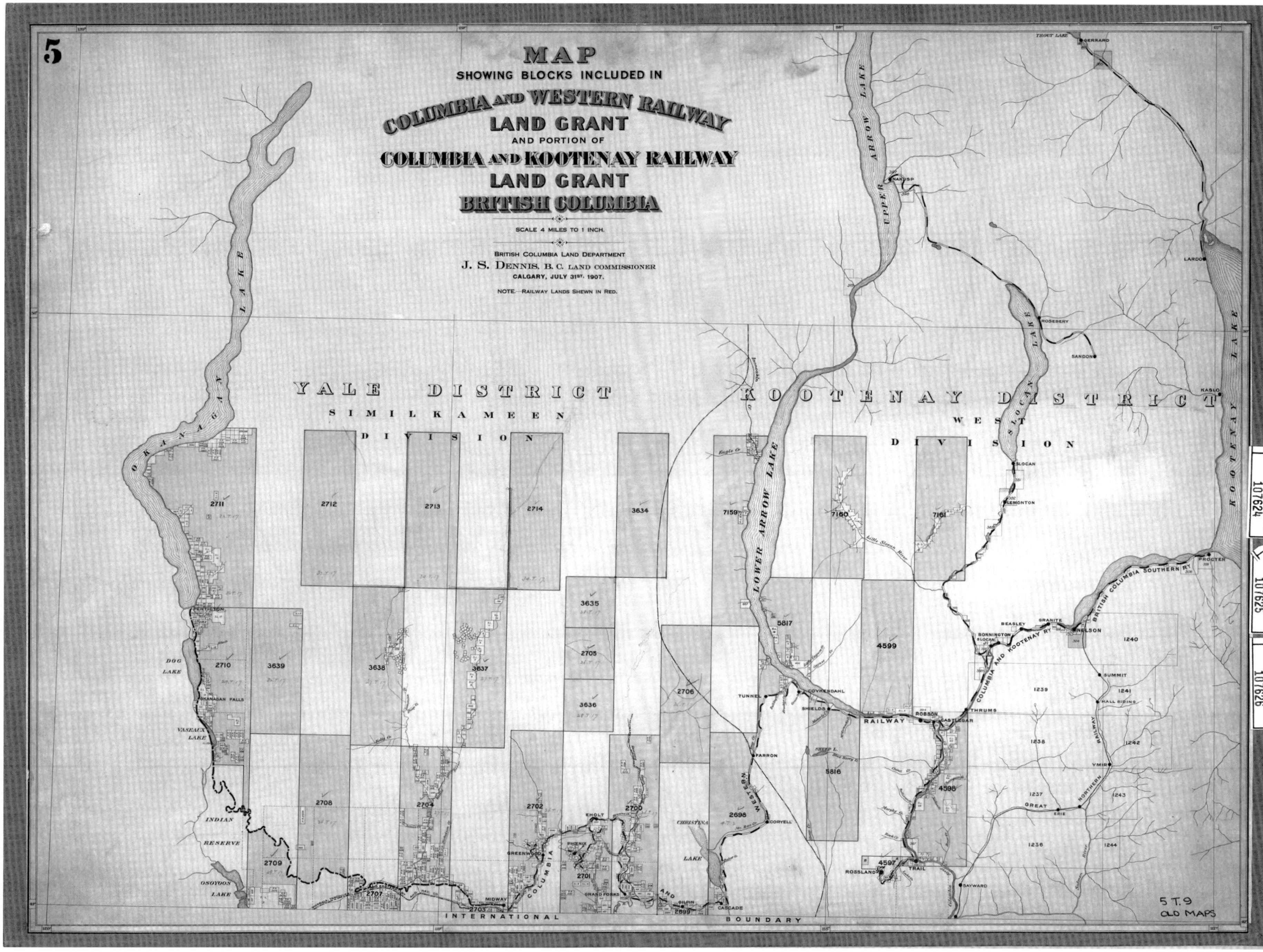

❖ Columbia and Western Railway land grant. The 1907 map (in comparison to the one produced in 1902 on pages 46–47) shows the additional blocks of land that the Columbia and Western received after the government officially cancelled the grants of Lots 4593 and 4594. The lots that the C & W did not survey were not relinquished until after the map was printed. One can see that Lot 2700 and Lot 3636 are not directly connected as the CPR had originally intended. There are other lots (like 2702 and 3637) that are also unconnected. Map 5TR9 old maps. Courtesy Land Title Survey Authority.

to make a survey to establish where the westerly limit of Lot 2700 intersected the railway right-of-way and how far north of the southeast corner of the block this point was located. From that point Swannell would survey the western boundary of Lot 2700 to the northwest corner of the lot, which was also the southeast corner of Lot 3636. There he would survey three adjoining lots to the north, one of the few places in the Columbia and Western land grant area that had a series of adjoining lots.

Swannell's season started in June. He had five men on his crew, including his good friend George Copley, who surveyed with him for the first time. Copley would later be Swannell's assistant from 1909 to 1914 in northern British Columbia. The cook was Nep Yuen, a Chinese man from Victoria who had worked on the construction of the original CPR railway and had been a cook on several field camps in remote areas of the province. Nep Yuen would also be a member of Swannell's survey crews in the northern part of the province from 1911 to 1914.

Swannell and his crew took the CPR from Vancouver to Sicamous. Then they travelled on the Shuswap & Okanagan rail line to Okanagan Landing at the head of Okanagan Lake, where they boarded a steamboat to Penticton. From Penticton they travelled by land to their initial survey point near Camp McKinney, a small, virtually abandoned mining community northeast of Osoyoos along the slopes of Mount Baldy. In a report to the surveyor general, Swannell noted that "the area available for cultivation is small, owing to scarcity of water, and difficulty of getting it on the land."

The work began with the two lot surveys northeast of Osoyoos. Swannell surveyed the west boundary of Lot 2709 south. The crew crossed Deep Creek, which Swannell described as "a deep box canyon, with perpendicular rock walls, 500 ft. [150 m] high. This canyon, which commences where the stream debouches into the Okanagan Valley, runs up without a break to a fork of the stream several miles above where the survey line crosses." Swannell again noted that a lack of water precluded cultivation, but he observed that it would form a good cattle range. He commented, "Many bands of wild horses were noticed here during the progress of the survey." About 12 kilometres along the survey, "the mountains descend precipitously to the valley of Osoyoos Lake—a sheer drop of over a quarter of a mile [over 400 m]." From here the survey line followed rolling country to the international boundary.

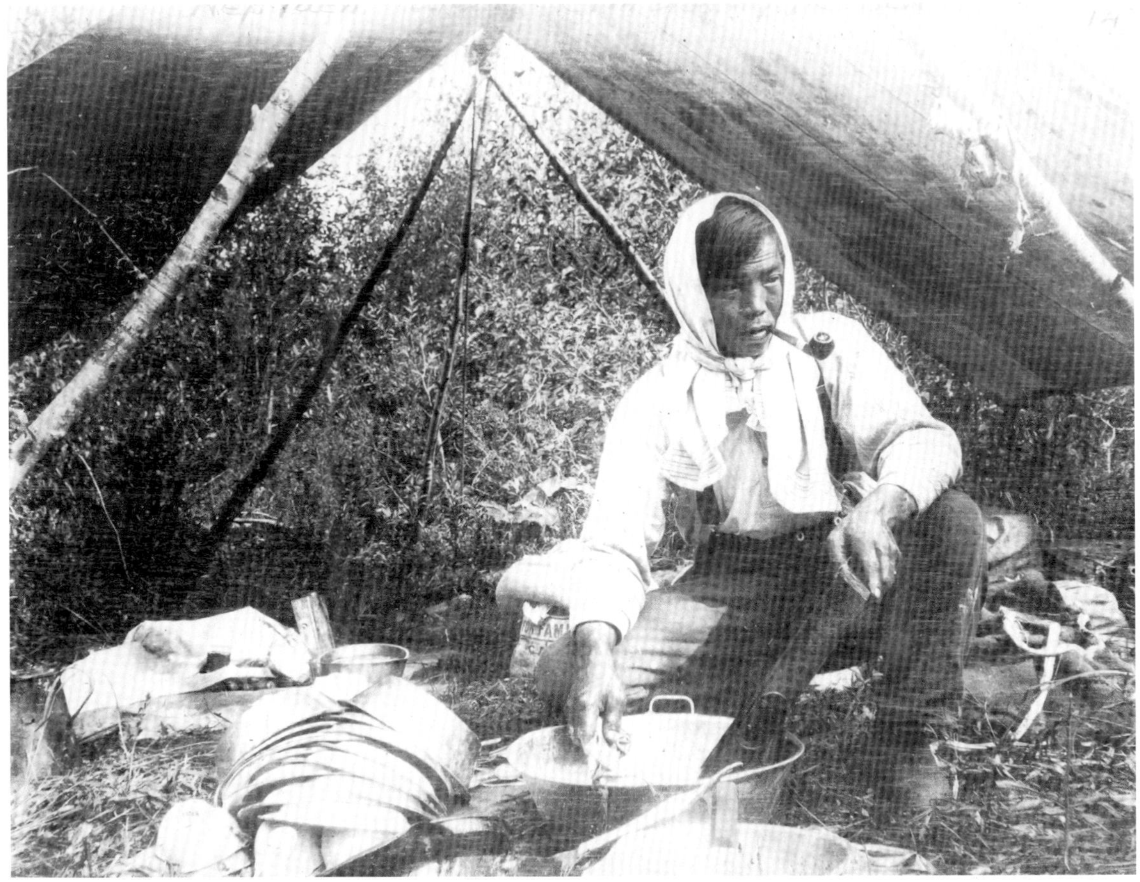

❖Nep Yuen making bread. Image MS-0392 vol 50, p 5 courtesy of Royal BC Museum, BC Archives.

The crew's next survey in the area was the east boundary of Lot 2710. This survey line started on land east of Oliver owned by J.P. McCuddy (who also operated a general store, first at Camp McKinney and later at Fairview) and headed north, reaching its highest point at 6.4 kilometres. From this location there was a descent of 600 metres to Sawmill Creek. The survey line then went through terrain that had a large amount of timber and bunchgrass. North of Shuttleworth Creek the survey line crossed a series

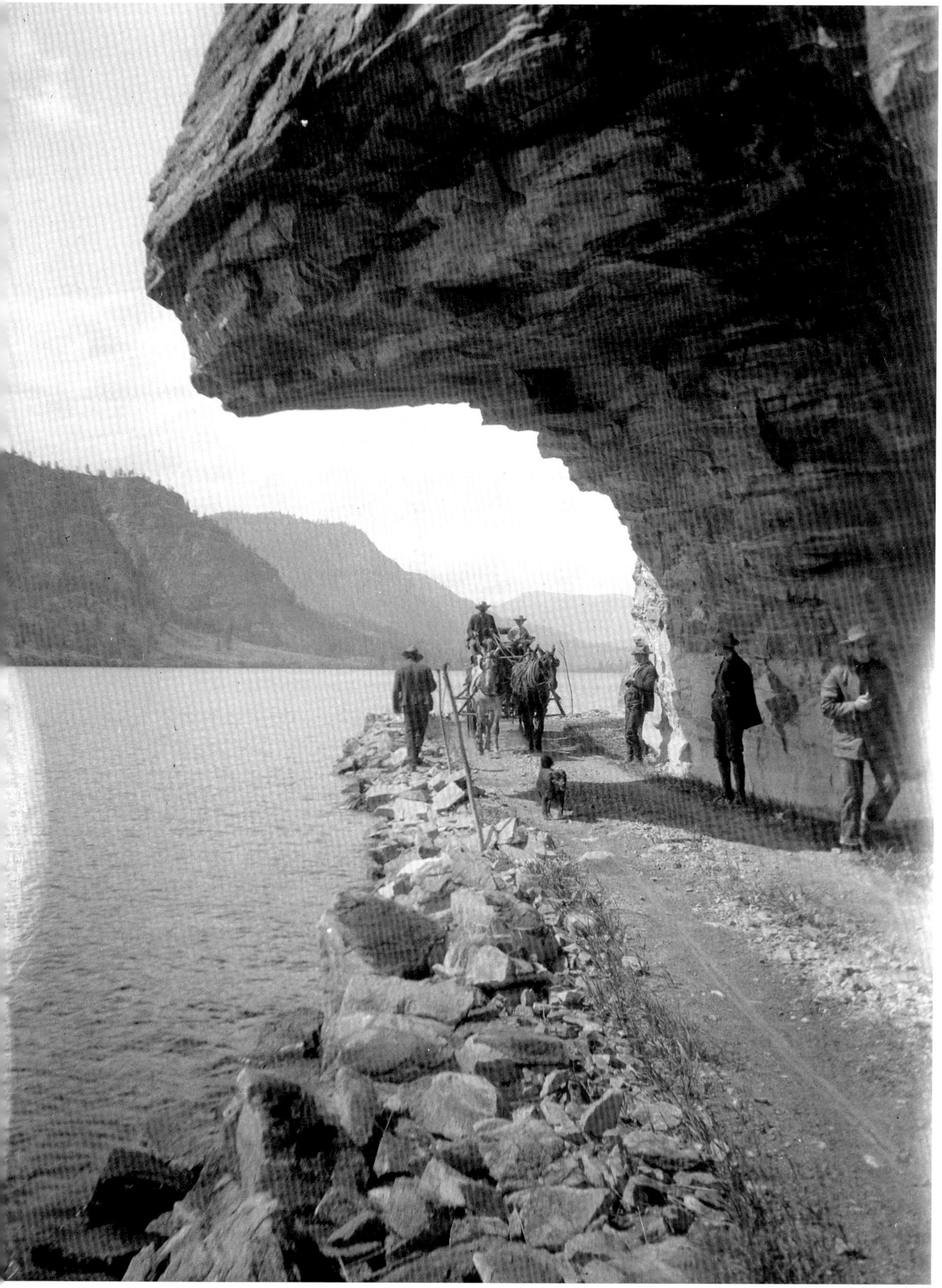

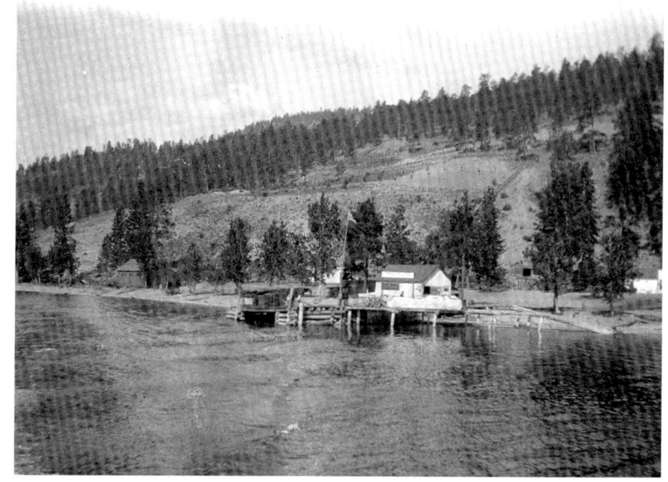

of benches until the 17 mile [27 km] post, where there was a long, steep descent to Ellis Creek. Swannell observed that "the whole southern slope of this valley is burnt as far as visible to the east, and for a couple of miles west."

After completing these surveys, Swannell and the crew returned to the CPR mainline at Sicamous and travelled east to Revelstoke. Then they went south to Arrowhead, where they boarded a sternwheeler that took them down the Arrow Lakes to the Columbia and Western Railway. They travelled west to Eholt, a divisional point on the railway.

In July and August, Swannell and his crew surveyed three lots (2705, 3635, and 3636) in the North Fork of the Kettle River (today known as the Granby River) drainage to the north of the mining town of Eholt. This area was not too far from where Swannell had worked in 1899 on the Columbia and Western Railway right-of-way. All three lots were six miles, or 9.6 kilometres, wide from east to west. Lots 3636 and 2705 were six miles from north to south, while Lot 3635 was 5.5 miles (8.8 kilometres).

The Columbia and Western Railway land grant was done in a hurry and not surveyed methodically, which caused some problems with the work. When Swannell arrived at the starting point, according to the CPR's instructions for this portion of the survey, he found that McLatchie had not yet established a

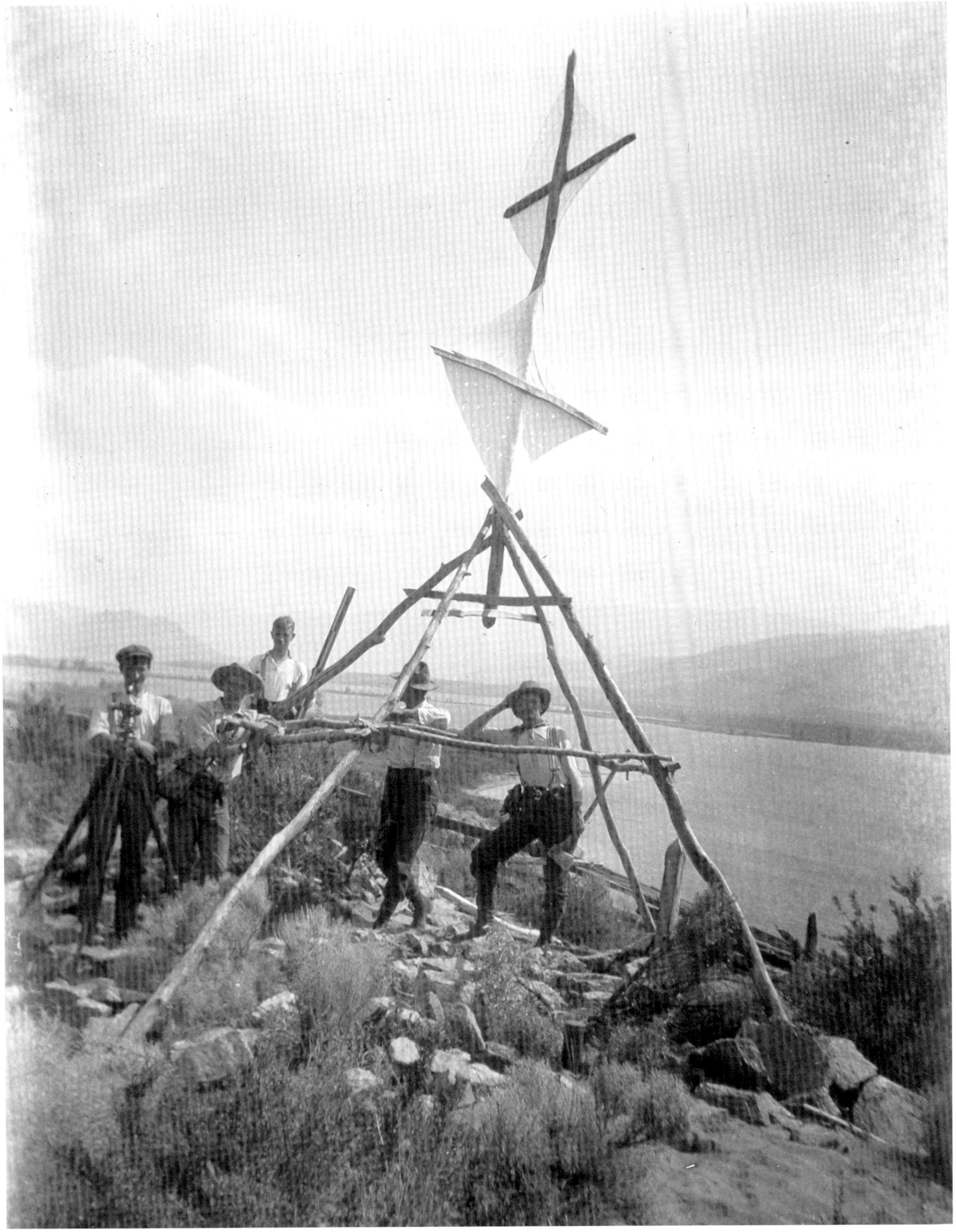

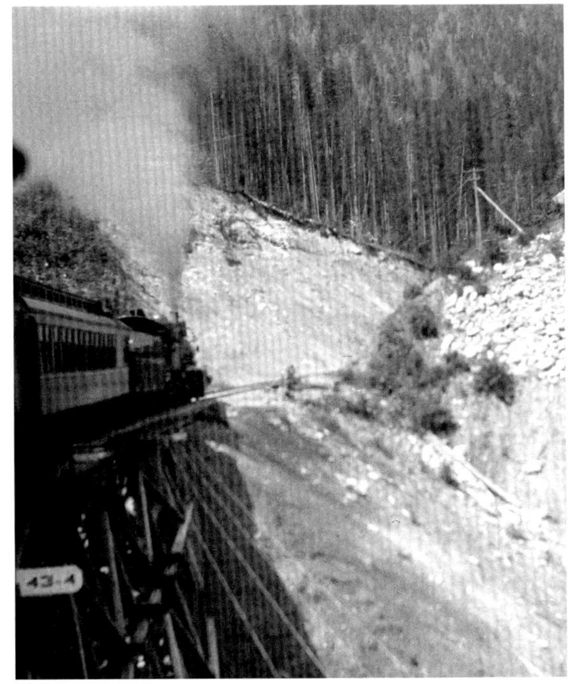

❖ On the notes for this photograph Swannell wrote that he took this train while travelling along the "Columbia and Western Railway near big [Bulldog] tunnel." Image I-57069 courtesy of Royal BC Museum, BC Archives.

❖ Survey marker above Osoyoos Lake. Swannell connected his survey of Lot 2709 with a US Geological Survey station at the international border above Osoyoos Lake. Image H-04701 courtesy of Royal BC Museum, BC Archives.

Opoosite Page

❖ Left: Okanagan survey party at Dog (Vaseux) Lake. This rock outcrop is a prominent location along Vaseux Lake. Image I-33611 courtesy of Royal BC Museum, BC Archives. For 2014 image see page 152.

❖ Right: Landing at Peachland. The sternwheelers travelling on Okanagan Lake regularly stopped at Peachland. Image I-33610 courtesy of Royal BC Museum, BC Archives.

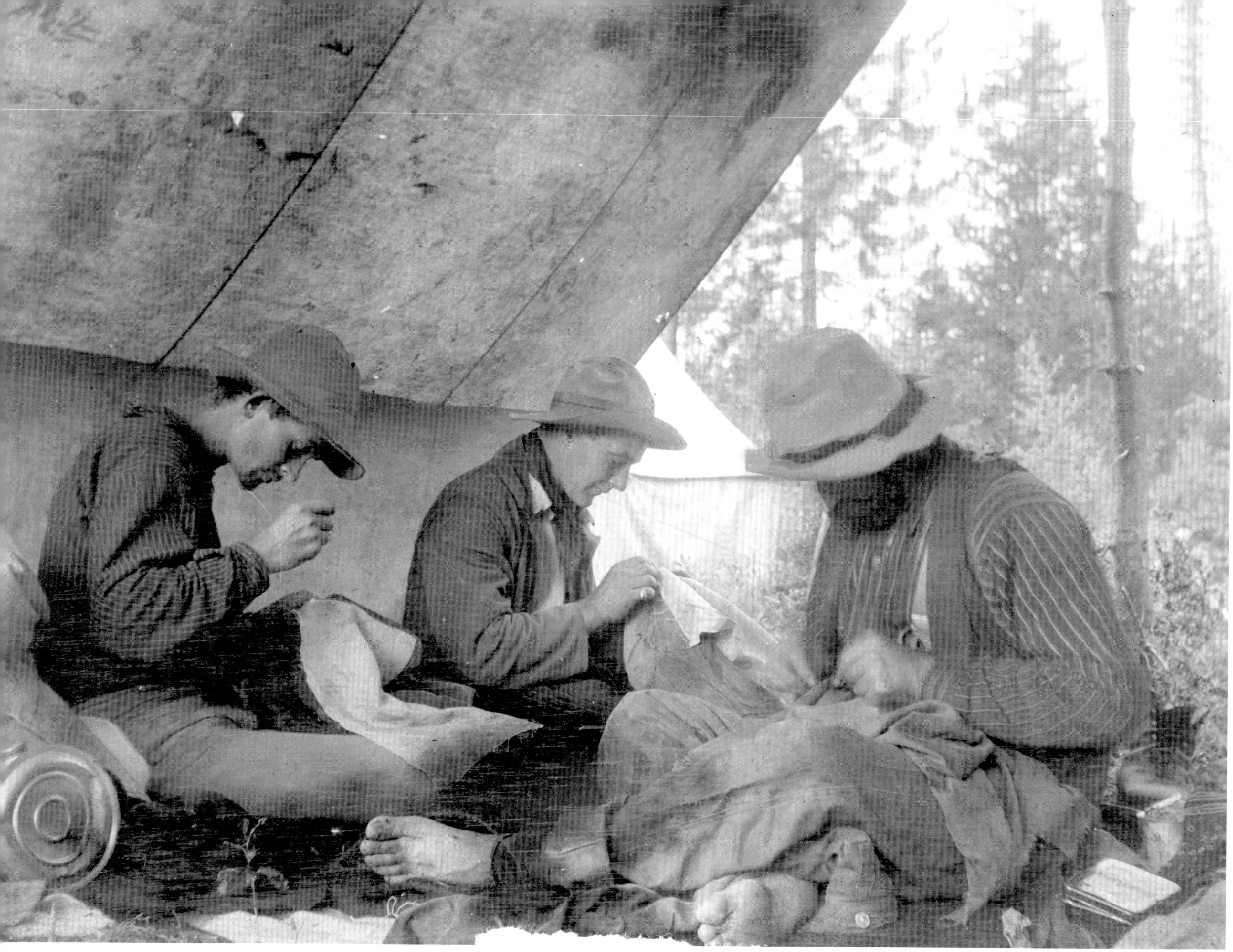

❖ Survey crew in camp at the Kettle River. The men are mending their clothes and shoes. Image I-33613 courtesy of Royal BC Museum, BC Archives.

surveying station there. Swannell decided to start his survey anyway. William Pearce explained the situation in a memorandum to J.S. Dennis. "Messrs. Gore & McGregor were written to and told in such case to establish 3635, 2705 and 3636 from the southeast angle of 2702. Two courses to be followed in so establishing presented themselves. One to run up the east boundary of 2702. The other to traverse 6 miles [9.6 km] east and then run north, assuming that the west limit of 2700 was 6 miles east of 2702, as was intended. They were urged to adopt the first course but as under their contract they had the privilege of adopting the second they elected to take it." Unfortunately Gore & McGregor made the wrong choice.

McLatchie arrived soon after Swannell started his survey. He had been delayed because there were errors

in both the lot surveys and his own work, and he wanted to make sure that everything was correct before establishing the point. In a letter to William Pearce on July 31, McLatchie wrote that he discovered that Lot 2700, which was supposed to be 6 miles (9.6 kilometres) in width, was actually 5 miles, 2109 feet (8.6 kilometres), while Lot 2701, which was also supposed to be 6 miles, had an additional width of 429 feet (130 metres). He asked Pearce, "What are you going to do with the boundary that Mr. Swannell is running? It seems to me that it will make a mess of the whole of his survey." If Swannell had surveyed up the west boundary of Lot 2702 he would have arrived at the actual southwest corner of Lot 3636.

However, the error was caught almost immediately and the solution was simple. Pearce noted, "Under the ruling of the Department of Lands and Works arising out of the dispute over the east boundary of 4590 at Sand and Lizard Creeks any mistake, if it passes the Department of Land and Works, must stand. Therefore, in spite of the mistake of 429 feet too much in easting the boundaries of 3635, 2705 and 3636 as run by Messrs. Gore & McGregor [the error] must stand once the returns are approved by the Lands and Works Department." This was before the 1914 court case that overturned this decision. For the Columbia and Western land grant the lot was being surveyed before the Crown grant was given, so the survey on the ground would take precedence once it was approved. In addition, the changed location did not affect the size of the lots nor any other properties, since these lots were surrounded by vacant Crown land.

In Gore and McGregor's contract Swannell was only to survey the outer boundary of the three lots. The east-west lines dividing Lots 3636 and 2705 and Lots 2705 and 3635 were to be marked by a surveying post, but the line itself was not to be run. The survey lines covered a wide variety of terrain, from the Granby River valley to mountains of the Midway Range of the Monashees, which are over 2,200 metres in elevation.

Swannell wrote a report on his surveying. He started at the southeast corner of Lot 3636, "a high summit, the sides and top of which have in most places been burnt over." He found that he was only able to survey the south boundary of Lot 3636 for just over a mile (1.6 kilometres), "inaccessible crags being encountered at the beginning of the second mile from the southeast corner." Next, Swannell and his crew proceeded north along the eastern boundary of the three lots. They reached the North Fork of the Kettle River soon after the 4-mile (6.4-kilometre) post and then climbed a steep side hill to the northeast corner. Swannell wrote, "The eastern boundary of this block is practically at the foot of a high mountain range running north and south whose summits (7000–8000) [2,100 m–2,400 m] [lie] along the western boundary of the block."

The men then continued north along the east boundary of Lot 2705. Swannell noted the steep mountain slopes to the west and the lack of merchantable timber. The most northerly lot, 3635, was also the steepest. The crew reached a rocky summit at about 6 kilometres, "a ridge of a very high bare conical peak 3/4 mile [1.2 km] west of line. The altitude here is about 7000 feet [2,100 m] and from here on the line could only be run by triangulation from arête to arête across rock-walled basins and rock slides." Swannell found that the northeast corner of Lot 3635 was inaccessible.

The north boundary of Lot 3635 went through steep, rugged terrain, which Swannell described.

"At 23.66 [chains, or .5 km] west a witness post was set on the main mountain ridge—here a jagged knife edge. Westward there is a sheer precipice of 800–1000 feet [240–300 m] to a rock-filled basin. The entire western slope of this mountain is very steep, broken by crags and rock slides. The descent to the river is over 3000 feet [900 m] in 2 miles [3.2 km], the only route down, discoverable being a torrent bed 1½ miles [2.5 km] north of line. The river in its course through this block is walled in on both sides by the steep mountain slopes."

Swannell was able to survey about 6 kilometres south down the west boundary of Lot 3635. After that he found that the "west boundary of Lots 3635, 2705

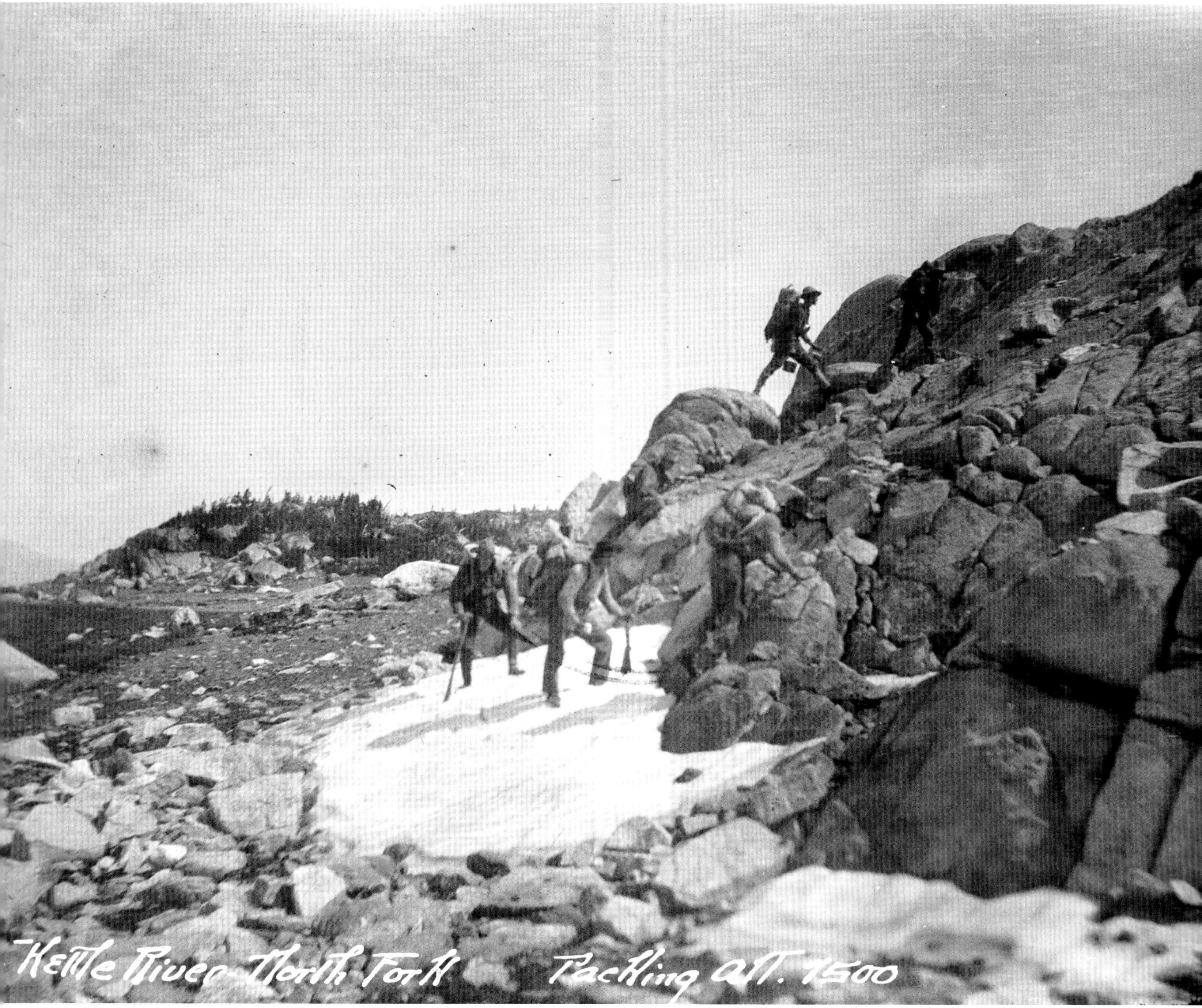
Kettle River North Fork
Packing alt. 1500

and 3636 cannot be run or defined on the ground as it traverses the summits of a broken mountain range, many of the ridges being inaccessible." Swannell did make a traverse survey to be sure that his work checked accurately. In his final sentence Swannell noted little economic potential for the land he surveyed. "I should estimate that only about 5% of the area embraced by these three blocks carries valuable timber, 95% of the total area is classifiable as 3rd class mountain land." Swannell and his crew completed their survey by the end of August.

Swannell was not the only surveyor who was unable to run his entire boundary line because of mountainous terrain. There were several lots whose boundaries could not be totally established. The Department of Lands and Works had to decide whether it would accept the surveys as sufficient. The CPR's contract provided for payment only when a surveyor's work was accepted by the BC government. Swannell made an affidavit for the sections that he was unable to complete and the reasons why. The Department of Lands and Works eventually accepted Swannell's survey of the three blocks. But there were a few lots that the Department rejected because it did not believe the surveyor made a sufficient effort to define the lot line in difficult terrain, or because there was insufficient surveying to validate the portion of the boundary that was completed.

In 1906 the CPR needed to complete the Columbia and Western Railway land grant. However, after the experiences of 1905, Gore & McGregor declined to bid for any of the remaining lots. The CPR contacted several other surveyors and none wanted to do this work. The company now faced a dilemma. Should they spend the money to complete the survey when much of the land they would receive appeared to have little or no economic value, or should they forfeit land to which they were entitled? Eventually the CPR found a few surveyors to complete some of the work, but in 1907 the BC government cancelled three remaining unsurveyed lots that were in mountainous terrain in the north part of the Columbia and Western land grant area.

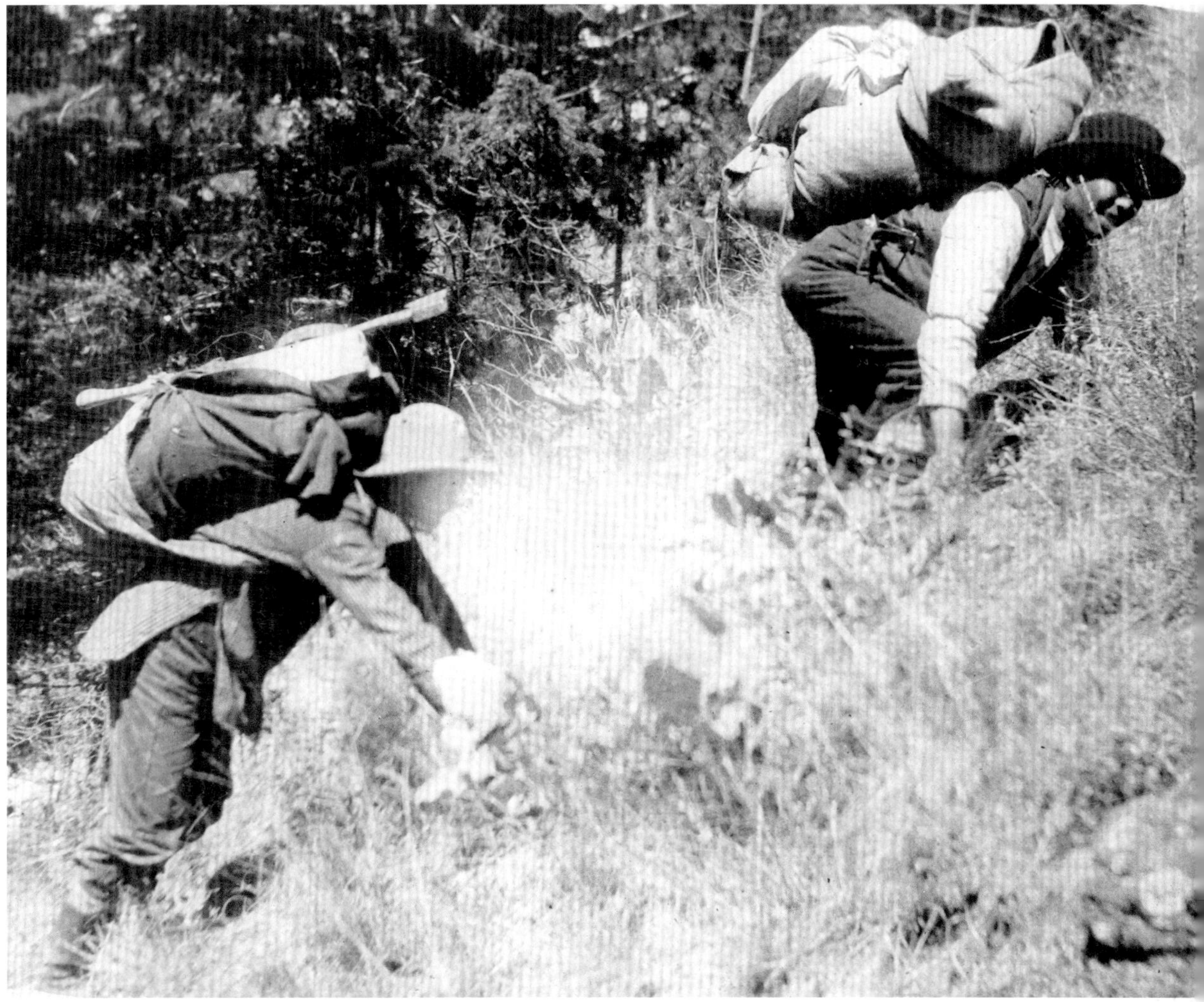

❖ Copley and Nep Yuen packing in the mountains. Image J-00320 courtesy of Royal BC Museum, BC Archives.

❖ Opposite page: Surveying high in the mountains. Image I-58396 courtesy of Royal BC Museum, BC Archives.

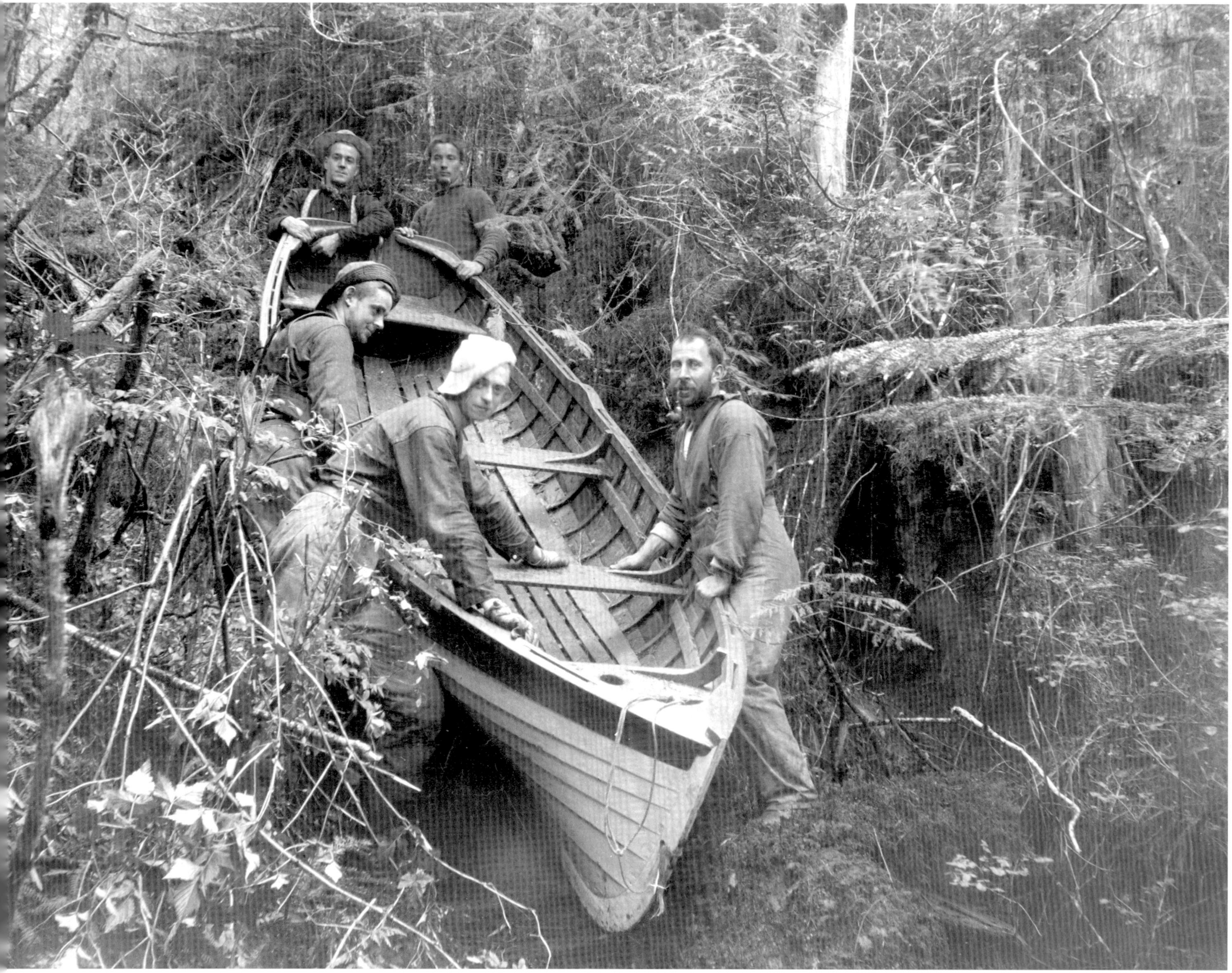

Swannell and his crew are portaging their rowboat to the four lakes on King Island near Rattenbury Point. There the men would survey Lot 212 between the lakes and the mountains on the east side. Image I-33614 courtesy of Royal BC Museum, BC Archives.

1906

Swannell began the field season by spending a month surveying timber licences in the Muir Creek area west of Sooke.

On April 15, Swannell left Victoria for his first surveying contract along the central coast region of BC. The *British Colonist* reported, "The *Tees* [a CPR steamer] will sail north again tonight, taking a large cargo, consisting for the most part of cannery stores." Both Swannell and J.H. McGregor had a crew to survey lots for the pulp and paper leases of the Bella Coola Development Company. Swannell took some photographs of places where they stopped en route to the Ocean Falls area.

Swannell and McGregor surveyed in the Ocean Falls area from May through July. Most of the lots were along the ocean shore and scattered throughout the Ocean Falls area, wherever the best timber for pulp was located.

While he was in the region, Swannell surveyed a proposed cannery site in Labouchere Channel for Brynild Brynildson. However, the development never occurred.

In the fall, Swannell returned to the Cariboo to survey more lots in the region. During this trip he met many well-known pioneer families. J.H. McGregor was also doing surveys in the area and occasionally the two of them met. Swannell left Vancouver on September 27 and travelled to Ashcroft, arriving there at 4:30 in the morning. He and his crew promptly left on the stagecoach, reaching 83 Mile House at 8:30 that evening. The next day the men left at 4 a.m., arriving at 7:30 p.m. at 150 Mile House, and then continuing to Soda Creek, which they reached at 2 a.m. Later that morning the men crossed the Fraser River on the ferry

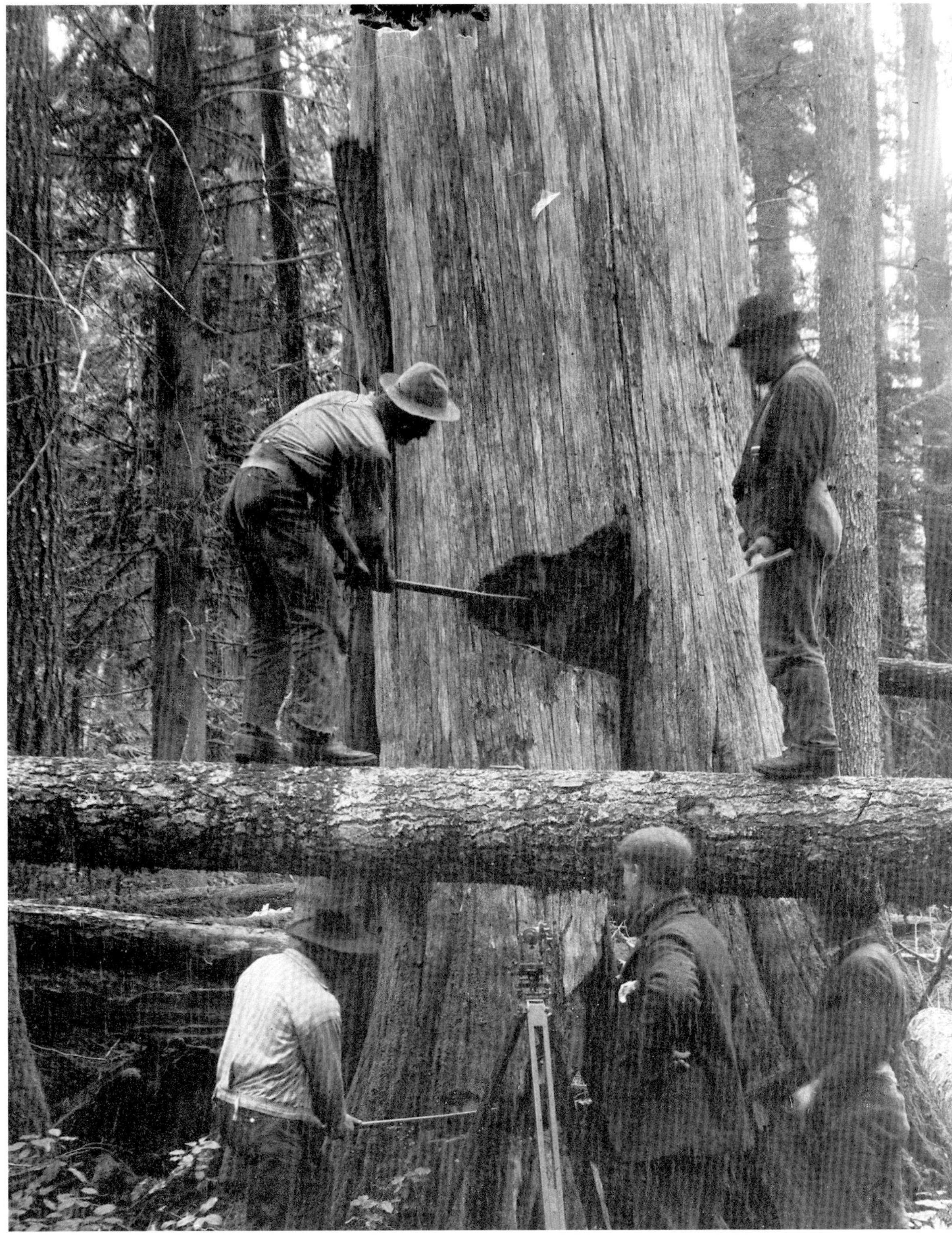

❖ Chopping a hole through a cedar on the line. This picture was taken during the Sooke timber survey. Image G-06439 courtesy of Royal BC Museum, BC Archives.

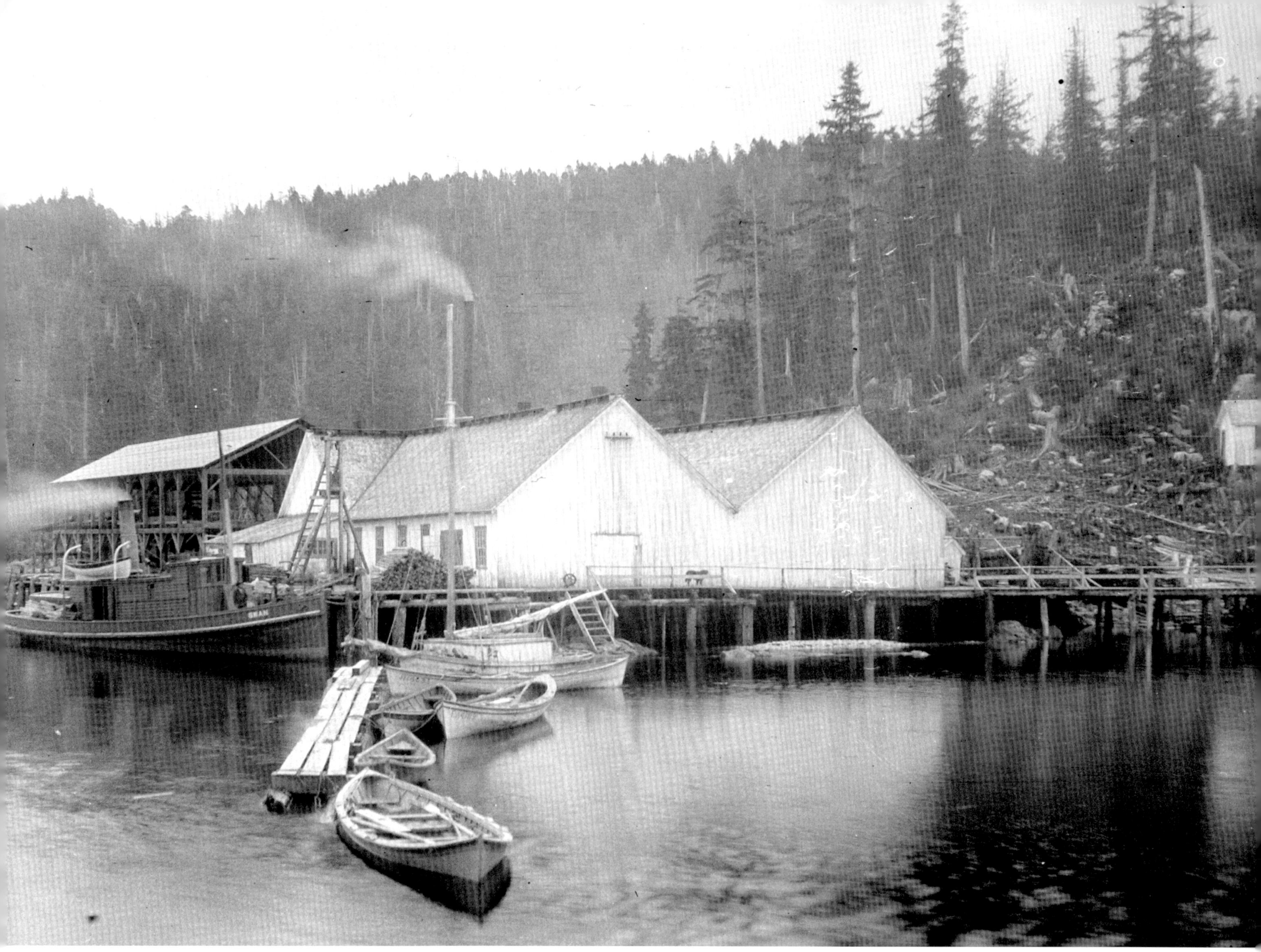

❖ The cannery at Namu began operating in 1893. It was one of the largest on the central coast and lasted until the latter part of the twentieth century. Swannell's photographs are among the oldest ones of this site. These two photographs, joined together, form a panorama view. Left: Image I-33619, right image G-06438 courtesy of Royal BC Museum, BC Archives.

and rode to Meldrum Creek, arriving after three long days of travelling. The men began surveying on October 1 but the next day there was "no work—cattle round up." The surveying resumed the following day, but heavy rain on October 4 curtailed surveying in the afternoon. After finishing this survey, Swannell went to Becher's near Riske Creek. During the next few days he made arrangements to do some future lot surveys in the area. His diary entry for October 17 provides a description of surveying in the fall in the Chilcotin. "Surveyed Lot 595—Thrift Meldrum, preemption 40 acres. Drove to Meldrum Cr., 16 [miles, or 25 km]—arrived after dark, very cold!" Swannell also surveyed a lot for Thrift's brother, Jack, and a lot that

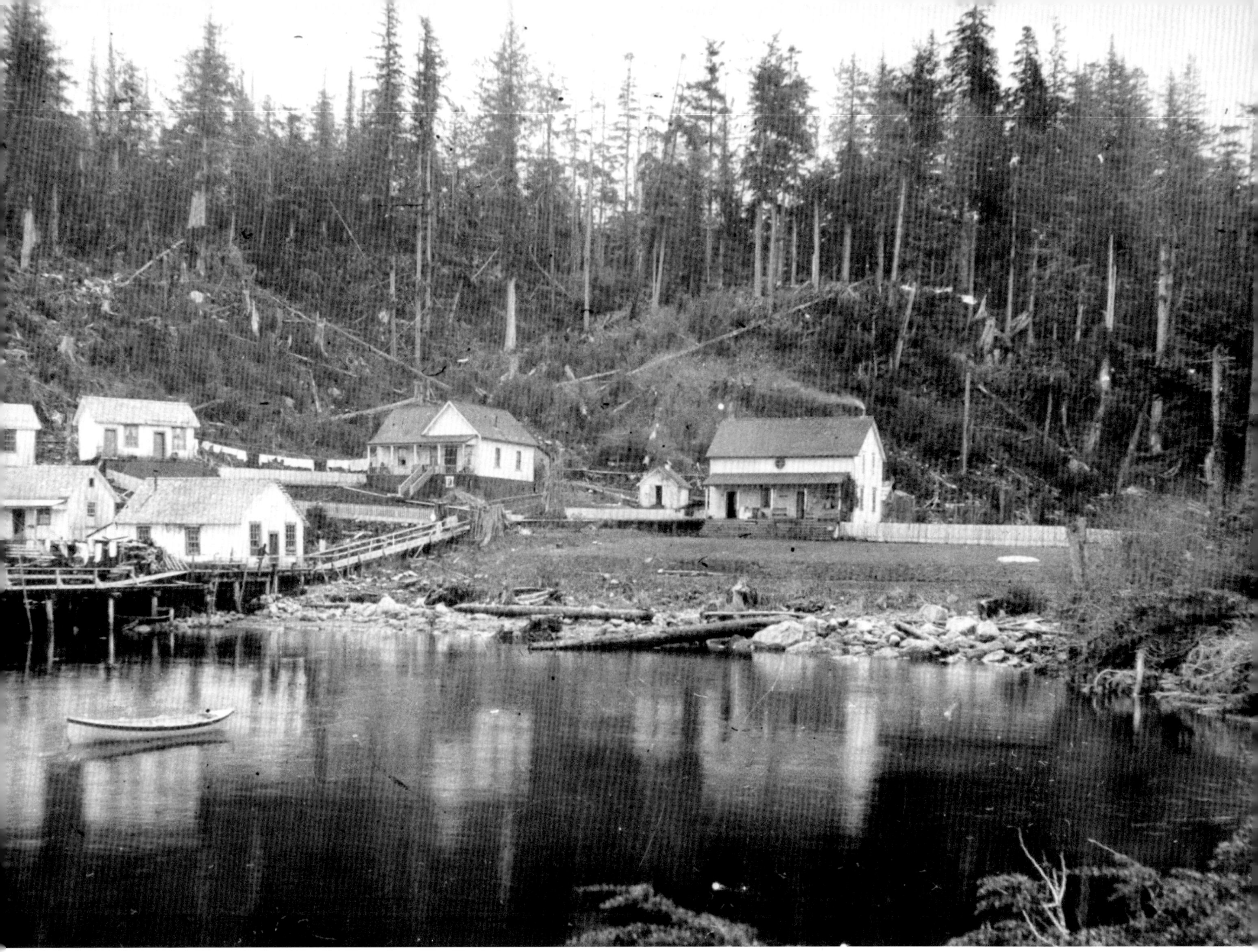

the two held in joint ownership in the area. On October 19, Jack Meldrum drove the men to Soda Creek. On the following day he spent the morning at Charles Foyle's home "fixing old Kentucky rifle" before taking the stage to 150 Mile House in the afternoon. There he met John Hargreaves and went to the Springfield Ranch, where the crew spent a week running levels for elevation for a water pipeline and ditch.

On November 1 Swannell "left 150 Mile House with McGregor and walked to the 137 [Mile House]. Temp. 0°F (-18°C). Reached the 122 (Forbes)." The next day they surveyed Lot 805 for Louis Crosina and had a dance in the evening. After finishing another lot for Crosina, the next day Swannell went to the

❖ Totem poles in Alert Bay. In the late nineteenth and early twentieth centuries, most commercial ships travelling up the British Columbia coast stopped at the First Nations village of Alert Bay on Cormorant Island, just off the east coast of Vancouver Island. While he was there Swannell took a photograph of this pair of iconic house posts made by the renowned and influential Kwakwak'wakw carver Charlie James between 1900 and 1905. Both poles have a thunderbird on top with a grizzly holding a human below. Later the house posts were moved to the First Nations community of Fort Rupert on Vancouver Island, where the famous photographer Edward Curtis incorporated them into his 1914 film, *In the Land of the Head Hunters*. In 1927 the posts were moved to Stanley Park. Today there is a replica of these poles in the park. Image G-06448 courtesy of Royal BC Museum, BC Archives.

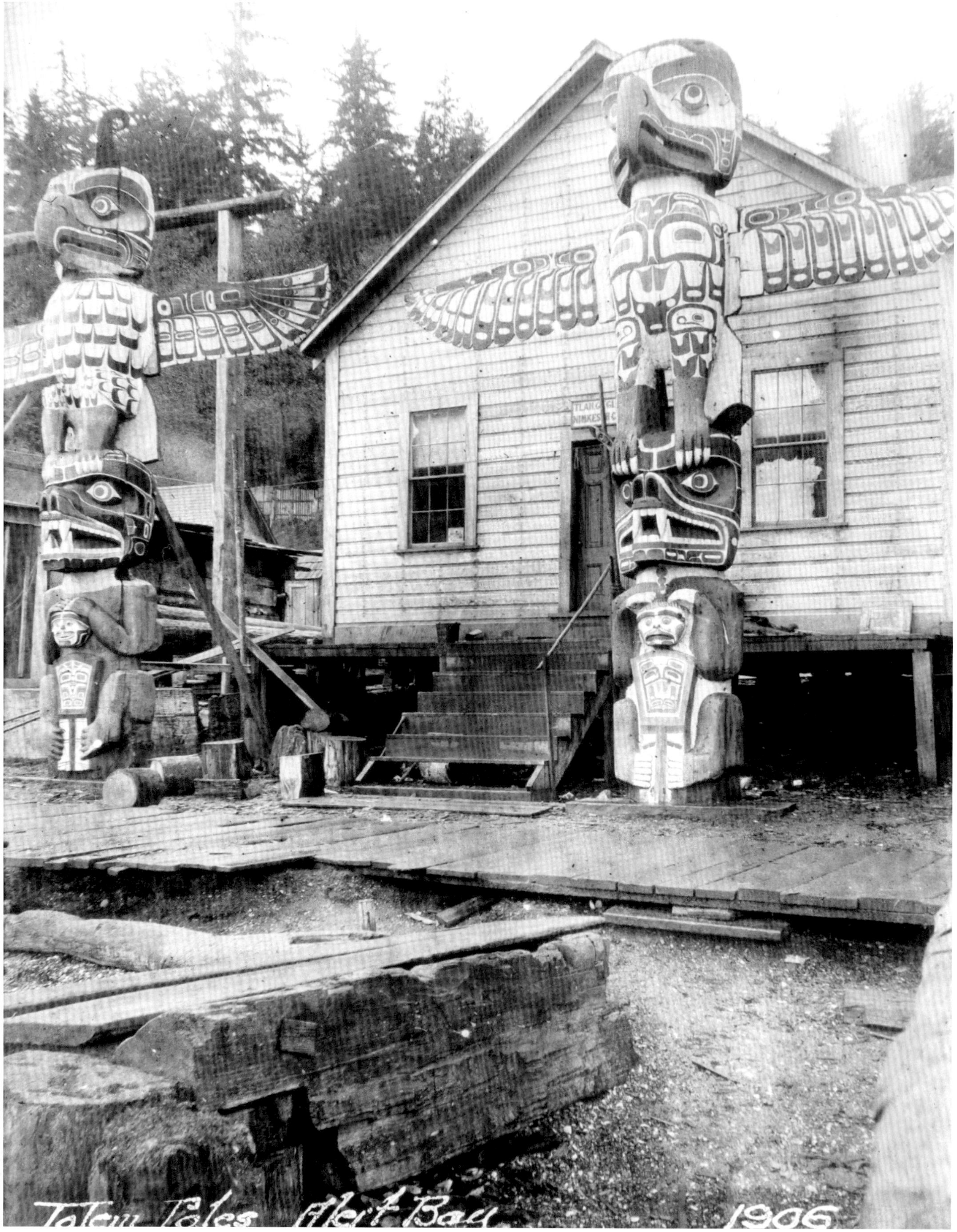

Opposite page

❖ Top: Beaver Cannery at Rivers Inlet. The SS *Tees* is docked at the cannery, which had been founded the previous year. Image I-33616 courtesy of Royal BC Museum, BC Archives.

❖ Bottom: Surveyors at Cousins Inlet. Image I-33615 courtesy of Royal BC Museum, BC Archives.

McKinlay Ranch at Lac la Hache. He surveyed a preemption for Peter Hamilton, Lot 1083, which included a school reserve and several other lots in the 111 Mile area. He finished Lot 1086 for Robert McLure in wet snow on November 14. It also snowed the following day when he surveyed Lot 1087 at Abel Lake. Next he surveyed Lot 1089 and 1090 for Moose Dixon near Lac La Hache. One day he was unable to work because of a snowstorm.

Before returning to Victoria, Swannell travelled 60 kilometres west to Springhouse. There he surveyed two lots for Antoine Boitanio. In his diary entry for the end of November he noted that he had driven 700 miles (1,100 kilometres) by wagon on this survey.

Swannell completed his survey at the end of November and returned to Victoria.

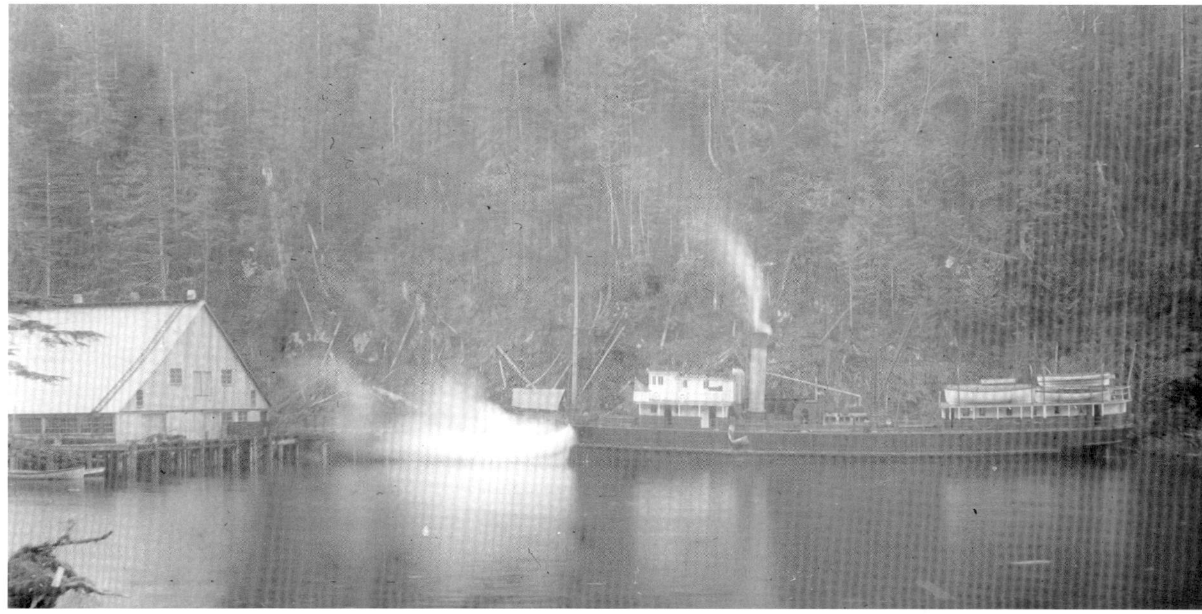

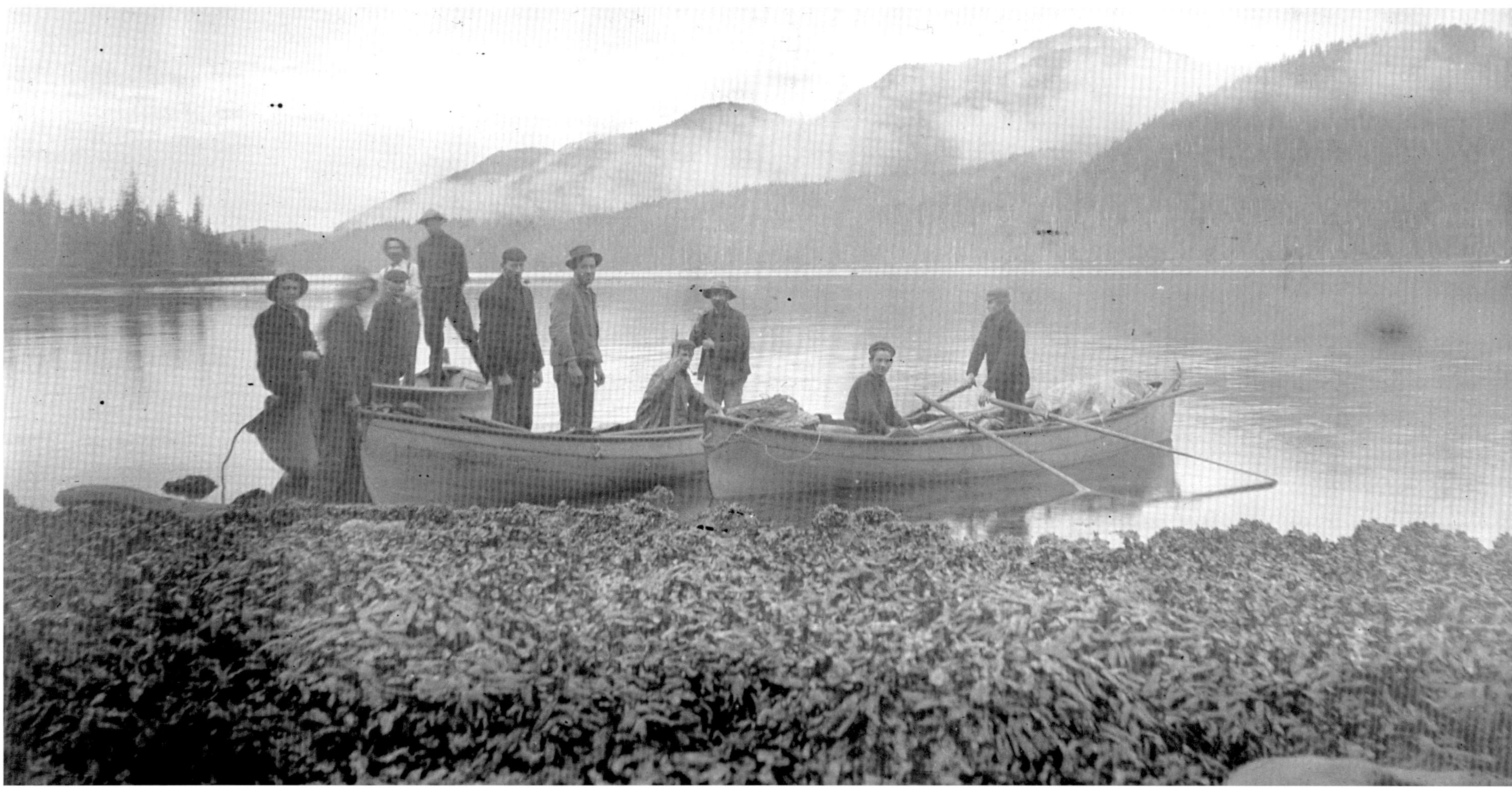

Bella Coola Development Company

In 1901 the BC government tried to encourage the growth of a pulp and paper industry in the province by offering leases with an annual rental of 2 cents per acre and low royalties to companies that would undertake to build and operate a pulp mill. The leases were offered for twenty-one years with a renewable option. The program lasted until 1903. Four companies became involved in this program and they received 354,399 acres [143,420 ha]. One of these, the Bella Coola Pulp and Paper Company, obtained 80,000 acres [32,375 ha] of leased land in the Ocean Falls area. The company was reorganized and became the Bella Coola Development Company in 1904. Ocean Falls, where the pulp and paper mill was located, was founded in 1906.

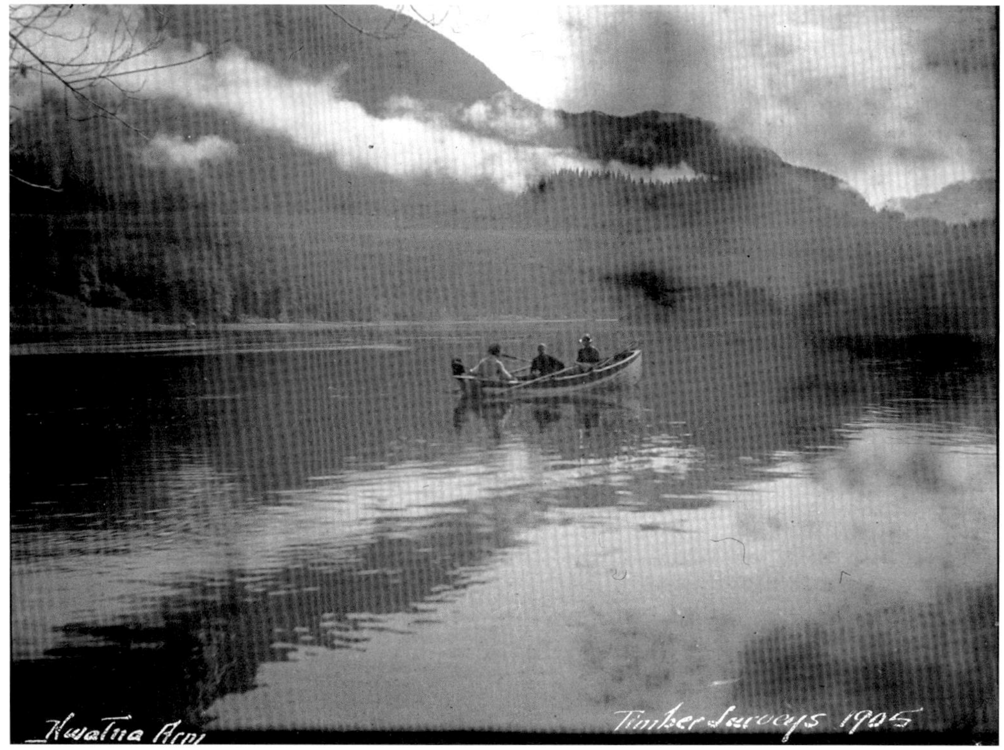

❖ In his notes for this photograph Swannell wrote that he took it at the head of Kwatna Inlet "after visit to Indian shack where barking [talking] sticks were found." Image I-57092 courtesy of Royal BC Museum, BC Archives.

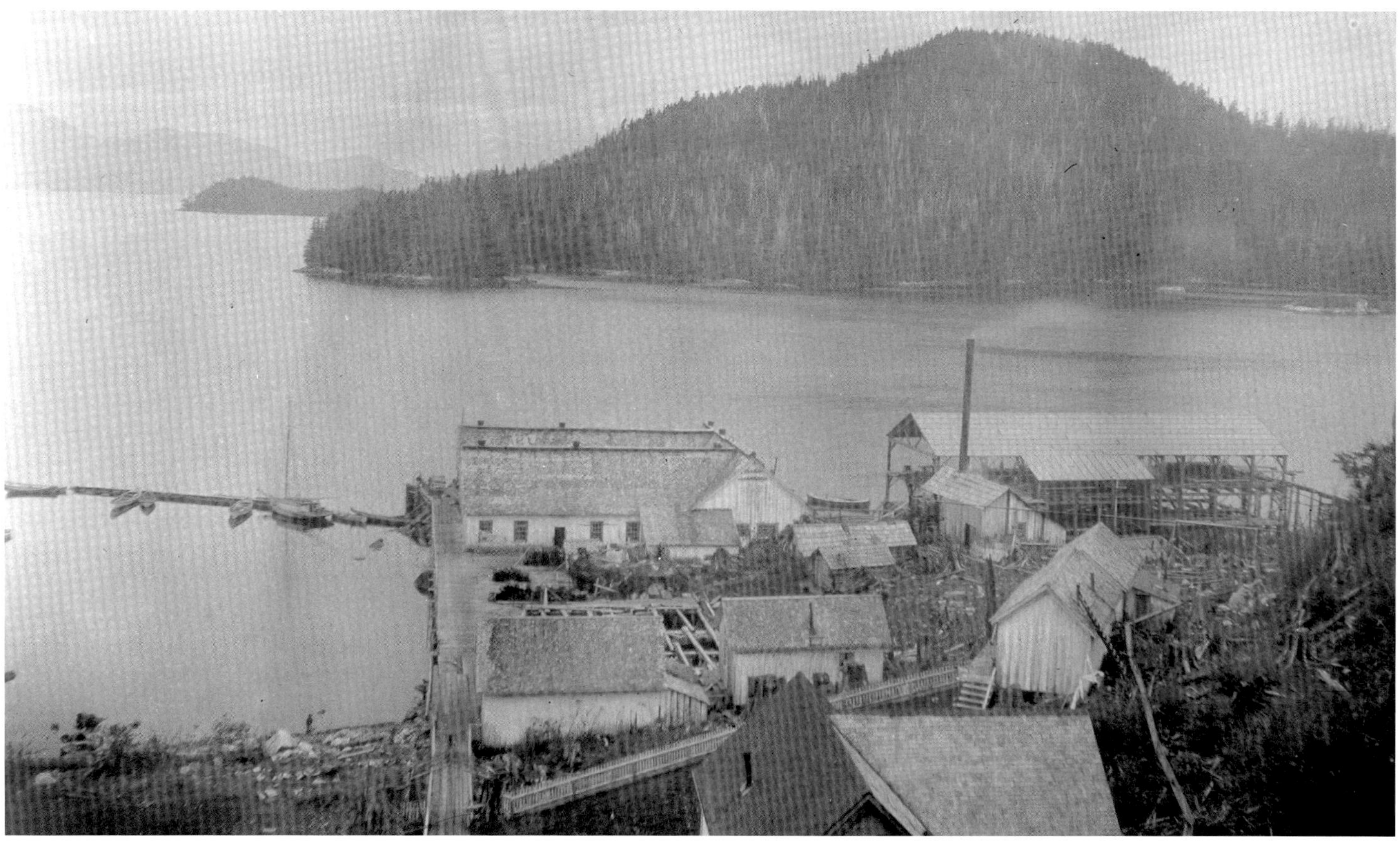

❖ Namu Cannery and sawmill. This is one of the few photographs of Namu taken from the land, and it shows the sawmill shortly after it was constructed. Image I-33618 courtesy of Royal BC Museum, BC Archives.

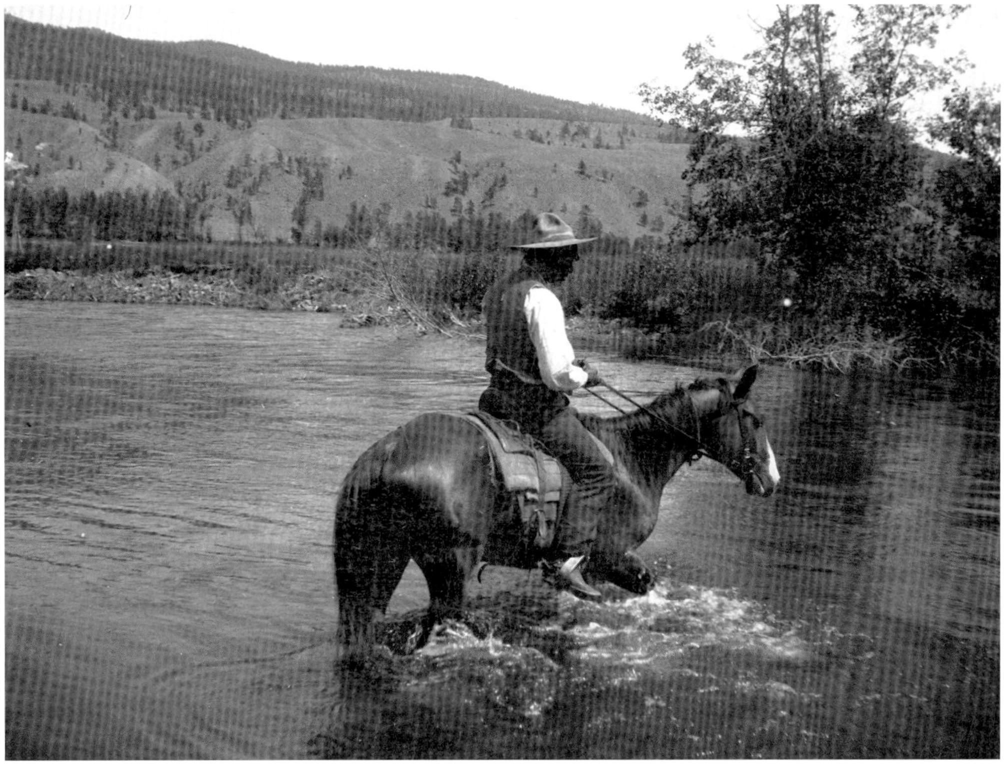

❖ Moose Dixon. It appears that Swannell did not have a camera on his trip to the Cariboo in 1906, but he took a picture of Moose Dixon when he met him again in 1907.Image I-57271 courtesy of Royal BC Museum, BC Archives.

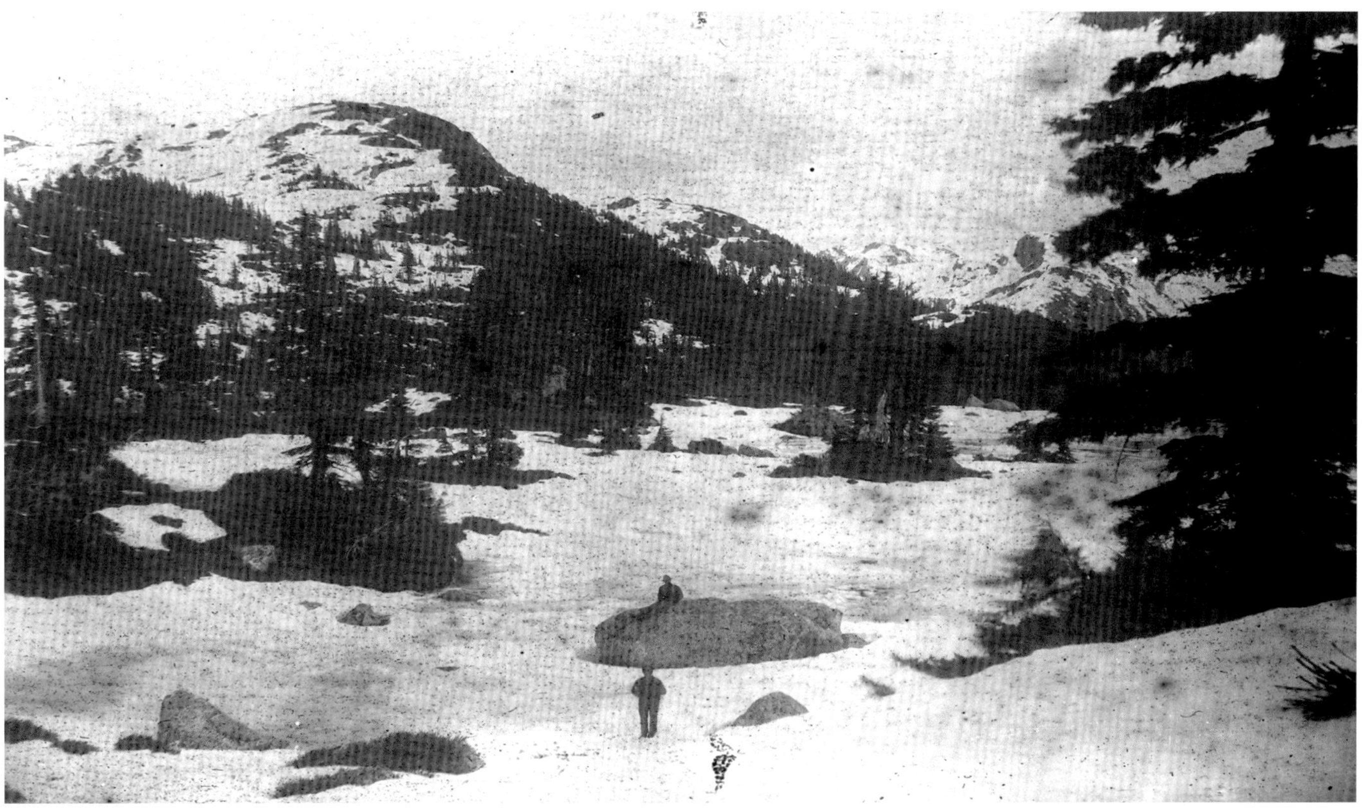

❖ Labouchere Channel. Swannell noted that they reached this summit 1 kilometre from the ocean, and he estimated that the mountains were 1150 metres high. Image I-57095 courtesy of Royal BC Museum, BC Archives.

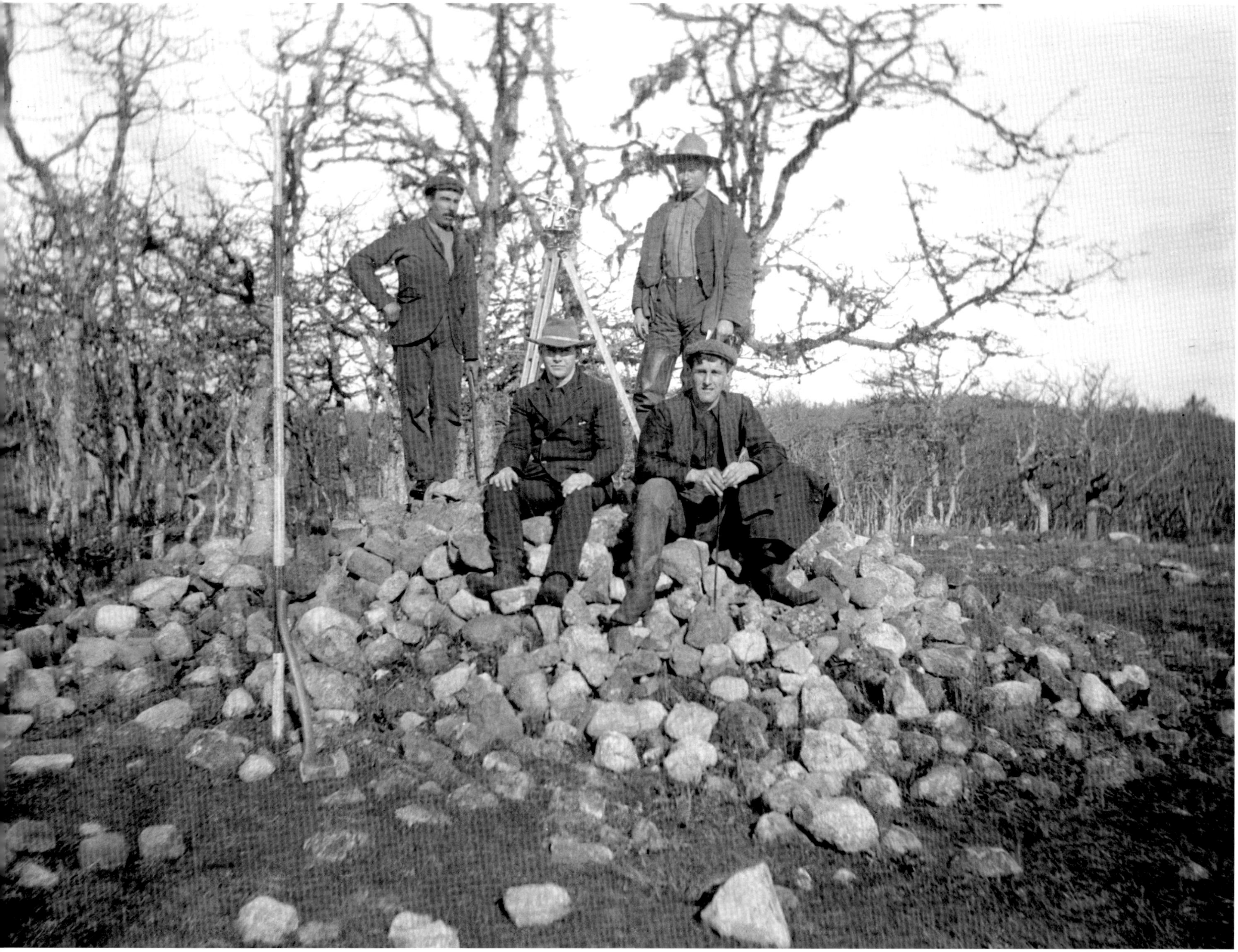

1907

Swannell spent the winter doing the office work for the lot surveys in the Cariboo, and he also made some surveys around Victoria. One of these was the subdivision of some farmland in the Uplands district where he found a First Nations gravesite.

On May 20 he, McGregor and a small crew left on the SS *Tees,* stopping at Bamfield before going on to Tofino for a survey at a nearby marble quarry. One evening they had wildlife for company. "All night at Sechart—4 whales on slip." The crew then took a boat to Port Alberni where they spent ten days subdividing a lot. From Port Alberni they travelled by stage to Nanaimo and then by train back to Victoria.

Gore & McGregor received a contract from the federal Department of Indian Affairs to survey several reserves during the summer of 1907, and Swannell was placed in charge. Most of these were in the Ashcroft area. In Victoria Swannell and Gore spent a day at the Department of Indian Affairs and a day at the Department of Lands preparing for the surveys.

Swannell, along with his crew of Wilfred Prevost (his surveying assistant), Eric Gordon, Alf O'Meara, J.F. Finlayson and his father, Fred Swannell, as cook, left Victoria on June 14. They took an evening train from Vancouver, arriving in Ashcroft at 1:30 a.m. The next day they went to the O'Hara Ranch in the Bonaparte River valley, where Swannell had surveyed in 1904. O'Hara wanted to lease an additional portion of the Bonaparte Reserve that adjoined his ranch on the east. On their first day of surveying, Swannell took an observation on the sun for a precise latitude and longitude for his work. The next day he took another observation, tested his surveying chain to make sure it was measuring correctly, and looked for the lot corner post that he planned to use to begin his survey. Although the O'Hara survey was small, it took a considerable amount of time. The land was within the Railway Belt (federal land), so it had been surveyed into one-mile sections. The boundaries of the Bonaparte Reserve and the section lines had been surveyed several years previously, so Swannell had to resurvey much of the nearby area. He then had to connect the boundaries of the land that O'Hara wanted with the reserve, the section lines, and the lot that he had surveyed in 1904. It took the remainder of June and the first week of July for the crew to complete this work.

After completing the survey on the Bonaparte Reserve on July 6, Swannell and his men returned to Ashcroft the next day. They took the train to Spatsum, farther down the Thompson River, where they began a survey of reserves for the Cook's Ferry First Nation, today a member community of the Nlaka'pamux Nation. There were two main purposes for this survey. Most of the reserves were along the Thompson River and some were having their boundaries altered for

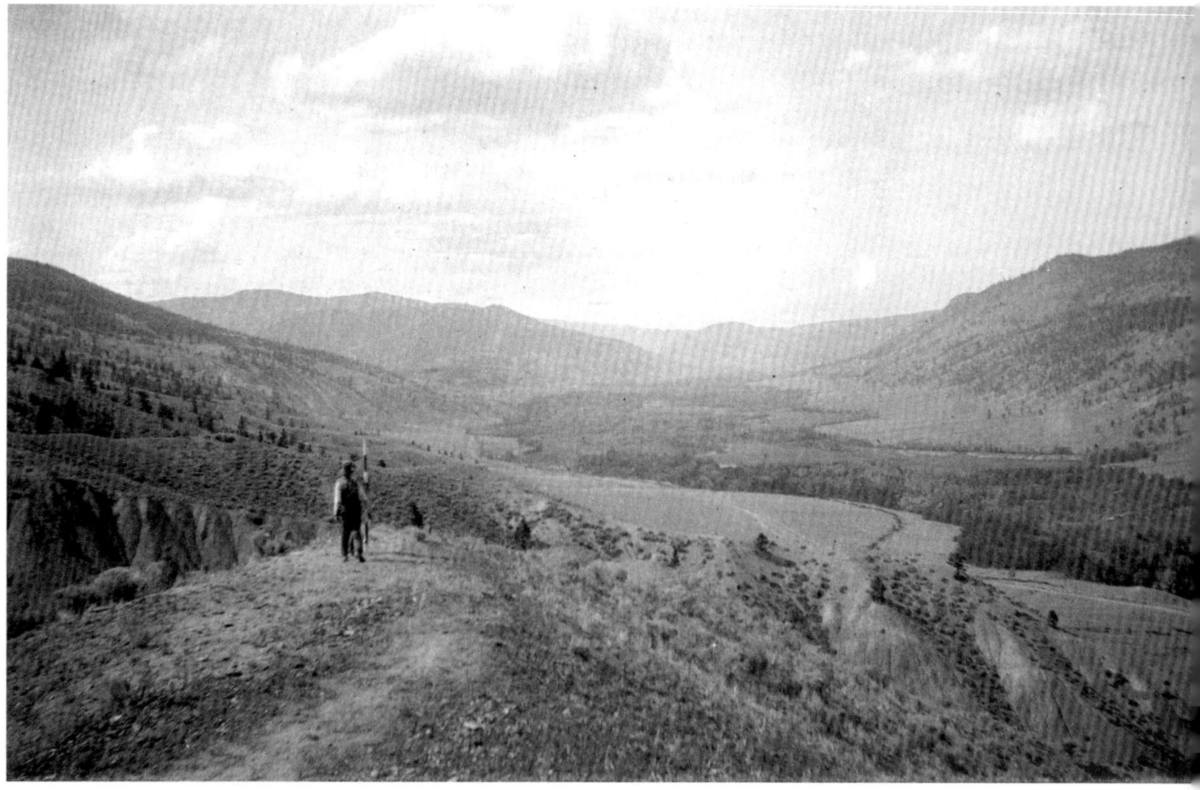

❖ Surveying above the Bonaparte valley. Image I-58437 courtesy of Royal BC Museum, BC Archives.

❖ Opposite page: Uplands subdivision party at a First Nations burial mound. While surveying Block 6 of the Uplands subdivision in suburban Victoria in early 1907, Swannell located this mound. There are many Native burial sites in the Uplands area and they were made by the people from the nearby Chekonein village, ancestors of the Songhees (Lekwungen) people. Each site contained a body buried in a shallow grave lined with stones and covered with dirt, rocks and boulders. Recent research has determined that some of the burial sites in the area are over one thousand years old. Swannell's picture is one of the oldest photographic records of one of these burial sites. Image I-33651 courtesy of Royal BC Museum, BC Archives.

right-of-way for the Canadian Pacific Railway. There was also one reserve along the river and four reserves in the remote Highland Valley about 40 kilometres to the southeast that had not yet been surveyed.

On the first evening, Swannell took an observation on Polaris to determine his location and he also searched for the original survey posts. He started by resurveying Indian Reserve (I.R.) #10 (Pokheitsk) and I.R. #11 (Spatsum) to include the new right-of-way additions. The resurvey of I.R. #11 included a "rectification of Ah Yep's boundary." Ah Yep had pre-empted land in the area before the Cook's Ferry reserves had been allotted. Originally the Indian Reserve commissioner had allocated the Nlaka'pamux Nation land belonging to Ah Yep, but this decision had been overturned. Swannell's survey clearly defined the boundary between Ah Yep's property and I.R. #11. Swannell also surveyed I.R. #11A (Spatsum), which had not previously been done. In addition he made a traverse of the railway line and connected it with the survey of these three reserves.

On July 18, Swannell and his crew travelled to Divide Lake (Fish Lake) in the Highland Valley on the height of land east of the Thompson River. At this time, Fred Swannell left and Alf O'Meara took over the cooking in addition to helping with the surveying.

The four reserves were not connected to each other so each was a separate survey. Before he started surveying in the Highland Valley, Swannell once again took an observation on Polaris. The crew made a traverse of Divide Lake as part of surveying I.R. #13 (Quil-lan-ton). Sunday, July 21, was a day off work for most of the men. Some went fishing at Divide Lake, where they caught several trout. Meanwhile, Swannell "sent [Eric] Gordon to Ashcroft for mail & medicine." Gordon returned on Tuesday with the provincial mineralogist, William Fleet Robertson, his assistant, Harold Nation, Yale MLA Stuart Henderson and Ashcroft physician Dr. George Sanson. Sanson was head of the Tamarac group, which had some mining claims in the Highland Valley that needed to be surveyed.

In the last week of July and the first week of August, Swannell and his crew finished I.R. #13 and I.R. #12 (Chilth Nux). For the next two weeks they sur-

Surveyors at O'Hara's ranch. O'Hara's ranch was on the Cariboo wagon road and adjoined the Bonaparte reserve. Daniel O'Hara was a friend of local rancher Charles Semlin, who was premier of British Columbia from 1898 to 1900. O'Hara was appointed to be the sergeant-at-arms for the British Columbia legislature for one session beginning in the summer of 1900. Image I-33687 courtesy of Royal BC Museum, BC Archives.

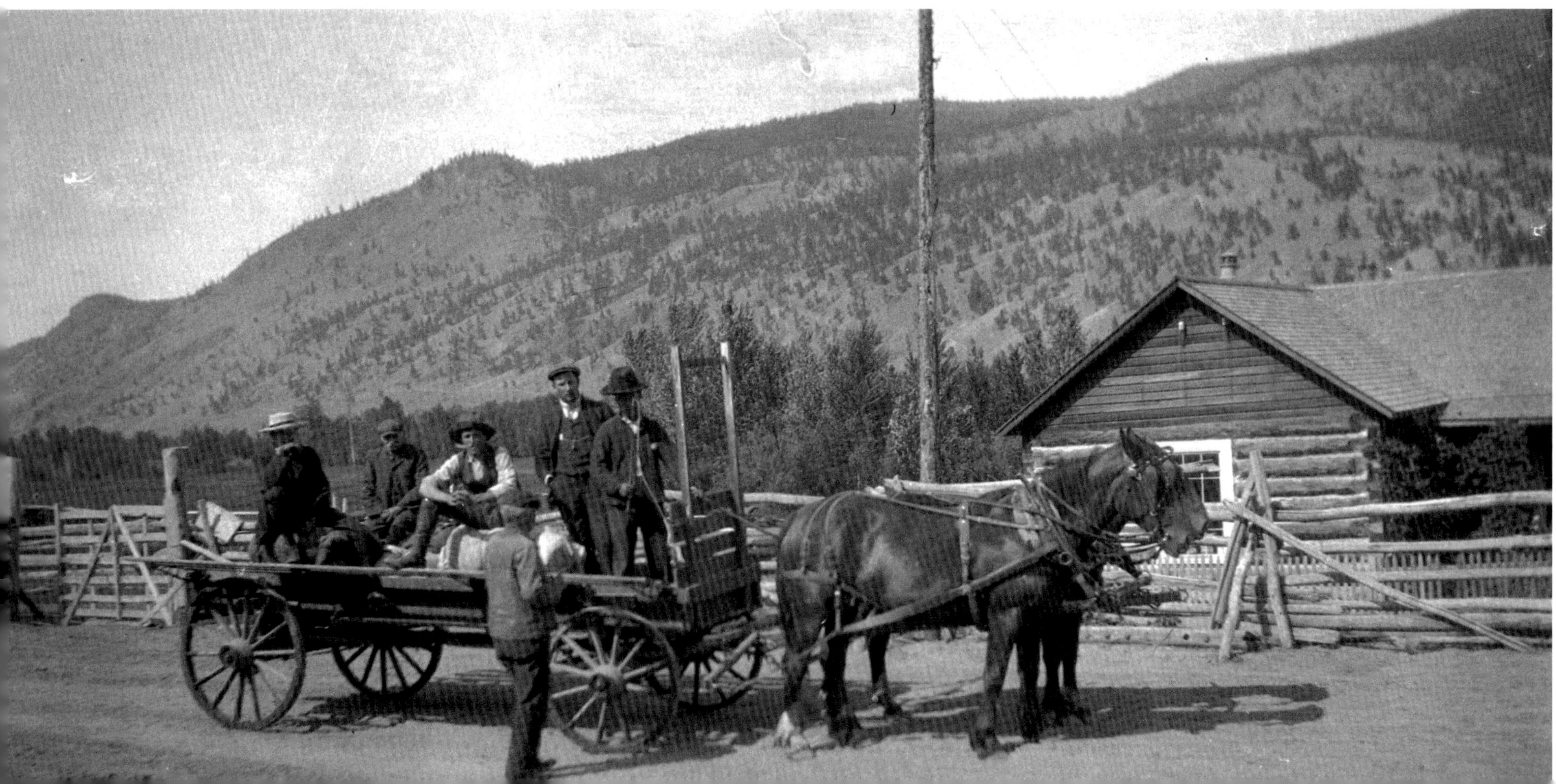

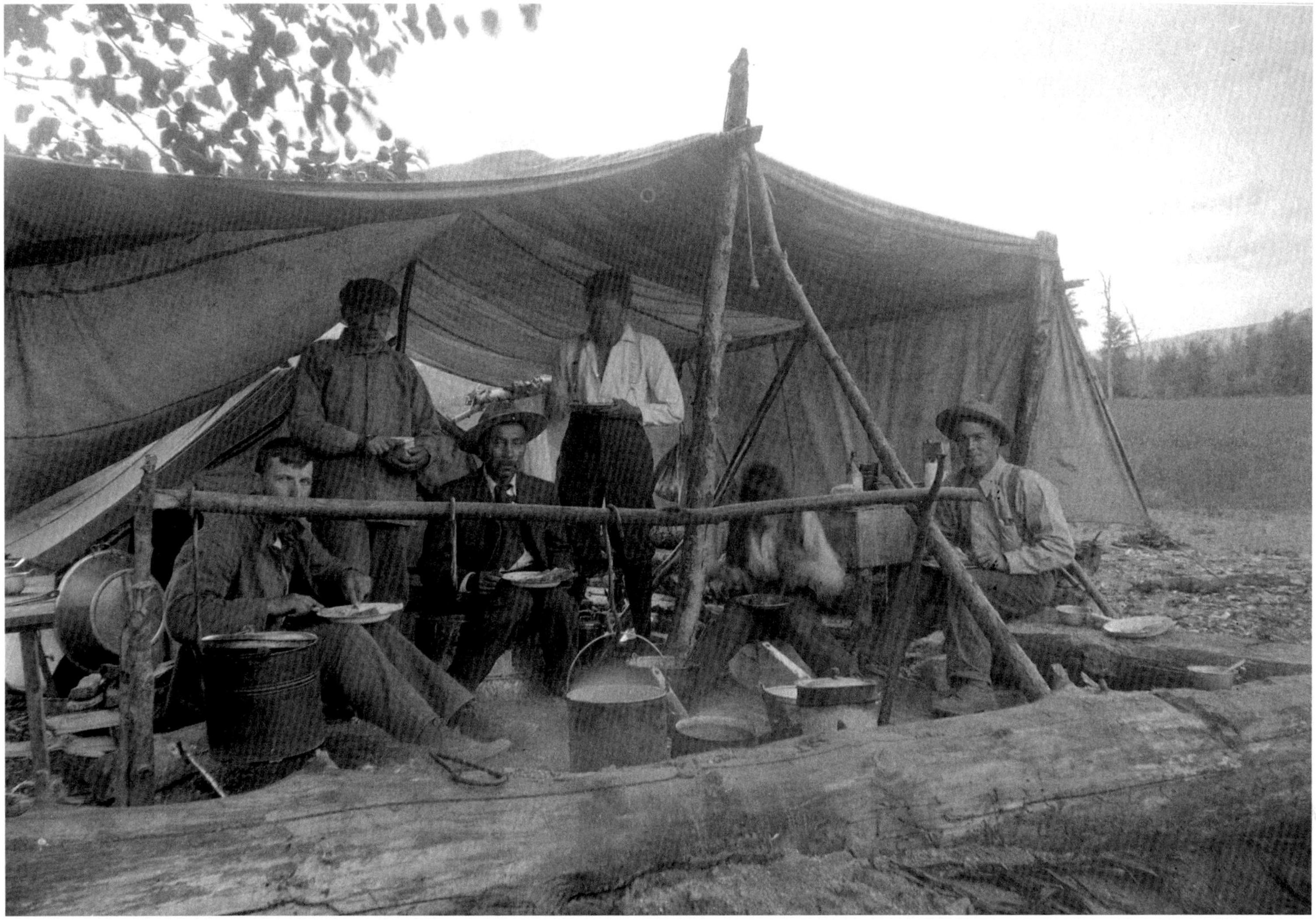

veyed the remaining two reserves, I.R. #14 (Enquotco) and I.R. #15 (Squetankilhats). During that time Swannell and Prevost took two days to go into Ashcroft to get more information about the mining claims to be surveyed for the Tamarac group.

On August 23 Swannell and his crew went to the Tamarac mine, where they spent six days surveying the three mining claims. The surveying proceeded smoothly, and on August 29, in a pouring rain, the men packed up and drove to Ashcroft. There they boarded the train to Spences Bridge, where Swannell hunted for old survey posts, adjusted his transit and took an observation on the sun. On the last day of August the crew started a resurvey of I.R. #4 (Spences Bridge).

❖ Secewepemc Chief Basil David (seated in the centre of the photograph) was a prominent First Nations leader. In 1906 he, along with Joe Capilano and Harley Isipaymilt, went to Ottawa and then to London, where he met King Edward VII to speak about the need to settle land claims in British Columbia. He was a signatory to the memorial written and presented to former prime minister Wilfrid Laurier when he visited Kamloops in 1910 and a similar memorial sent to the minister of the Interior, Frank Oliver, in 1911. Chief David was one of a group of First Nations leaders who made a presentation to the BC premier, McBride, in Victoria in 1911 and to the prime minister, Robert Borden, in Ottawa in 1912. On both occasions the prominent ethnologist James Teit served as his interpreter. The presentation to Borden led to the establishment of the Royal Commission on Indian Affairs (McKenna–McBride Commission) later that year. Chief David spoke to the McKenna–McBride Commission in 1914 and travelled again to Ottawa in 1916 to make a presentation to Borden protesting against the policies of the Department of Indian Affairs. When a First Nations political organization, the Allied Tribes of British Columbia, was formed the same year, Chief David was chosen to be on its original executive committee. Chief David was photographed on several occasions, but the two pictures that Swannell took are the only informal ones of this well-known advocate for the rights of First Nations people in British Columbia. Image I-33682 courtesy of Royal BC Museum, BC Archives.

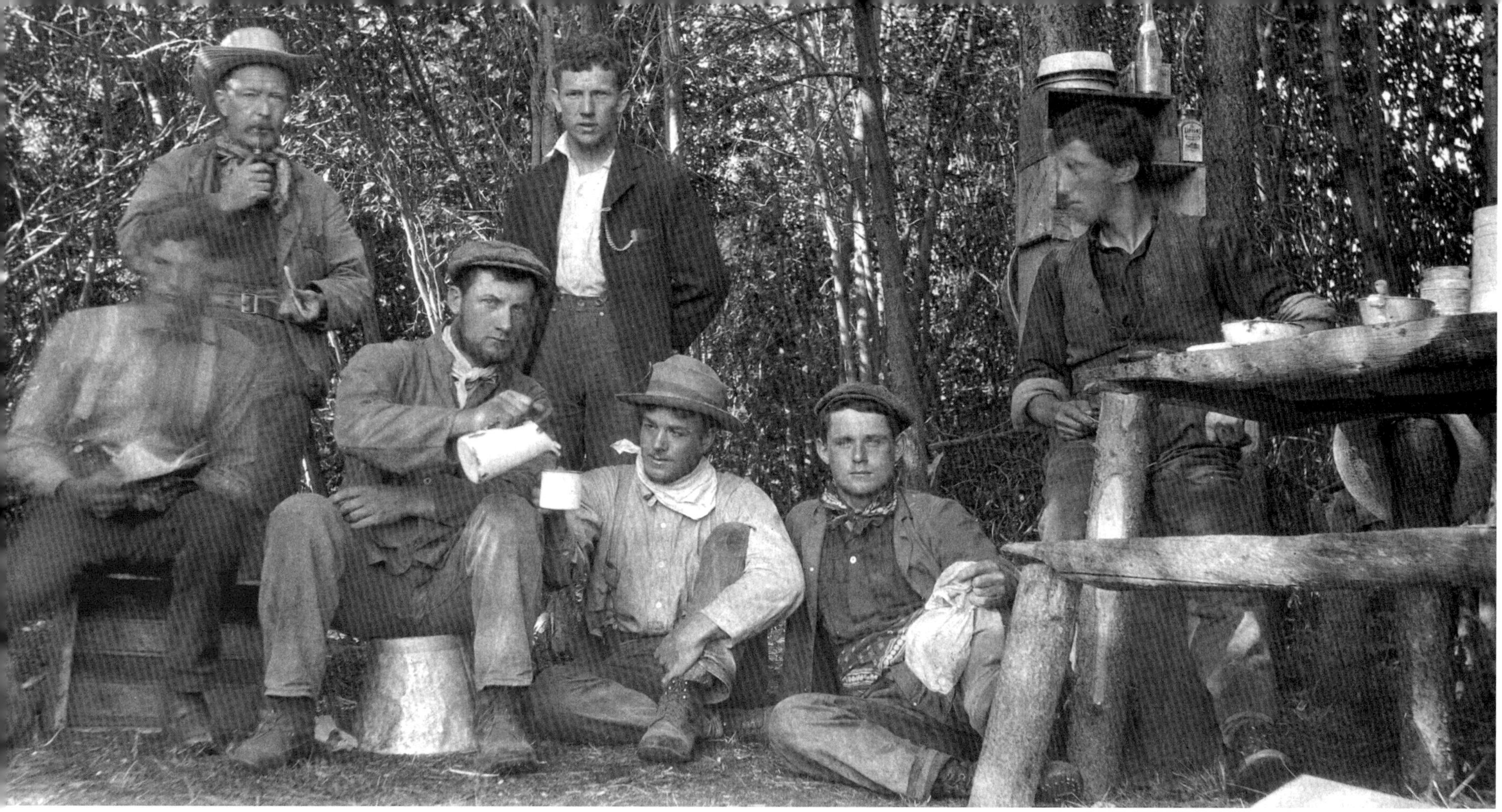

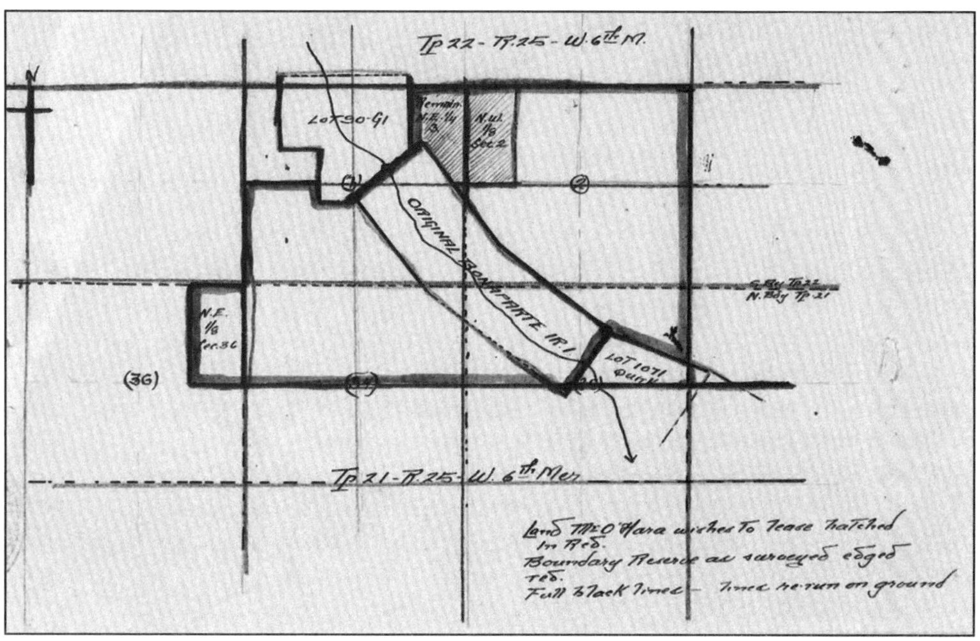

Swannell's map shows O'Hara's ranch (Lot 90), G1, the Bonaparte Reserve and the Cariboo wagon road. Swannell's sketch map shows all of the lines that he needed to resurvey for the land that O'Hara wanted to lease. The land that O'Hara applied for in 1904 is to the west of his lot.

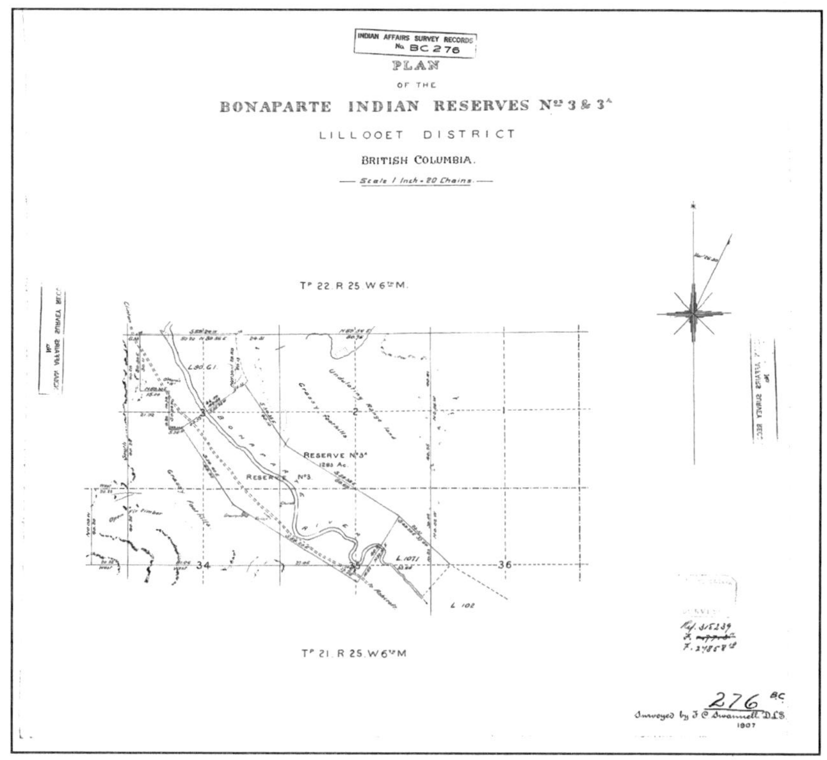

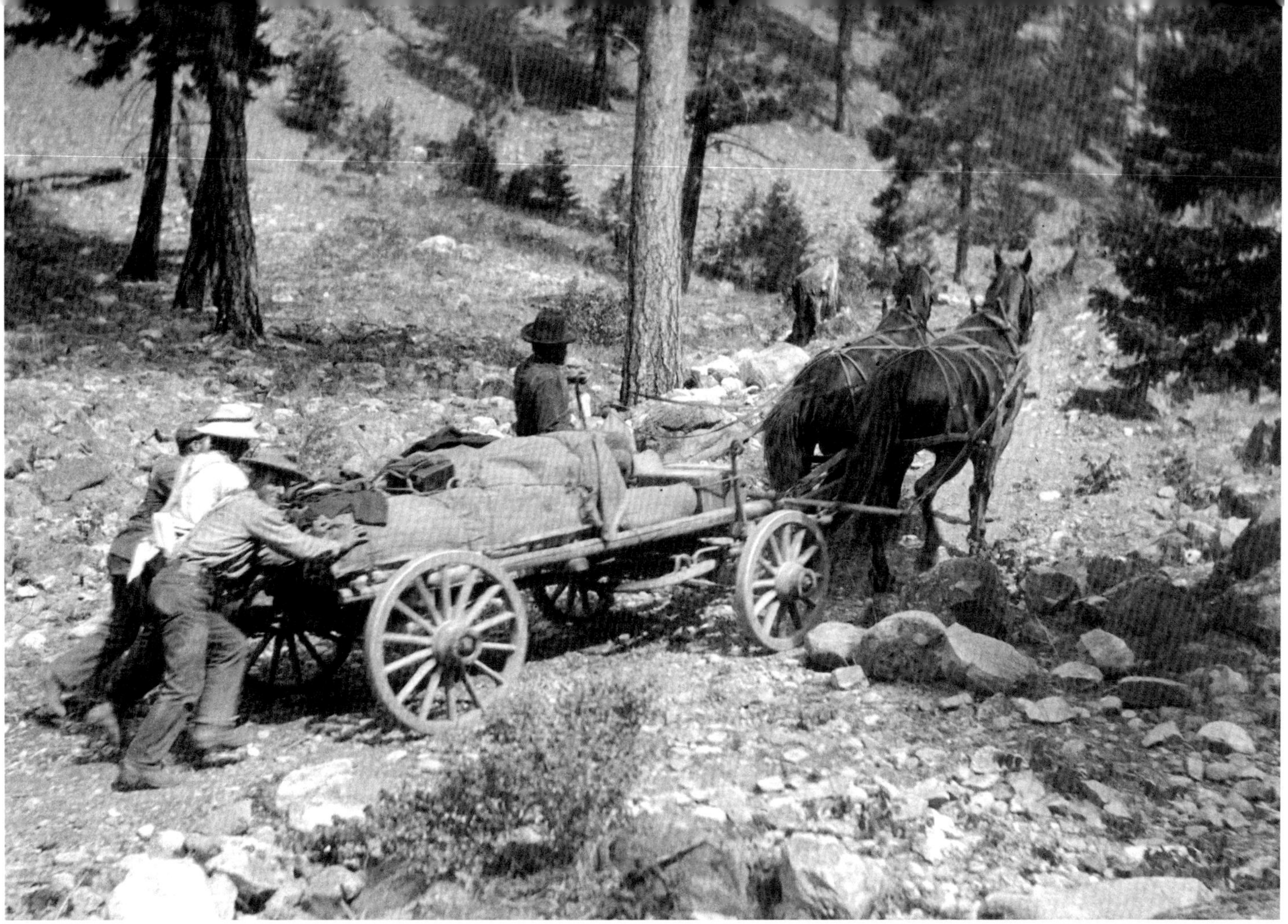

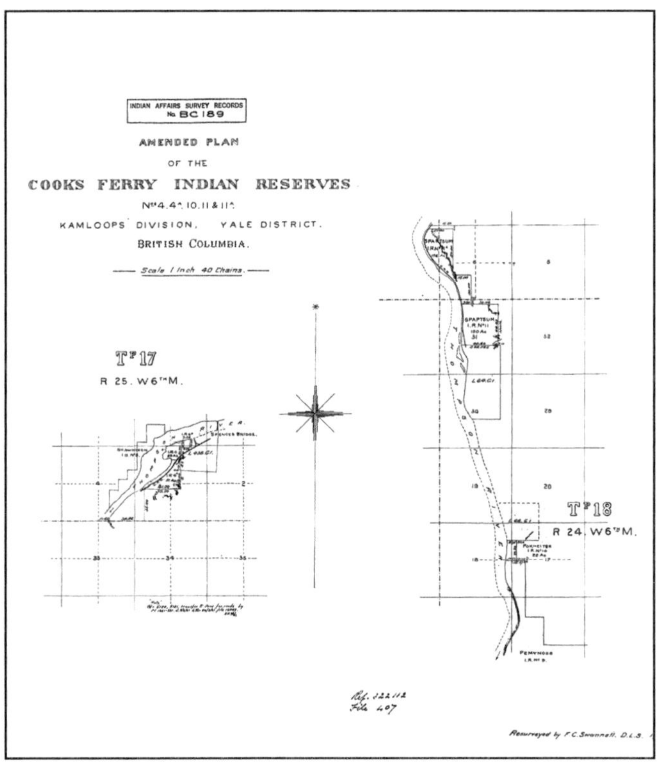

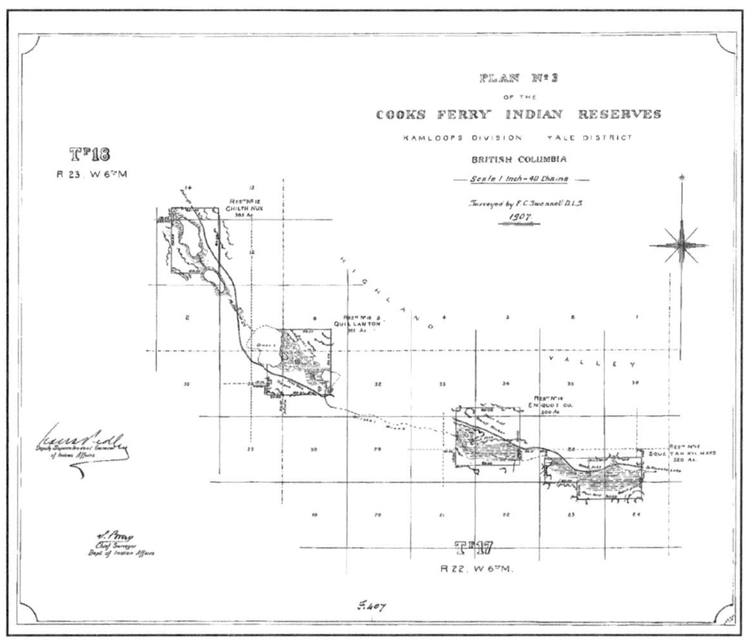

Opposite page

❖ Top: Robertson and Nation with Swannell's crew. William Fleet Robertson (standing back left) was the provincial mineralogist from 1897 to 1925. Harold Nation (back centre) was the assistant provincial mineralogist for many years. Almost every summer Robertson and Nation visited and examined mineral areas of British Columbia, and they usually took a camera with them. One of the locales that Robertson and Nation visited in 1907 was the mining claims in the Highland Valley. On the back right is Alf O'Meara, who was a member of Swannell's crew in 1907, 1909 and 1910. In the front left is Eric Gordon, a member of Swannell's crew in 1906 and 1907; J.F. Finlayson (pouring drink); and Wilfred Prevost, who was on Swannell's crew in 1906 and 1907. Frank Swannell is at the bottom right. Bureau of Mines photograph. Image I-56891 courtesy of Royal BC Museum, BC Archives.

❖ Bottom left: Bonaparte Reserve sketch. Image n0011246 courtesy Land Title Survey Authority.

❖ Bottom right: Bonaparte Reserve map. Image BC276, courtesy Natural Resources Canada.

This page

❖ Top: Helping a freight wagon uphill en route to Divide Lake. Image I-58409 courtesy of Royal BC Museum, BC Archives.

❖ Bottom left: Cook's Ferry Indian Reserves. This map shows the reserves that Swannell surveyed or resurveyed along the Thompson River. Image BC189, courtesy Natural Resources Canada.

❖ Bottom right: Cook's Ferry Indian Reserves. This map shows the four reserves that Swannell surveyed in the Highland valley. Image 10899, courtesy Natural Resources Canada.

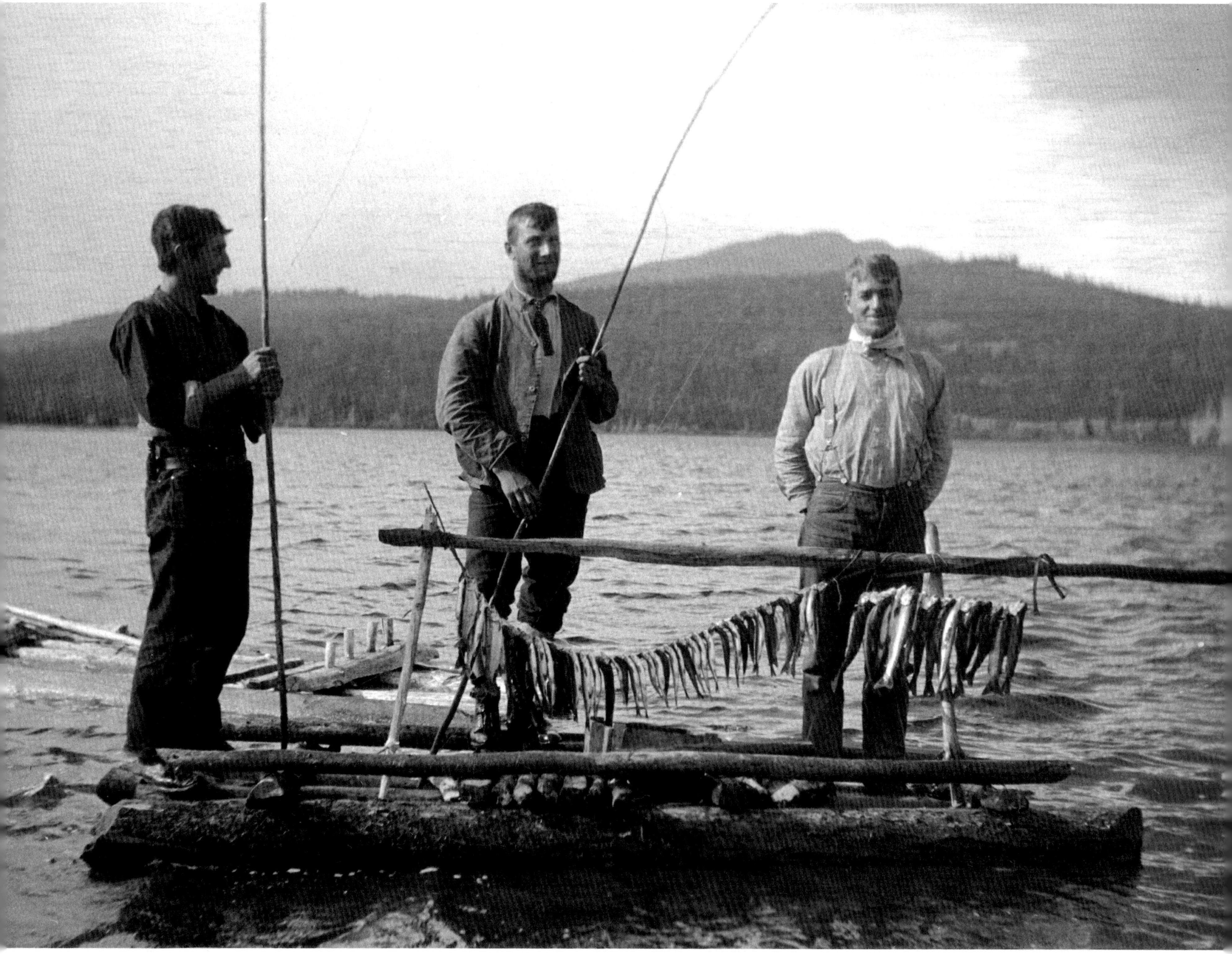

❖ Alf O'Meara (left), J.F. Finlayson (centre) and Wilfred Prevost (right) after a successful outing at Divide (Fish) Lake. Image I-58429 courtesy of Royal BC Museum, BC Archives.

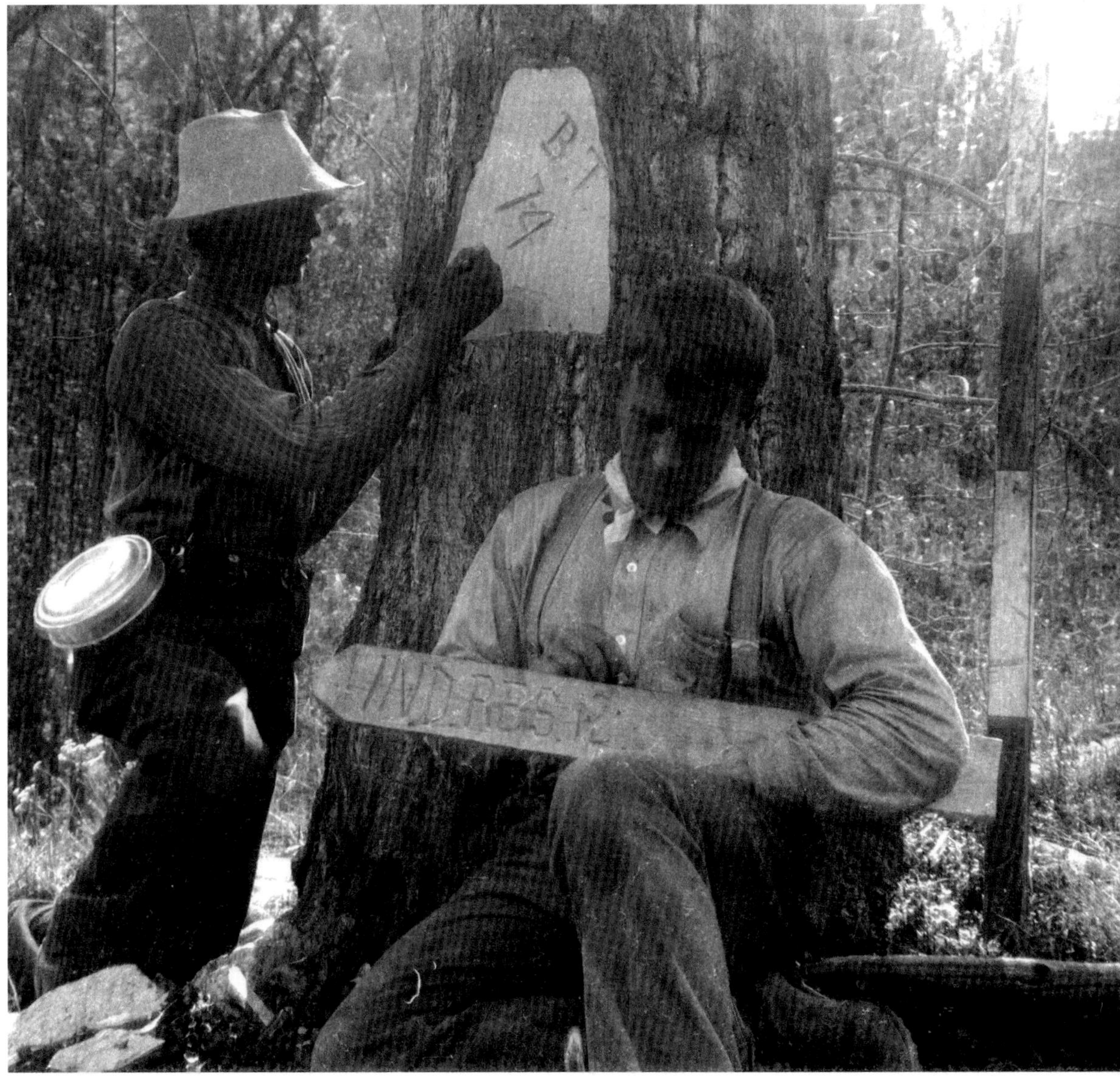

❖ Surveyors setting out boundary markers. One person has scribed the information on the survey post that is going into the ground while another is writing the information for the bearing tree. Image I-58278 courtesy of Royal BC Museum, BC Archives.

The work at Spences Bridge took a few days and included a traverse along the Thompson River.

Swannell's final reserve survey was I.R. #17 (Chilliwack) which he did from September 7 to 9. Then the crew travelled on the SS *Beaver* to New Westminster, and the following day they took the SS *R.P. Rithet* to Victoria.

Swannell was only in Victoria for a few days before leaving for Cranbrook to testify at the third trial involving John Watt's Flathead petroleum claims. On the evening of September 16, he took an overnight trip to Vancouver on the SS *Princess May*. Since Swannell was travelling directly to Cranbrook it was easier and quicker to go through the United States. From Vancouver he travelled on the Great Northern railway route to Everett, then proceeded to Spokane and finally arrived in Cranbrook. It took thirty-one hours to make the journey, but that was still quicker than the CPR route. Swannell spent the entire day of September 19 and the morning of September 20 in court. When the trial was finished, he went to Nelson to spend a day with his father and stepbrother. From Nelson he took a train to Spokane and then a Northern Pacific train to Seattle. At Seattle he travelled on the SS *Chippewa* back to Victoria.

For the remainder of the fall Swannell did some small surveys on Vancouver Island. The notes, calculations and maps for the Indian reserve surveys took a considerable amount of time. There was also the paperwork that needed to be done for a federal government survey.

By the end of 1907 Swannell had worked for Gore & McGregor for eight years. A teenager when he started with the survey company, Swannell was now in his mid-twenties, married and with two children. He had spent three years articling with Gore, learning from one of BC's most qualified surveyors and developing the skills that would make Swannell one of the province's outstanding men in his field. He had spent five years in charge of surveying projects, learning all the aspects of the work—surveying, handling people, managing logistics and finances, mapping and calculations—all while spending months at a time in remote areas of the province. He had made mistakes in his work, but from these experiences he developed a thoroughness and precision in his surveying. Swannell had kept a record of his work through his photographs, letters to Ada and some journals. In the succeeding years his record, both written and photographic, would become even more detailed.

By 1907, under the leadership of Premier Richard McBride, British Columbia was emerging from years of debt and entering into an era of prosperity. It was the beginning of the golden age of surveying in the

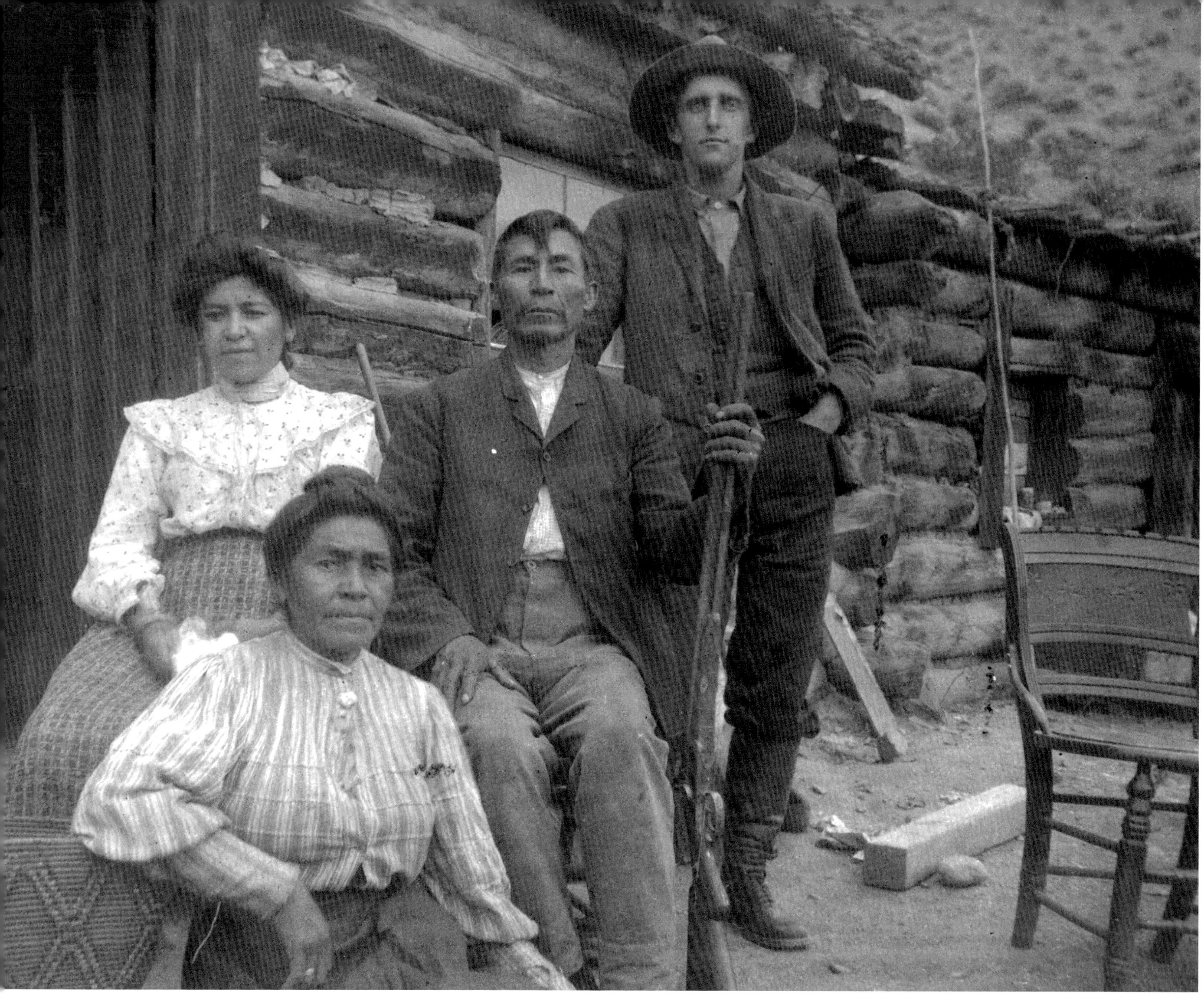

province. British Columbia was the only province in western Canada that had control of its own land (except for the Railway Belt and Peace River Block, which it gave to the federal government as part of the terms for the construction of the Canadian Pacific Railway) but the provincial government was also responsible for the costs of surveying its land. Mountainous terrain along with scattered economic and land settlement made surveying much more difficult than on the open Prairies, limiting the cash-strapped government

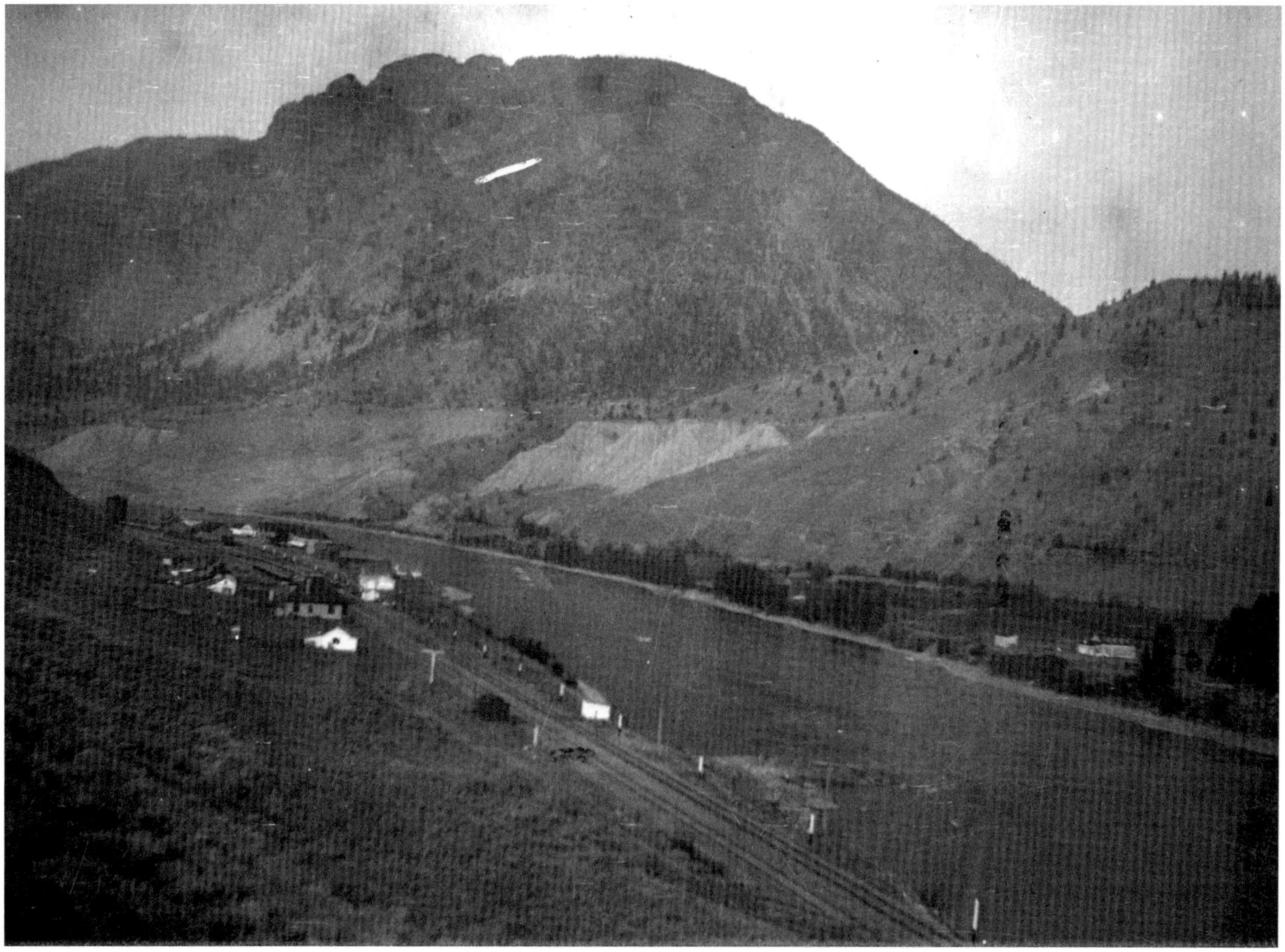

❖ Spences Bridge. In 1907 there was no bridge across the river at this location and the CN railway tracks did not exist. Image I-33688 courtesy of Royal BC Museum, BC Archives. For 2104 image see page 151.

❖ Opposite page: John Tedlenitsa and family, Divide Lake. Tedlenitsa was a well-known Nlaka'pamux leader and chief of the Pekaist First Nations band in the early twentieth century. His first political involvement was in 1908 when he was one of the signatories of the "Prayer of Indian Chiefs," a four-page petition to A.W. Vowell, Superintendent of Indian Affairs in British Columbia, written by James Teit. Like Chief David he was involved in many important political events, advocating for better conditions for First Nations people in British Columbia. He also spoke to the Honourable Richard McBride in Victoria and the Right Honourable Robert Borden in Ottawa, was a signatory to the memorials to Wilfrid Laurier and Frank Oliver and made a presentation to the McKenna–McBride Commission. Chief Tedlenitsa travelled to Ottawa a second time as part of the 1916 delegation that spoke to Borden. At the British Columbia Indian Conference held in Vancouver in 1916, Chief Tedlenitsa was appointed a member of one of the committees created by the conference delegates. The two photographs that Swannell took are the only known pictures that show Tedlenitsa with his family at home. The traditional basket and the chair provide an interesting contrast in this photograph. Image I-33681 courtesy of Royal BC Museum, BC Archives.

to only the basic necessities of surveying. In 1906 the British Columbia government spent less than $7,000 on surveying, but by 1911 expenditures increased astronomically to $448,885 as the government and its surveying personnel tried to keep pace with the province's prosperity and the backlog of work from past years. From 1907 to 1911 the number of lots surveyed annually increased from 1,830 to 7,312, while the amount of acreage more than tripled.

The BC government had several large projects that needed to be surveyed, most of them in the central and northern part of the province. The Grand Trunk Pacific and Canadian Northern, two new transcontinental railways, were going to begin construction. The Grand Trunk's line would traverse a large area of agricultural land in the Nechako, Burns Lake and Bulkley valley areas, and this land (along with the Peace River) was being billed as the last agricultural frontier in North America. The government needed to survey a large number of lots to have ready for the anticipated settlers.

Swannell was now confident of his surveying skills and his ability to work in the field. He wanted to be in charge of his own surveying and to travel to even more remote areas beyond the southern part of the province. The next year he would start working on contract for the BC government and begin surveying northern British Columbia.

Mining in the Highland Valley

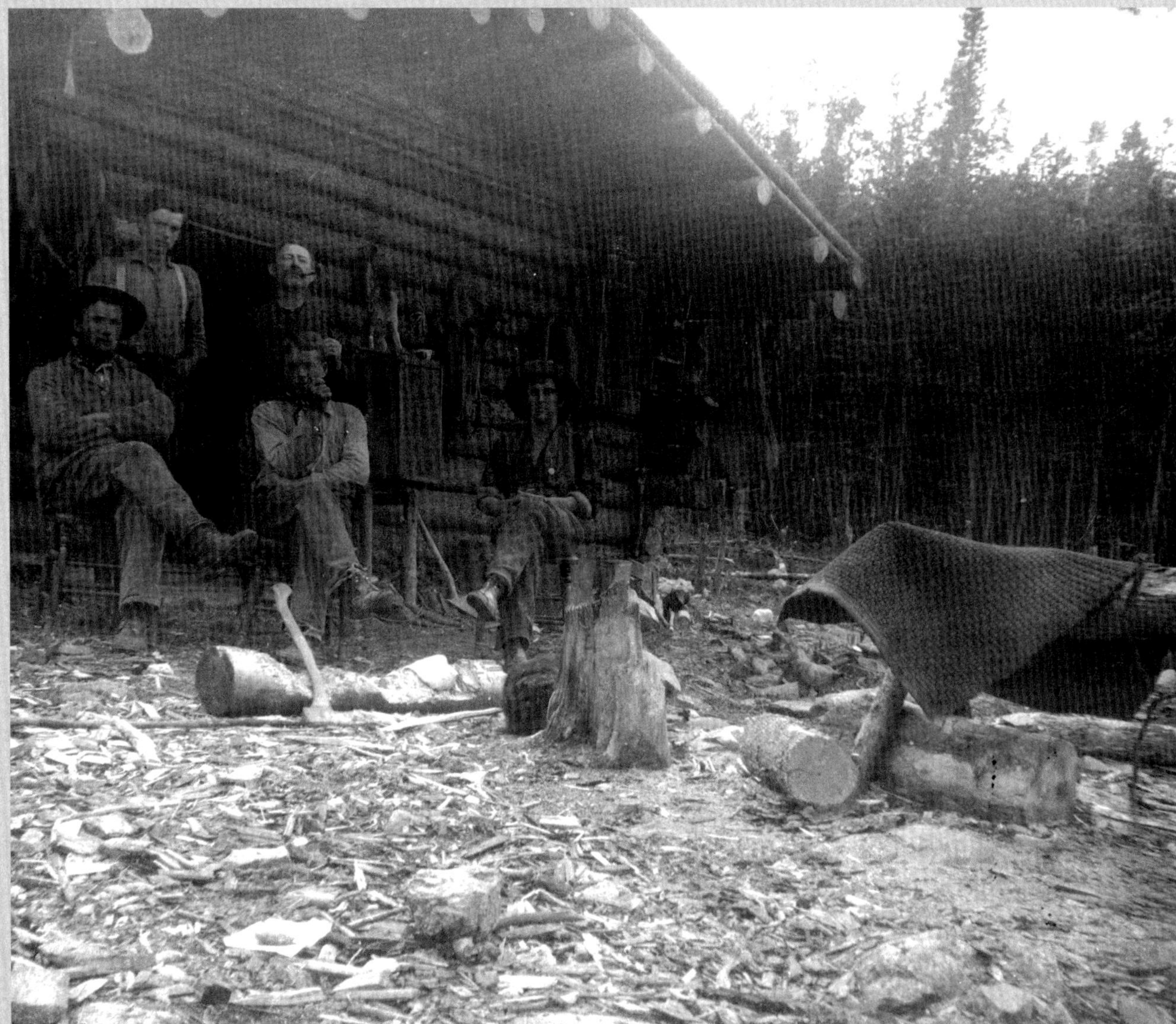

❖ Surveyors at Tamarac Mine. Image I-33685 courtesy of Royal BC Museum, BC Archives.

George Sanson was a pioneer doctor in the Ashcroft region for many years. He was also an unsuccessful candidate for the British Columbia legislature and was involved in the copper mining claims in the Highland Valley as head of the Tamarac group. George Ward, owner of the Grand Central Hotel in Ashcroft, was one of the members. Lack of suitable transportation routes had inhibited the development of mining in the Highland Valley, but in early 1907 a wagon road was constructed from Ashcroft, on the CPR line, to the valley. In his 1907 government report G.C. Tunstall, gold commissioner for the Yale

District, wrote: "The Highland Valley mines in the Ashcroft Division are fulfilling the most favourable expectations of the parties interested in them. A large outlay has been expended in development work that has been amply justified by the results." Robertson and Nation used the new road to inspect the mining claims of the area. Henderson probably accompanied the two mineralogists, not only because he was MLA for the area, but because he also had invested in one of the mining groups in the Highland Valley. (Henderson was one of the premier lawyers in British Columbia. In 1919 he successfully defended Simon Gunanoot, the famous First Nations fugitive, against the charge of two murders in northern BC.)

In his report, the provincial mineralogist noted seven main sites of activity and wrote about the Tamarac group: "The Tamarac group, consisting of the Tamarac, Shamrock, King, Duke, Billy, Muir Fraction, May L. and Star mineral claims, is situated at an altitude of 5,200 feet [1,600 m], about one and a half miles [2.4 km] to the northwest of the wagon road at Fish Lakes, and is owned by Dr. Sanson and others, of Ashcroft, who have built a branch road up to the property and erected a very good cabin. The development consists of three or four shafts, each sunk about 25 feet [7.5 m] deep, and a number of open cuts." After briefly describing his observations of the quartz veins, he noted that Robertson commented, "the mineral occurs in bunches of varying size in the quartz vein matter, and the selected ore assayed high in copper."

Sanson had three claims that he needed surveyed, the Tamarac, Star and Shamrock. Since Swannell was already in the area he must have made arrangements to have these properties surveyed.

In 1915 the Tamarac group became part of the Highland Mining and Development organization based out of Spokane. In the 1960s the Highland Valley became the site of a large open-pit copper mine that is still in production in 2014.

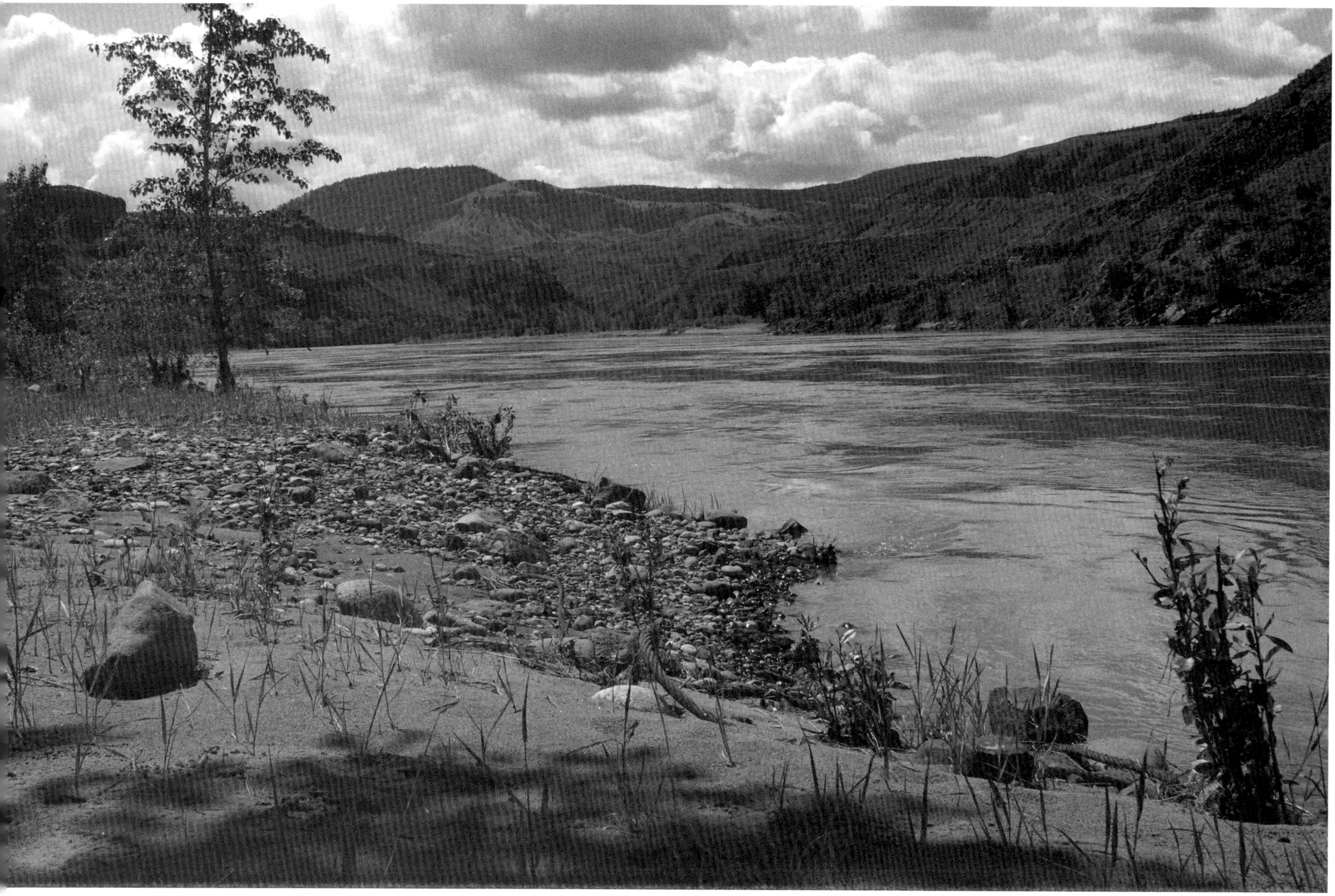

❖ On the opposite bank, in the centre of the picture, the gravel bar is visible where Churn Creek enters the Fraser River. Downstream there is much more vegetation along the river bottom than in 1904. In the foreground are pieces of cable from the ferry. For comparison, see archive photo I-33358 on page 101.

SOUTHERN BRITISH COLUMBIA IN 2014

Ten years ago, as I was completing *Surveying Northern British Columbia,* I went on my first trip to visit places where Frank Swannell surveyed and take some pictures at locales that he photographed. During parts of several intervening summers I have followed Swannell's routes, and in the process (thanks to many people) I have had the opportunity to visit parts of the province that few people see. As I finished the final Swannell book this year, I took trips to search for some of the settings of the photographs used in *Surveying Southern British Columbia.*

British Columbia has changed considerably in the century since Swannell began taking pictures. The photographs that Swannell took before he started working for the BC government are not documented as precisely as his later pictures, so it is much more difficult to find the location for some of his photographs, particularly the backcountry scenery. Highways, bridges and rail lines cross large portions of the landscape that Swannell saw. At the same time, many places that Swannell photographed have disappeared. Several of his pictures in the Kootenays were taken at locations that are now flooded by reservoirs. His pictures are a reminder of what BC looked like at that time, providing a contrast to today's scenery. Despite these difficulties I was able to find and rephotograph some of Swannell's locations. As always, seeing the same setting through the lens of my camera was very satisfying.

In mid-June I took a trip to the Cariboo where Swannell had surveyed in 1901, 1904 and 1906. Kris Andrews was my host at Williams Lake. Her father, Gerry Andrews, had articled with Swannell and she shared some of her childhood memories of Frank. We spent a glorious day travelling southwest from Williams Lake past the Springhouse Ranch where Swannell surveyed in 1906 to Dog Creek, where Swannell and his crew arrived in the spring of 1904 on the weekly stage from Clinton. The scenery was spectacular and there were many birds as we travelled across the high plateau to the steep terrain above the Fraser River. From Dog Creek we drove to the present crossing of the Fraser River not far from the old Churn Creek ferry. Kris knew the location of the ferry, and we walked the last section down to the Fraser River. There we found the place where Swannell took his photograph in April 1904. Pieces of the old cable still remained at the site.

In the afternoon we drove past the Gang Ranch, where Swannell took some pictures in 1904, to the

❖ The main change to the landscape at Farwell (Salmon) Canyon since Swannell's visit is the bridge, which is higher and upstream from the one at this site in 1904. For comparison, see archive photo I-57591 on page 102.

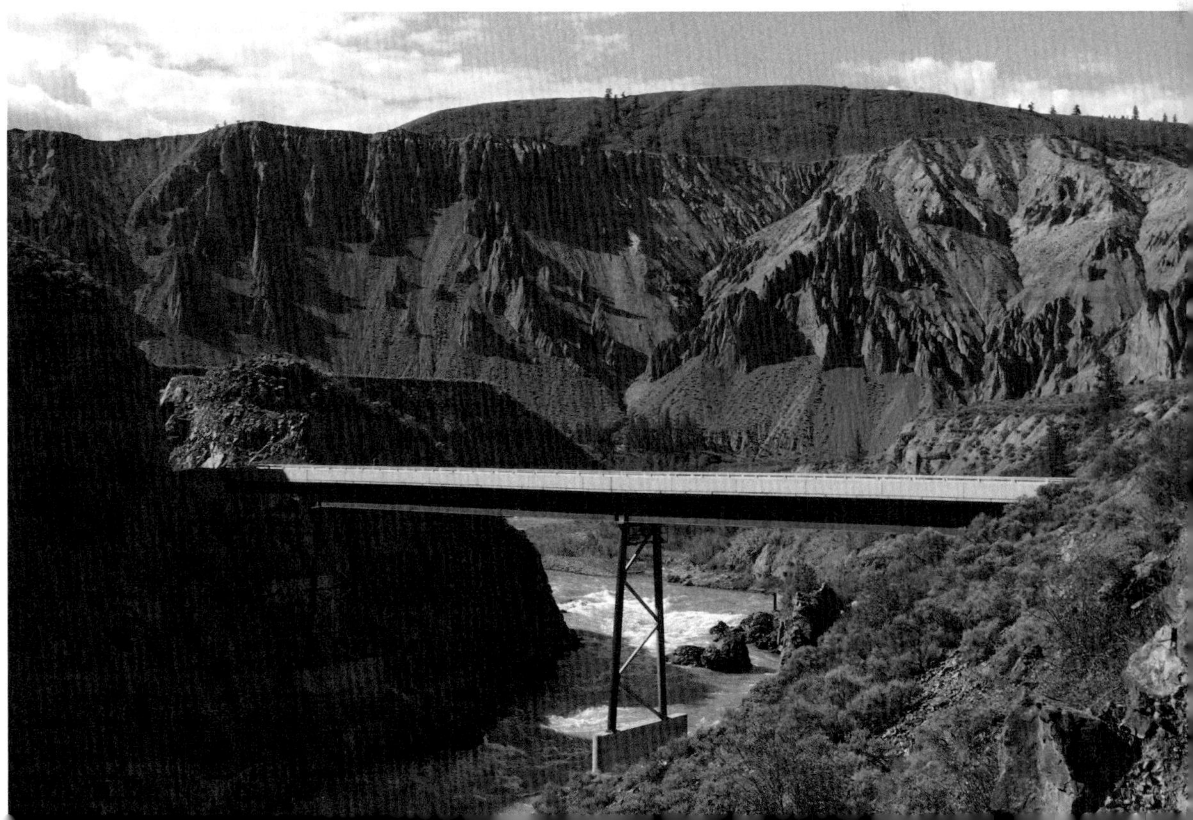

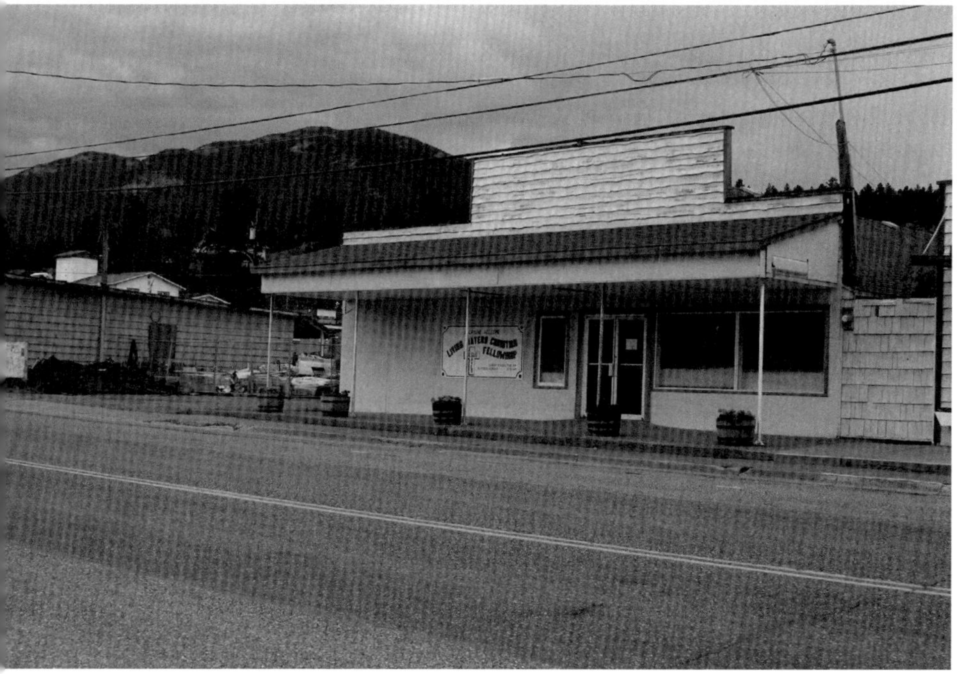

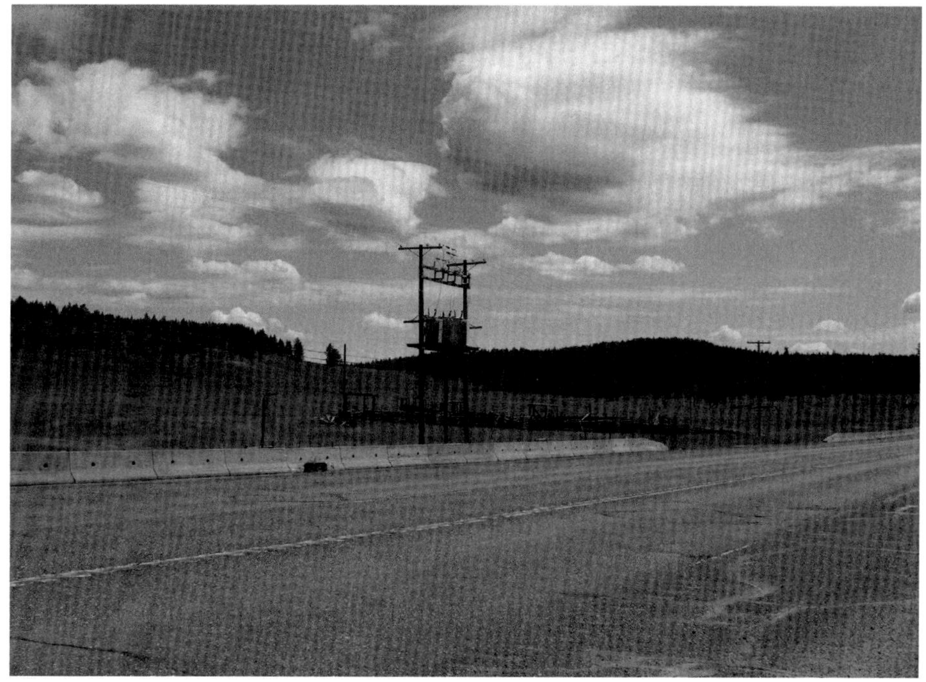

❖ Above left: The building that is at the same location in Clinton today is very similar in style to the one that Swannell photographed in 1904. For comparison see archive photo I-33678. on page 99.

❖ Above right: There are only a few structures to mark the remains of the once thriving ranch and roadhouse at 111 Mile House. The creek crosses under the highway almost immediately in front. For comparison see archive photo I-33353 on page 34.

❖ Right: The change to the foreground, because of the gravel pit and the loss of the vegetation behind the open hillside, provides a dramatic contrast to the 1904 landscape along this section of Williams Lake. For comparison see archive photo I-33348 on page 104.

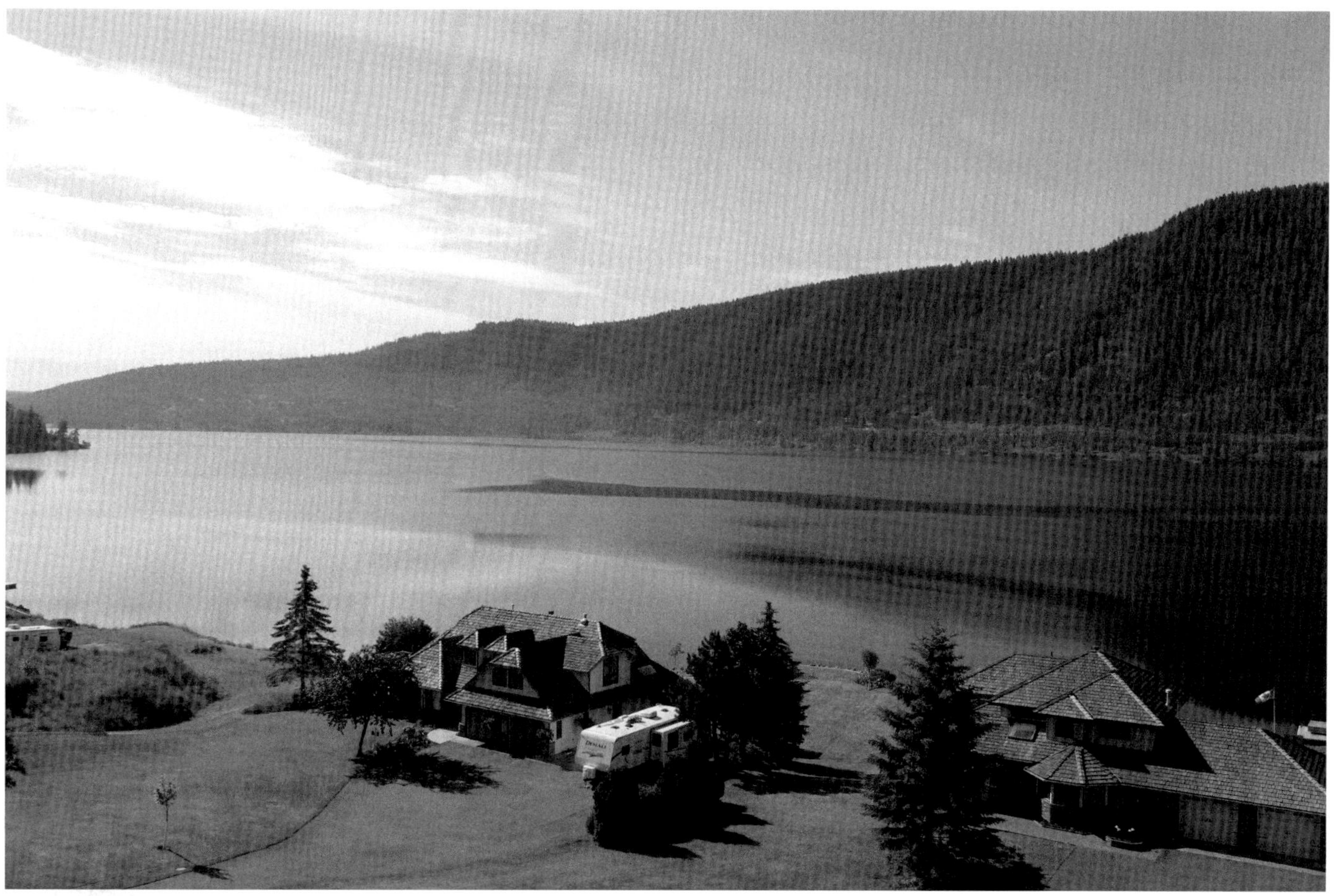

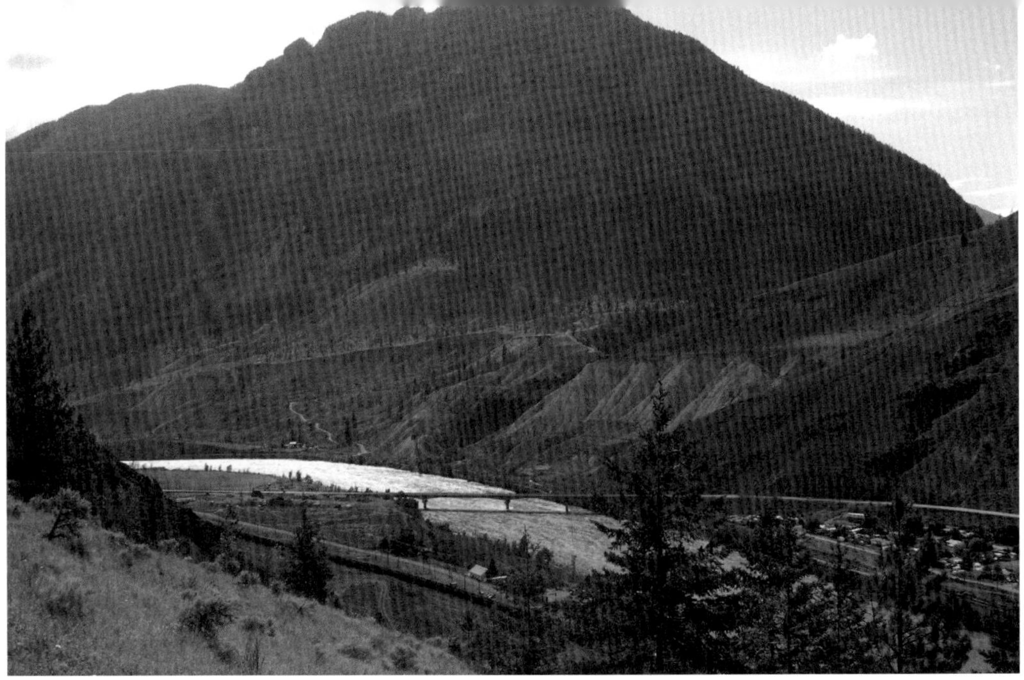

crossing of the Chilcotin River. In Swannell's photograph he calls the place Salmon Canyon, but today it is known as Farwell Canyon. Because of erosion and flooding it was impossible to reach the exact place where Swannell took his picture.

The next morning, following Kris's directions, I found the locale above Williams Lake where Swannell took a photograph in 1904. Since that time the place was used for a gravel pit, and it has now been transformed into some nice homes. Much of the foreground for the picture is gone along with the place where Swannell was standing, but I was able to take a photograph near the locale.

On my way back to Burnaby I stopped at 111 Mile House. In 1901, when Swannell visited, there was a ranch and roadhouse but they are gone. Highway 97 follows the Cariboo Wagon Road at this location. While taking the picture I realized that in the foreground of Swannell's picture there is evidence of the narrowing of the wagon road as it crosses a small creek. I looked at the place beside the highway where I was standing and saw the same small creek in front of me. It was easy to use the mountains in the background to find where Swannell took a picture while he stopped in Clinton in 1904. I also found the place on the southern approach to Clinton where Swannell took his interesting diamond-shaped picture of the community in 1901, but vegetation growth prevented me from taking a clear photograph. At Spences Bridge I hiked up the hillside above the town and found a place that provided an approximate location to Swannell's photograph in 1907. From this vantage I could see the changes that had occurred in the Thompson River valley in the century since that time.

In July I took a trip to the Kootenays. I had previously been there in 2010 searching for the locations of some of Swannell's photographs, so I knew where some of the places would be. Also, in 2012 I had been to the North Fork of the Flathead River valley and hiked to the top of Mount Gerry Andrews with a group of people as part of the Mount Gerry Andrews ceremony. Although I did not find any specific Swannell locations, I was able to see the Kishinena and Sage Creek valleys where he did his oil claim surveys in 1904 and the broad river valley that Swannell followed when he and his crew travelled from the Great Northern rail line in Montana to this area.

At Nelson I stopped at the museum and showed staff the three pictures that Swannell took there in 1902. They immediately recognized the location of one picture. It had been taken at Central School. The present-day building, constructed in 1908, still exists, and replaced an earlier school on the same site. Perhaps Swannell went one day to the school that his niece and nephew attended and took a picture of downtown Nelson from an upper story window. Vegetation in the foreground precluded a clear picture of Nelson today, but the lake and background are immediately recognizable. The museum staff were also able to determine the general location of a second photograph. I went to the Rosemont section of the city

❖ Above: When Swannell took his photograph in 1907, neither the CN railway tracks nor the highway bridge existed. Most of the settlement of Spences Bridge was on the near side of the river along the CPR railway line. For comparison see archive photo I-33688 on page 145.

❖ Below: The view from Central School toward downtown Nelson. The trees in the foreground were not present in 1902. There is also considerably more vegetation on Elephant Mountain in the background. For comparison see archive photo I-58036 on page 51.

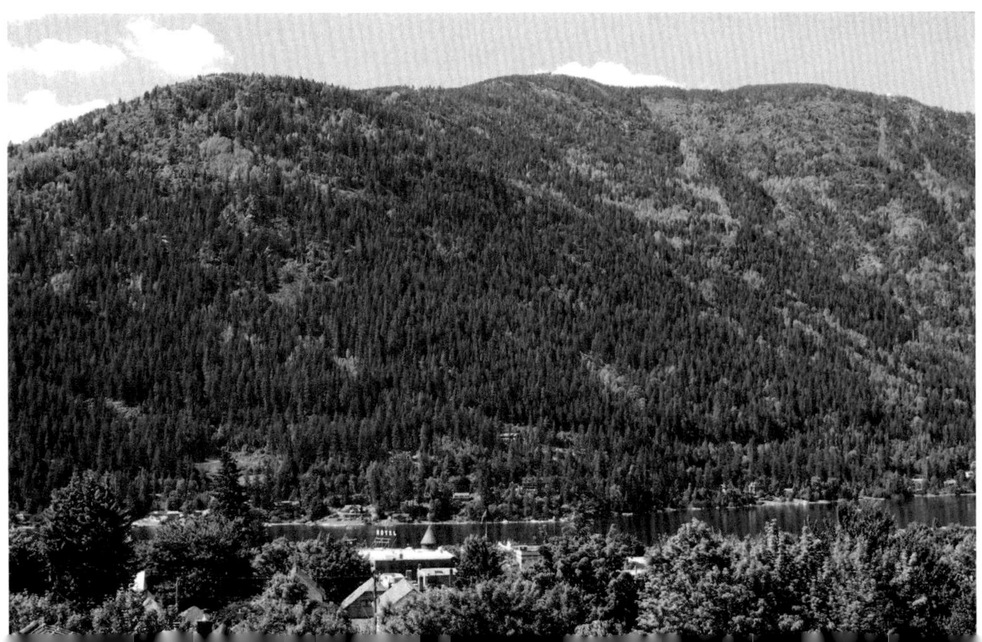

❖ Right: The CPR still maintains a large rail yard in Cranbrook. For comparison see archive photo I-58258 on page 87.

❖ Below: Dog (Vaseux) Lake. The outcropping that created an overhang on the original road was removed to make the highway. For comparison see archive photo I-33611 on page 122.

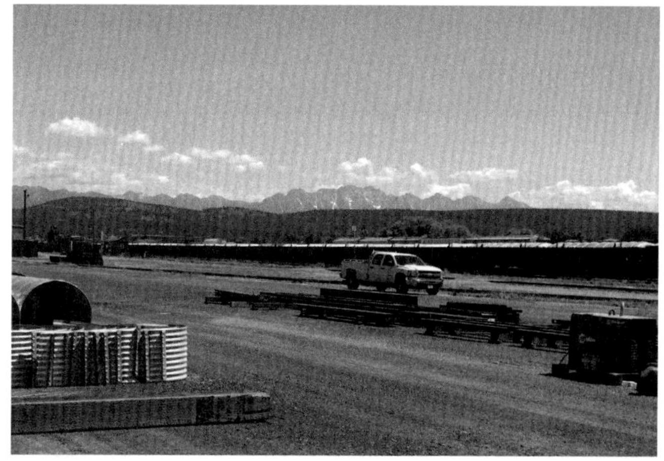

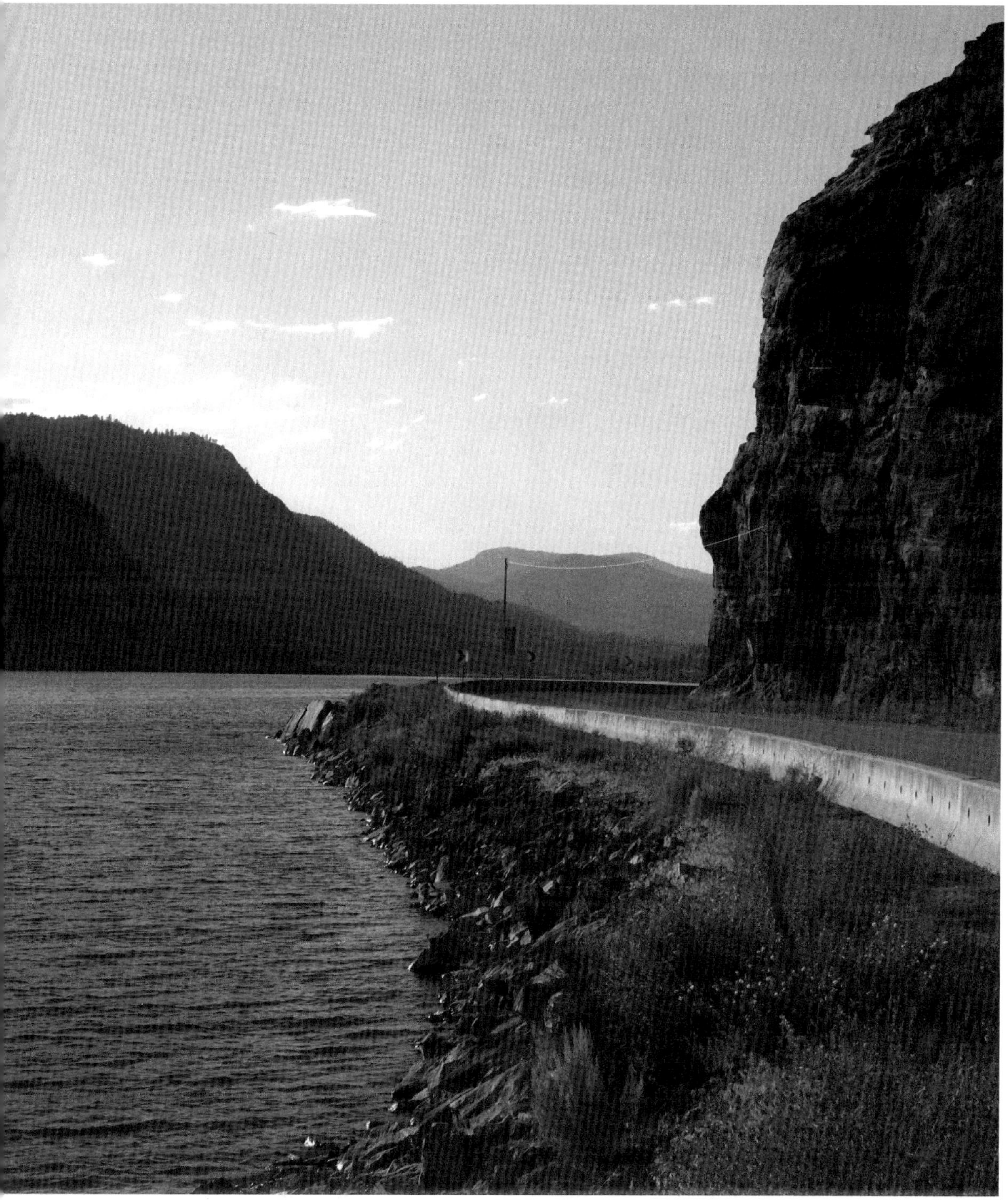

and found an open viewpoint that was close to where Swannell took his picture. It was interesting to learn that Swannell's three photographs each showed a different part of Nelson and what it looked like at the beginning of the twentieth century.

From Nelson I drove to Balfour and took the ferry across Kootenay Lake. Swannell would have travelled down the lake to Kootenay Landing. While crossing the lake I saw some of the same scenery that Swannell would have viewed and observed the Pilot Bay smelter that he would have seen in operation. Travelling to Creston along the east shore of Kootenay Lake I passed the small community of Sirdar. Near here the BC Southern rail line headed west to Kootenay Landing where the train and boat met.

In the East Kootenay region I went to Wardner. When I was there in 2010 I met local historian Chris Graf, who along with his wife, wrote a book about the history of the area. The Grafs had used one of Swannell's photographs in their book, and Chris recognized the locations for all of Swannell's pictures in the community. The rock outcrop at the western edge of the railway bridge still remains and is quite prominent. I took a picture similar to Swannell's, and while I was in Wardner I briefly visited Chris again. At Fort Steele I took a picture at almost the same location where Swannell photographed the bridge across the Kootenay River that existed in 1902. At nearby Wildhorse Creek I took a re-photograph of the place where the bridge crossed this creek also in 1902. Swannell's

photographs of three prominent bridges on or near the Kootenay River are an important historical record for this area.

On the return trip I stopped at the rail yard in Cranbrook where Swannell took two pictures in 1903. It must have been a cloudy day at that time for the Rocky Mountains in the background are not visible in his picture.

My final picture was in the Okanagan at Vaseux Lake. In 1905, en route to his surveying of Columbia and Western land subsidy lots near Osoyoos, Swannell photographed the stage coach on which they were riding passing under the big rock outcrop over the lake. It's an excellent photograph of what this place looked like over a century ago.

Once again I am appreciative to the people who helped me find places where Swannell took his pictures. It was nice to stand at the locales and see what some of the places in southern British Columbia look like a century later.

❖ Top: Water from Lake Koocanusa, the reservoir behind Libby Dam in Montana sometimes backs up the Kootenay River at Wardner. When I was here in August 2010 the river was flowing. The rock outcrop that is clearly visible in Swannell's photograph is noticeable in this picture. The island in the river in 1903 is barely above water, and shows the outline of the side channel where the Crow's Nest Pass Lumber Company stored its logs. For comparison see archive photo I-59699 on page 88.

❖ Below right: The bridge across the Kootenay River at Fort Steele in 1902 was lower and slightly upstream from the current bridge. Evidence of the old road down to the river still remains along the hillside above the tracks. The rail line which runs north to Golden did not exist in 1902. The Kootenay River channel has changed noticeably in the past century. For comparison see archive photo I-33937 on page 61.

❖ Below left: The Wild Horse Creek channel below the bridge has changed and it was impossible to take a picture at the same location as Swannell. The snow and glacial remnants barely visible on the mountains in the background of Swannell's photograph have virtually disappeared. For comparison see archive photo I-33946 on page 62.

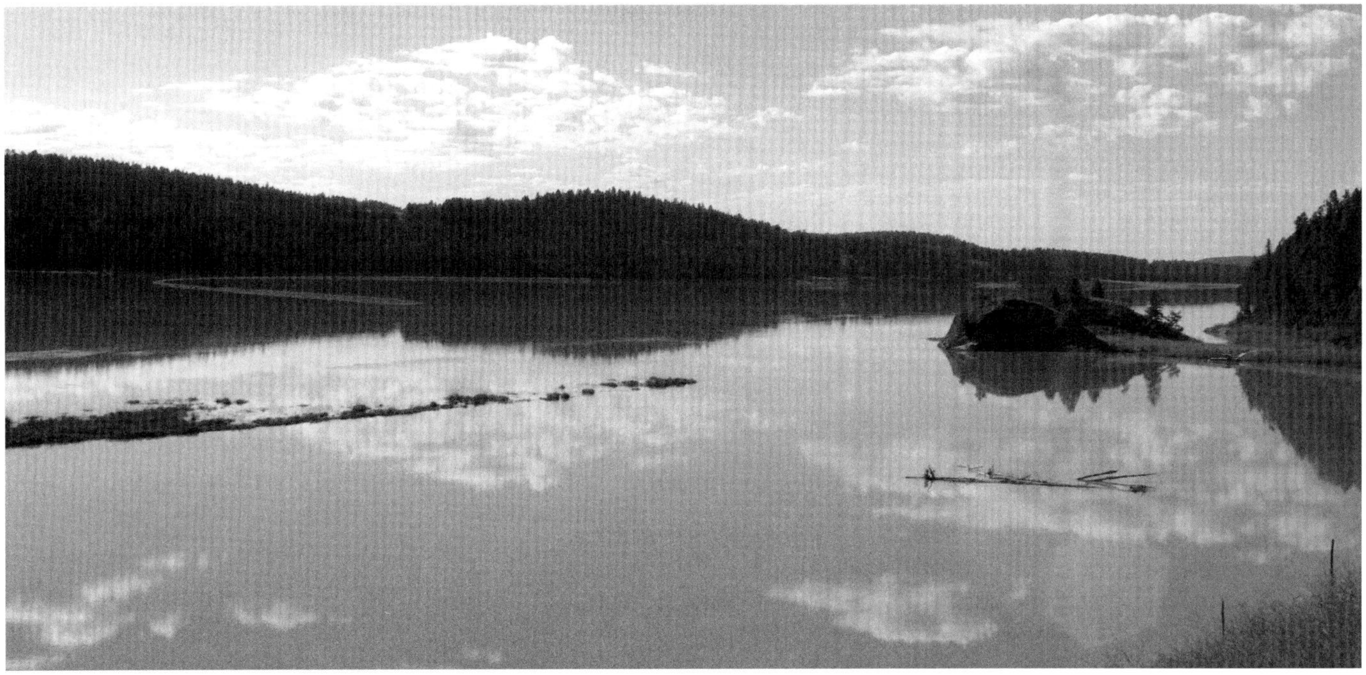

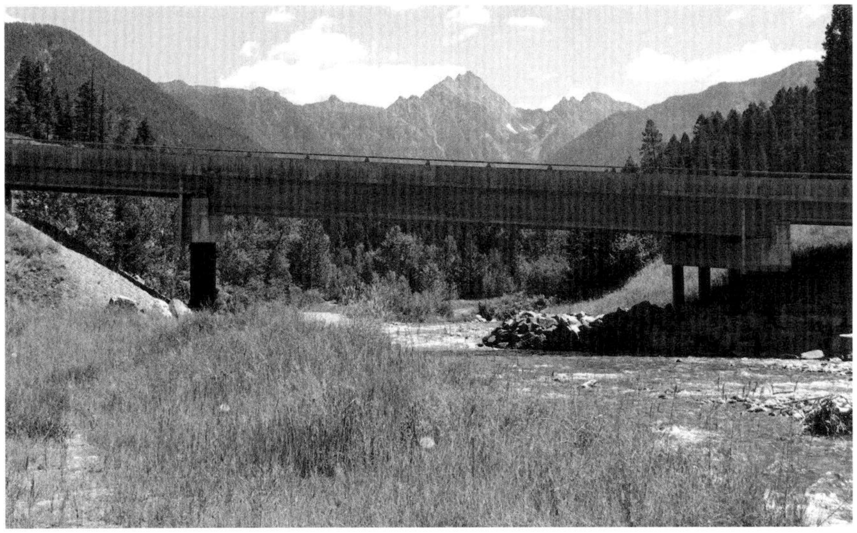

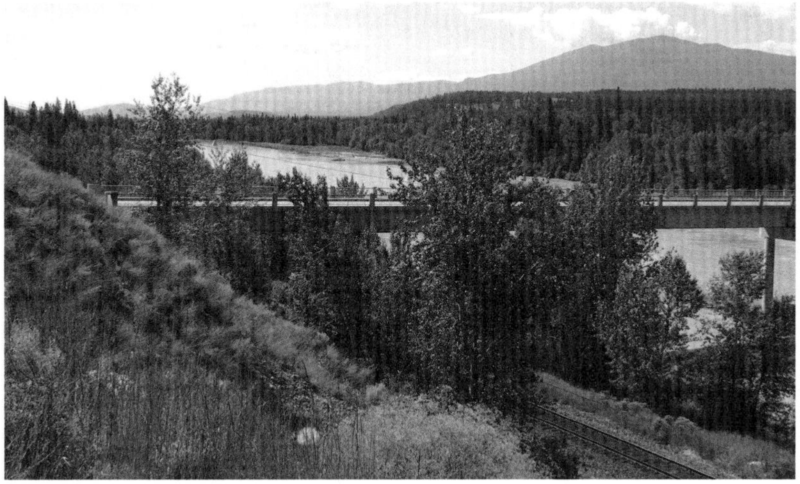

Acknowledgements

I have been fortunate to have a considerable amount of assistance for all of the books that I have written, and *Surveying Southern British Columbia* is no exception.

Several organizations and individuals have provided a financial contribution toward the acquisition of copies of the photographs used in the book, and I sincerely appreciate their support. The Historical and Biographical Committee of the Association of BC Land Surveyors and the Career Awareness Committee of the same organization both provided assistance and have provided support throughout the books that I have written. I would also like to thank Bill Chapman from Chapman Land Surveying for his generous contribution. I first met Bill at the Mount Gerry Andrews celebration in 2012 and am pleased that he has continued our association. In addition, I would like to express my thanks to Vici Johnstone, from Caitlin Press, and the BC Archives for their financial assistance.

Robert Allen, who has shared his expertise as a BC Land Surveyor by answering my surveying questions in previous books, provided his insights into the surveying techniques used in the early twentieth century, particularly for the surveying chapter. Calvin Woelke, from the Land Title Survey Authority, located some of the historical maps and field books that I used. Janet Mason, Provincial Toponymist, and Louis (Buddy) Syrette helped me find the file records for the lots granted by the BC government as part of the railway subsidies and information relating to the Reserve surveys that Frank Swannell did in 1907. The staff of the Glenbow Museum provided prompt assistance and service during my three-day blitz of their Canadian Pacific Railway files. And once again the front desk staff at the BC Archives cheerfully and patiently handled my many requests. Staff at the Victoria branch of the BC Law Library showed me how to locate historical law cases which provided invaluable information about the two court cases related to surveys that Swannell did. Michael Carter from the BC Archives assisted me with the Pinkerton files, an unusual research source. My thanks goes to all of you for your assistance.

I would like to express my appreciation to Kris Andrews for hosting me on my trip to the Cariboo and Chilcotin and for taking me on a day trip to find some of the locations where Swannell surveyed in 1901 and 1904. It was a spectacular spring day and wonderful to visit places that Swannell mentioned and photographed. I would also like to thank Mary Andrews for providing me a place to stay while I was in Victoria, and Robin and Max Phillips for hosting me while I was in Calgary. Kris, Mary and Robin all knew Swannell, so it was nice to continue their connection with him.

My thanks goes to Vici Johnstone of Caitlin Press for her willingness to publish the final book in the Frank Swannell series. I produced the first Swannell book with Caitlin Press (publisher Cynthia Wilson died in 2005) and it is nice to be reconnected with this organization. I hope that we will produce many more books together.

Finally, I would like to thank my wife, Linda, who has been of continual assistance throughout all the books that I have written in the past ten years. Your support and the many ways that you help mean a lot.

—Jay Sherwood, July 2014

Index

Bold indicates photograph

E

F

G

H

I

J

K

L

Resources

Books

Cail, Robert E. *Land, Man and the Law: The disposal of crown lands in British Columbia, 1871-1913.* Vancouver: UBC Press, 1974.

Early Land Surveyors of British Columbia. Compiled and edited by John Whittaker. Victoria: Corporation of Land Surveyors of the Province of British Columbia, 1990.

Graf, Constance. *Reflections on the Kootenay: Wardner BC 1897-1997*, by Constance and Christopher Graf. Wardner, BC, c1997.

Harbord, Heather. *Texada Tapestry.* Madeira Park: Harbour Publishing, 2011.

Hayes, Derek. *British Columbia: A New Historical Atlas*. Vancouver: Douglas & McIntyre, 2012.

Headrick, Daniel R. *The Invisible Weapon: Telecommunications and International Politics 1851-1945.* New York: Oxford University Press, 1991.

Keddie, Grant. *Songhees Pictorial: A History of the Songhees People as Seen by Outsiders, 1790–1912.* Victoria: Royal BC Museum Press, 2003.

The L.S. Group: British Columbia's First Land Surveyors. Saanich: The Association of British Columbia Land Surveyors, 2007.

Roy, Patricia E. *Boundless Optimism: Richard McBride's British Columbia.* Vancouver: UBC Press, 2012.

Turner, Robert D. and David S. Wilkie. *The Skyline Limited: The Kaslo and Slocan Railway*. Victoria: Sono Nis, 1993.

Turner, Robert D. and David S. Wilkie. *Steam Along the Boundary: Canadian Pacific, Great Northern and the Great Boundary Copper Boom*. Winlaw: Sono Nis Press, 2007.

Turner, Robert D. *Sternwheelers and Steam Tugs: An Illustrated History of the Canadian Pacific Railway's British Columbia Lake and River System.* Victoria: Sono Nis, 1984.

Turner, Robert D. and and David S. Wilkie. *West of the Great Divide: An illustrated history of the canadian Pacific Railway in British Columbia 1880-1986.* Winlaw: Sono Nis, 2003.

Williams, David Ricardo. *Call in Pinkerton's: American Detectives at Work for Canada.* Toronto: Ddurn Press, 1998.

Articles

Baker et al. v. Smart et al. Leckie et al. v. Watt et al. *The British Columbia Reports.* Volume 12, pp. 129-148. Victoria: The Law Society of British Columbia, 1906.

Darlington, Esther. Dr. George Sanson. *B.C. Historical News*, Fall 1997, pp. 2-5.

Seippel Lumber Company v. Herchmer et al. *The British Columbia Reports*. Volume 19, pp. 436-441. Victoria: The Law Society of British Columbia, 1915.

Veazey, Phyllis. John Hendry and the Vancouver, Westminster and Yukon Railway: it would put us on easy street. *BC Studies*, Autumn 1983, pp. 44-63.

Newspapers

British Colonist (Victoria)

No Bonus for the Paint Works. March 14, 1905 – p.3.

Interviewed the Indian Agent. May 20, 1905 – p.5.

Songhees Refuse. May 31, 1905 – p. 5.

The B.C. Oil News. March 16, 1913 – p.4.

To Hear Appeal in Kootenay Case. June 20, 1908 – p.7.

Budget Debate is Concluded. February 27, 1913 – p. 14.

Prospector (Fort Steele)

CPR in the Flathead. April 23, 1904 – p. 1.

Archives

BC Archives

GR0056 – British Columbia Provincial Police correspondence, Volume 3

GR441 – Vol. 16/2 File 40/01 T.S. Gore to Premier April 27, 1901

MS392 – Frank Swannell papers

MS2259 – BC Land Surveyors file

MS98002-017 – Frank Swannell photograph collection

BC Land Title Survey Office – surveyor field books and Crown grants

Glenbow – Canadian Pacific Railway, Land Settlement and Development fonds

M2269-1963 Oil descriptions and maps, 1904-1905

M2269-2056 British Columbia Southern Railway Company, Lot 4590

M2269-2075 Land Department papers, 1902-1910
M2269-2136 Columbia and Western Railway lands, T.S. Gore and J.H. McGregor
M2269-2139 British Columbia Southern Railway lands, 1901-1904
M2269-2141 Columbia and Western Railway lands, 1905-1906
M2269-2142 Columbia and Western Railway lands, 1905-1907
M2269-2230 British Columbia Southern Railway Company, Lot 4590
Minnesota Historical Society—Vancouver, Westminster, Northern and Yukon Railway.
Manuscript ID 6539858
BC Law Library
Baker v. Smart, Leckie et al. v. Watt et al. Judgment April 20, 1906. BC Supreme Court. 1906
Carswell BC44, 2MMC373
In re Coal Mines Act: Re: Baker, Leckie, Watt et al. BC Supreme Court. Judgment November 4,
1907. 1907 Carswell BC32, 2MMC530
Seippel Lumber Co. v. Herchmer. BC Supreme Court. Judgment February 16, 1914. 1914
Carswell 297, 7WWR 333

BC Government Documents

BC Sessional Papers, 1902. *Annual Report of the Minister of Mines, 1901.*
BC Sessional Papers, 1904. Report of the Minister of Mines. Southeast Kootenay District.
Report of the Provincial Mineralogist, H73-95.
BC Sessional Papers, 1908. Report of the Minister of Mines. Ashcroft Mining District: notes by
the Provincial Mineralogist, L134-136.
Petition No. 5—1899 Vancouver, Northern and Yukon Railway Co.
Petition No. 3—1900 Vancouver, Northern and Yukon Railway Co.
Columbia and Western Railway Subsidy—1903.
Report of Select Committee re Columbia and Western Railway Land Subsidy—1903.
Journals of the Legislative Assembly—February 26, 1913.

Federal Sessional Papers

1908 Annual Report of the Department of Indian Affairs.
1943 Annual Report of the Department of Indian Affairs. Schedule of Indian Reserves in the Dominion of Canada. Part 2. Reserves in the Province of British Columbia. Recompiled and Corrected up to March 31, 1943.

Internet

"Underwater cables and Canadian Innovation," Royal BC Museum blog, July 5, 2011
www.royalbcmuseum.bc.ca
"History of the Atlantic Cable & Undersea Communications: Bamfield Cable Station"
www.atlantic-cable.com
"Extracts of interest to British Columbia land surveyors from British Columbia law reports:
Seippel Lumber Company v. Herchmer et al." www.abcls.ca